# THE BEAUTY OF GEOMETRY

## TWELVE ESSAYS

H. S. M. Coxeter

DOVER PUBLICATIONS, INC.
Mineola, New York

*in memoriam NICO 1951–1967*

*Bibliographical Note*

This Dover edition, first published in 1999, is an unabridged
republication of the work originally published as *Twelve Geometric
Essays* in 1968 by Southern Illinois University Press, Carbondale
and Edwardsville, Ill., and Feffer and Simons, London and
Amsterdam.

*Library of Congress Cataloging-in-Publication Data*

Coxeter, H. S. M. (Harold Scott Macdonald), 1907–
    [Twelve geometric essays. English]
    The beauty of geometry : twelve essays / H.S.M. Coxeter. —
Dover ed.
         p.    cm.
    Originally published: Twelve geometric essays. Carbondale
and Edwardsville, Ill. : Southern Illinois University Press, 1968.
    Includes bibliographical references and index.
    ISBN 0-486-40919-8 (pbk.)
    1. Geometry.   I. Title.
QA443.C6713      1999
516—dc21                                                    99-35678
                                                                CIP

Manufactured in the United States of America
Dover Publications, Inc., 31 East 2nd Street, Mineola, N.Y. 11501

# PREFACE

I am grateful to G. Kimball Plochmann and the Southern Illinois University Press for offering to reproduce a dozen of my published papers, especially as this offer enables me to correct a considerable number of small errors. I hope that some of the twelve chapters may encourage the further study of elementary geometry, not only for its own sake but also for its applications to analysis (Chapter 1), number theory (Chapter 2), group theory (Chapters 3, 5, 6, 7), graph theory (Chapters 6 and 7), communication theory (Chapter 9) and relativity theory (Chapter 11). The order of the chapters is not always chronological, but has been arranged so that, with one exception, every reference in one chapter to another is to an *earlier* chapter.

The first of all my published works was a brief note that my school teacher, Alan Robson, submitted in my name to the *Mathematical Gazette* (**13**, 1926, p. 205). It invited any reader to give an elementary verification of the following three results (which had arisen from my investigation of four-dimensional figures):

$$\int_0^{\pi/2} \text{arc cos} \, \frac{\cos x}{1 + 2 \cos x} \, dx = \frac{5\pi^2}{24},$$

$$\int_0^{\pi/3} \text{arc cos} \, \frac{1}{1 + 2 \cos x} \, dx = \frac{\pi^2}{8},$$

$$\int_0^{\pi/3} \text{arc cos} \, \frac{1 - \cos x}{2 \cos x} \, dx = \frac{11\pi^2}{72}.$$

The only reader who responded was Professor G. H. Hardy (then at Oxford), who admitted that he could never resist the challenge of a definite integral. He sent me three letters (totaling fourteen pages) and a seven-page manuscript. I can still recall the thrill of receiving these during my second month as a freshman in Cambridge. The first four pages of the manuscript contain an ingenious (but essentially elementary) verification of all three results.

The rest, and the letters, extended the results to a wider theory, and gave me references to N. I. Lobatschefsky and L. J. Rogers. About six years later, Professor H. F. Baker suggested that I should try to express the related integral

$$\int \frac{\text{arc tan} \sqrt{x}}{(a - x)\sqrt{b - x}} \, dx$$

as a power series. This idea enabled me to complete an elementary account of the wider theory, and then Hardy recommended it for publication in the *Oxford Quarterly Journal*. That paper is now reprinted (without change) as Chapter 1. It has been fruitfully extended by Johannes Böhm (*Zu Coxeters Integrationsmethode in gekrümmten Räumen*, Math. Nachrichten **27**, 1964, pp. 179–214).

In Acts 12:4, we read that the imprisoned Peter was guarded by "four quaternions." William Rowan Hamilton, in 1843, adopted this name *quaternion* for a hypercomplex number $a+bi+cj+dk$, where the symbols $i$, $j$, $k$ satisfy the relations

$$i^2 = j^2 = k^2 = ijk = -1.$$

These relations pleased him so much that he scratched them on a bridge near Dublin. Although any three quaternions satisfy the "associative law" $AB \cdot C = A \cdot BC$ (which allowed him to write $ijk$ without risk of confusion), two quaternions may fail to satisfy the "commutative law" $AB = BA$ (for instance, $ij = -ji$). Just as a complex number $a+bi$ is a combination of two real numbers $a$ and $b$, and a quaternion

$$a + bi + (c + di)\, j$$

is a combination of two complex numbers $a + bi$ and $c + di$, it is also possible to obtain a significant algebra by combining *two quaternions* (though *not*, as in the Bible, four quaternions). Such an algebra was discovered in that same year 1843 by Hamilton's friend J. T. Graves. A few years later, Arthur Cayley pointed out that three elements of the algebra of Graves may fail to satisfy the associative law! Some authors use the name "Cayley numbers," others "Cayley-Graves numbers." Hamilton (in 1848) proposed the non-controversial term *octaves*. If I had read F. van der Blij's *History of the octaves* (Simon Stevin **34**, 1961, pp. 106–125) before writing Chapter 2, I would have entitled it "Integral octaves."

The papers of Schläfli, which are referred to in these first two chapters, can be seen conveniently in his *Gesammelte mathematische Abhandlungen* (Vol. 1, 1950, pp. 167–392; Vol. 2, 1953, pp. 156–270).

Chapter 3 develops an idea and a notation which will reappear in Chapters 4, 5 and 10. The idea is the use of a set of mirrors (that is, a kaleidoscope) to derive the whole of a regular polytope (or, more generally, a "uniform" polytope) from a single one of its vertices. (A 1-dimensional polytope is a line segment, a 2-dimensional polytope is a polygon, a 3-dimensional polytope is a polyhedron, and so on. An $n$-dimensional polytope is said to be *regular* if it has a unique *center* from which all the vertices are at the same distance $_0R$, all the edges at the same distance $_1R$, all the plane faces at the same distance $_2R$, ... and all the *facets* or "cells" at the same distance $_{n-1}R$. When $n < 3$, a regular polytope is also called *uniform*. When $n \geq 3$, a polytope is said to be uniform if its facets are uniform and its vertices are all "equivalent." This means that its symmetry group is big enough to transform one vertex into the other vertices in turn.)

The notation consists of a "graph" whose vertices or "dots" or "nodes" represent the mirrors while its edges or "links" or "branches" indicate the angles between pairs of mirrors. For instance, an isolated node represents a single mirror, and such a node with a ring around it represents the line segment (or "one-dimensional polytope") whose vertices consist of a point (not lying on the mirror) and its image. Two nodes joined by a branch marked $p$ represent two mirrors inclined at an angle $\pi/p$. When one of the two nodes (say the first) is ringed, the graph represents the regular $p$-gon whose $p$ vertices consist of a point on one mirror (the second) and all its images in the dihedral kaleidoscope. When both nodes are ringed, the graph represents the regular $2p$-gon whose $2p$ vertices consist of a point equidistant from the two mirrors and all its images. For further details see Chapters 5 and 11 of my *Regular polytopes* (Macmillan, New York, 1963) and Chapter 5 of W. W. R. Ball, *Mathematical recreations and essays* (Macmillan, London and New York, 1967). For a discussion of polytopes that are not necessarily uniform, see Branko Grünbaum, *Convex polytopes* (Wiley, New York, 1967).

Chapter 4 describes an application of real projective geometry to the study of convex polyhedra whose faces are centrally symmetric. One infinite family of such polyhedra consists of the *polar zonohedra* (Chilton and Coxeter, *Amer. Math. Monthly* **70**, 1963, pp. 946–951). Three members of that family occur among the five *isozonohedra*, whose faces are congruent rhombi. Two of these three occur also among the five *parallelohedra*, which can be repeated by translation to fill the whole space without overlapping, and thus play a role in the geometry of numbers.

Chapter 5 generalizes the notion of a regular polyhedron in a manner that arose from some conversations in 1926 with my school friend John F. Petrie. According to the revised definition (which applies also to tessellations), a polyhedron is *regular* if it possesses two special symmetry operations: one which cyclically permutes the edges that are sides of one face, and one which cyclically permutes the edges that come together at one vertex, this vertex belonging to the "one face." (The last clause is necessary because otherwise a square pyramid would be regular!)

Chapter 6 was an invited address to the American Mathematical Society in 1948. The central idea is due to F. W. Levi, who considered an "even" graph whose nodes are colored alternately red and blue. These two colors should have been changed to white and black, respectively, for perfect agreement with the figures (which have been redrawn by my former student, Miss Nancy Myers). The special graphs called *cages* were first recognized by W. T. Tutte, who considers them further in his book, *Connectivity in Graphs* (University of Toronto Press, 1966). The cycle of three "Hessenberg triangles", discussed on pp. 433 and 434, is more properly called a cycle of three *Graves Triangles;* see Coxeter, *Projective Geometry* (Blaisdell, Waltham, Mass., 1967), p. 39.

Chapter 7 was one of six lectures offered during my tour of the Western States as a roving lecturer for the Mathematical Association of America in 1957. It partially answers a question raised many years earlier by Richard Brauer: Is there any "natural" geometric representation for the five simple groups of Mathieu? The rest of the answer appeared later in a sequel by J. A. Todd (*On representations of the Mathieu groups as collineation groups*, J. London Math. Soc., **34**, 1959, pp. 406–416).

Chapter 8 is related to the work of L. Fejes Tóth, *Regular figures* (Macmillan, New York, 1964). As it was printed in Hungary, some British and American readers may be puzzled at first by the decimal point appearing as a comma, and by the functions

$$\text{cosh, sinh, tanh, tan, cot}$$

appearing as

$$\text{ch,} \quad \text{sh,} \quad \text{th,} \quad \text{tg,} \quad \text{ctg.}$$

Chapter 9 was an invited address for the American Mathematical Society's Symposium on *Convexity* in Seattle. Its notation is related to that of Chapter 1 by the formula

$$F_4(\alpha) = \frac{2}{\pi^2} S\left(\frac{\pi}{2} - \alpha, \frac{\pi}{3}, \frac{\pi}{6}\right).$$

It deals with $n$-dimensional sphere-packing, a subject which surprisingly has an application to the theory of communication. This was pointed out by G. R. Lang, who has devised a program for evaluating the function $F_n(\alpha)$ quite rapidly.

Chapter 10 was an invited address to the International Congress of Mathematicians in Amsterdam, 1958. It exhibits the multiply-connected regular polyhedra $\{\frac{5}{2}, 5\}$ and $\{5, \frac{5}{2}\}$ of Kepler and Poinsot as the simplest members of families of non-Euclidean tessellations $\{m/2, m\}$ and $\{m, m/2\}$ ($m = 5, 7, 9, \ldots$), which are spherical or hyperbolic according as $m = 5$ or $m > 5$. There are analogous "star honeycombs" in hyperbolic 4-space. This investigation proceeds by comparing the contents of various figures in hyperbolic spaces by "cutting and pasting."

Chapter 11 was requested by Banesh Hoffmann for his collection of essays (*Perspectives in Geometry and Relativity*) in honor of Václav Hlavatý. It applies projective geometry to relativistic cosmology so as to solve an almost "practical" problem without using differential calculus or tensors.

Chapter 12 was commissioned by the George Washington University in the form of a pair of lectures given one morning and afternoon in May 1964. Besides summarizing many of the preceding chapters, this includes (on page 258) the amusing formula

$$v_{,} = c\sqrt{\{A\alpha,\ B\beta\}},$$

which expresses a velocity as the square root of a cross ratio.

Toronto, Canada                                          H. S. M. Coxeter
September 22, 1967

# CONTENTS

# THE BEAUTY OF
# GEOMETRY

**Explanatory Note**
Italic foot folios on all pages represent
the original pagination of reprinted articles.

# 1

# THE FUNCTIONS OF
# SCHLÄFLI AND LOBATSCHEFSKY

Reprinted, with the editors' permission,
from *The Quarterly Journal of Mathematics*,
Oxford Series, Vol. 6, No. 21, March, 1935.

# 1. Geometrical introduction

EVERY polygon can be divided into right-angled triangles. A convenient way of doing this is by joining any point to all the vertices and drawing perpendiculars from this same point on to all the sides. The number of triangles is then twice the number of sides. In calculating the area some of the triangles may have to be considered negative. But if the polygon is regular, we can simply take the special point to be the centre, and so make all the triangles equal.

Analogously, every polyhedron can be divided into *double-rectangular tetrahedra*.* To do this we join any point $A$ to the right-angled triangles that make up the faces, taking the special point in each face to be the foot of the perpendicular from $A$. The number of tetrahedra is then four times the number of edges. In calculating the volume some of the tetrahedra may have to be considered negative. But, if the polyhedron is regular, we can simply take $A$ to be the centre, and so make all the tetrahedra equal.†

The vertices of a double-rectangular tetrahedron can be called $A$, $B$, $C$, $D$ in such a way that the three edges $AB$, $BC$, $CD$ are all perpendicular. The opposite faces 1, 2, 3, 4 are then such that the dihedral angles (1 3), (1 4), (2 4) are right angles. In Euclidean space the volume is simply $\frac{1}{6}AB \cdot BC \cdot CD$. In spherical (or elliptic) space Schläfli‡ considered the volume as a function of the dihedral angles

$$\alpha = (1\,2), \qquad \beta = (2\,3), \qquad \gamma = (3\,4).$$

In hyperbolic space Lobatschefsky§ gave an explicit formula for the

---

* W. A. Wythoff, 'The rule of Neper in the four-dimensional space': *K. Akad. van Wet. te Amsterdam, Proc. Sect. of Sci.*, 9 (1906), 529–34.

† For the generalization to $n$ dimensions, see L. Schläfli, 'On the multiple integral $\int\int \dots \int dx\, dy \dots dz$ whose limits are $p_1 = a_1 x + b_1 y + \dots + h_1 z > 0$, $p_2 > 0$, ..., $p_n > 0$, and $x^2 + y^2 + \dots + z^2 < 1$': *Quart. J. of Math.* 3 (1860), 54–68, 97–108. Hereafter we shall refer to this work as **S**. ‡ S, 97.

§ N. I. Lobatschefsky, *Imaginäre Geometrie und ihre Anwendung auf einige Integrale*, translated into German by H. Liebmann (Leipzig, 1904). Hereafter we shall refer to this work as **L**.

volume in terms of four angles, of which three are these same dihedral angles (although he apparently failed to recognize $\beta$), while the fourth is related to them trigonometrically.

In the present paper we reconcile the work of these two great men. In fact, we define Schläfli's function by means of a series, and show that, when the angles are such as to make the tetrahedron hyperbolic, Lobatschefsky's formula follows. (Actually, we find it convenient to replace Schläfli's $\alpha$ and $\gamma$ by their complements.)

In § 8 we apply our results to a special problem of some interest.

## 2. Analytical introduction

Lobatschefsky expressed most of his results in terms of the function

$$L(x) = - \int_0^x \log\cos x \, dx \qquad (-\tfrac{1}{2}\pi \leqslant x \leqslant \tfrac{1}{2}\pi). \qquad (2.1)$$

Abel[*] and Rogers[†] have considered the related functions

$$\psi(x) = - \int_0^x \frac{\log(1-x)}{x} \, dx = \sum_1^\infty \frac{x^n}{n^2} \quad (|x| \leqslant 1), \qquad (2.2)$$

$$R(x) = -\tfrac{1}{2} \int_0^x \left( \frac{\log(1-x)}{x} + \frac{\log x}{1-x} \right) dx \qquad (2.3)$$

$$= \psi(x) + \tfrac{1}{2}\log x \log(1-x) \qquad (0 \leqslant x \leqslant 1).$$

For the extreme values of $x$, the term $\tfrac{1}{2}\log x \log(1-x)$ is replaced by its limit, so that $R(0) = 0, \qquad R(1) = \tfrac{1}{6}\pi^2.$

Rogers gives three other cases in which $R(x)$ is commensurable with $\pi^2$: $\quad R(\tfrac{1}{2}) = \tfrac{1}{12}\pi^2, \qquad R(\sigma) = \tfrac{1}{10}\pi^2, \qquad R(\sigma^2) = \tfrac{1}{15}\pi^2,$

where $\qquad\qquad\qquad \sigma = \tfrac{1}{2}(\sqrt{5}-1). \qquad (2.4)$

Putting $x = \sigma^2$, $y = \sigma$ in his (11),[‡] and $x = \sigma^3$ in his (12), we get

$$\tfrac{1}{6}\pi^2 = R(\sigma^3) + R(\tfrac{1}{2}\sigma^2) + \tfrac{1}{12}\pi^2,$$

$$2R(\sigma^3) = R(\sigma^6) + 2R(\tfrac{1}{2}\sigma^2).$$

Hence $\qquad\qquad\qquad 4R(\sigma^3) - R(\sigma^6) = \tfrac{1}{6}\pi^2.$

[*] N. H. Abel, *Œuvres complètes*, 2 (Christiania, 1881), 189–93. See also W. Spence, *Essay on the Theory of the Various Orders of Logarithmic Transcendents* (London, 1820); J. Bertrand, *Calcul intégral* (Paris, 1870), 216–18.

[†] L. J. Rogers, *Proc. London Math. Soc.* (2), 4 (1907), 169–89. We have changed Rogers's $L$ into $R$ to avoid confusion with Lobatschefsky's $L$.

[‡] Loc. cit., 173.

In § 4 of the present paper, we verify this result by a method that gives also some further identities, of which the most striking are*

$$3R(\sigma^4)+3R(\sigma^6)-R(\sigma^{12}) = \tfrac{2}{15}\pi^2,$$

$$15R(\sigma^4)+2R(\sigma^{10})-R(\sigma^{20}) = \tfrac{7}{15}\pi^2,$$

$$\sum_1^\infty \frac{\sigma^n}{n^2}\cos\tfrac{2}{5}n\pi = \tfrac{1}{100}\pi^2,$$

$$\sum_1^\infty \left(\frac{\cos x}{\cos y}\right)^n \frac{\cos nx\cos ny-\cos^2\tfrac{1}{2}n\pi}{n^2} = \tfrac{1}{2}(\tfrac{1}{2}\pi-x)^2$$

$$(0 \leqslant x \leqslant \pi;\ \cos^2 x \leqslant \cos^2 y).$$

Incidentally, we express the integral

$$\int_0^{\cdot} \frac{\arctan\sqrt{x}\,dx}{(a-x)\sqrt{(b-x)}} \qquad (x \leqslant b < a) \qquad (2.5)$$

as a series, and show how to evaluate it in a number of special cases.

In § 5, we find†    $L(\tfrac{1}{12}\pi) = \tfrac{2}{3}L(\tfrac{1}{4}\pi)-\tfrac{1}{4}L(\tfrac{1}{3}\pi),$

$$L(\tfrac{4}{15}\pi)-L(\tfrac{1}{15}\pi) = L(\tfrac{2}{5}\pi)-\tfrac{1}{3}L(\tfrac{1}{5}\pi)-\tfrac{2}{15}\pi\log 2.$$

## 3. The series for $S(\alpha, \beta, \gamma)$

We shall make use of the elementary identities

$$\sum_1^\infty \frac{\cos 2nx}{n^2} = (\tfrac{1}{2}\pi-x)^2-\tfrac{1}{12}\pi^2 \qquad (0 \leqslant x \leqslant \pi), \qquad (3.11)$$

$$\sum_1^\infty \frac{y^n}{n}\cos 2nx = -\tfrac{1}{2}\log(1-2y\cos 2x+y^2) \quad (-1 \leqslant y \leqslant 1), \qquad (3.12)$$

$$\sum_1^\infty \frac{y^n}{n}\sin 2nx = \arctan\left(\frac{1+y}{1-y}\tan x\right)-x \qquad (0 \leqslant x \leqslant \pi;\ -1 \leqslant y \leqslant 1), \qquad (3.13)$$

where the 'arctan' is supposed to lie between 0 and $\pi$. We define

$$S(\alpha,\beta,\gamma) = \sum_1^\infty \frac{(-X)^n}{n^2}(\cos 2n\alpha-\cos 2n\beta+\cos 2n\gamma-1)-\alpha^2+\beta^2-\gamma^2, \qquad (3.2)$$

where      $X = \dfrac{\sin\alpha\sin\gamma-D}{\sin\alpha\sin\gamma+D},$      (3.21)

$$D = \sqrt{(\cos^2\alpha\cos^2\gamma-\cos^2\beta)} \qquad (3.22)$$

(positive, zero, or of the same sign as $i\cos\beta$),

$$0 \leqslant \alpha \leqslant \tfrac{1}{2}\pi, \qquad 0 \leqslant \beta \leqslant \pi, \qquad 0 \leqslant \gamma \leqslant \tfrac{1}{2}\pi. \qquad (3.23)$$

* (4.84), (4.85), (4.82), (4.22).            † (5.91), (5.92).

The series obviously converges, since $|X| \leqslant 1$. It is proper to associate this function with the name of Schläfli, since, as we shall see in § 6, it is related to his $f(\alpha, \beta, \gamma)$ by the formula

$$S(\alpha, \beta, \gamma) = \tfrac{1}{2}\pi^2 f(\tfrac{1}{2}\pi - \alpha, \beta, \tfrac{1}{2}\pi - \gamma).$$

We see at once that $\quad S(\gamma, \beta, \alpha) = S(\alpha, \beta, \gamma).$ \hfill (3.31)

When $\alpha = 0$, we have $X = -1$, so that, by (3.11),

$$S(0, \beta, \gamma) = \sum_1^\infty \frac{-\cos 2n\beta + \cos 2n\gamma}{n^2} + \beta^2 - \gamma^2$$

$$= -(\tfrac{1}{2}\pi - \beta)^2 + (\tfrac{1}{2}\pi - \gamma)^2 + \beta^2 - \gamma^2$$

$$= \pi(\beta - \gamma). \tag{3.32}$$

When $\qquad \sin^2\alpha + \sin^2\gamma = \sin^2\beta,$ \hfill (3.4)

we have $D = \sin\alpha \sin\gamma$ and $X = 0$, so that in this case

$$S(\alpha, \beta, \gamma) = -\alpha^2 + \beta^2 - \gamma^2. \tag{3.41}*$$

When $\qquad \cos\alpha \cos\gamma = \cos\beta$ \hfill (3.5)

(which implies $\beta \leqslant \tfrac{1}{2}\pi$), we have $D = 0$ and $X = 1$, so that, by (3.11) (with $\tfrac{1}{2}\pi - x = \alpha, \beta, \gamma, 0$ in turn),

$$S(\alpha, \beta, \gamma) = 0. \tag{3.51}$$

In particular, $\qquad S(\tfrac{1}{4}\pi, \tfrac{1}{3}\pi, \tfrac{1}{4}\pi) = 0.$ \hfill (3.52)†

## 4. The case $\cos^2\alpha \cos^2\gamma \geqslant \cos^2\beta$

When $\cos^2\alpha \cos^2\gamma \geqslant \cos^2\beta$, it is clear that $D$, $X$, and $S(\alpha, \beta, \gamma)$ are real. In this case, differentiating (3.2), we have

$$dS(\alpha, \beta, \gamma) = \sum_1^\infty \frac{(-X)^n}{n}(\cos 2n\alpha - \cos 2n\beta + \cos 2n\gamma - 1)d\log X -$$

$$- 2\sum_1^\infty \frac{(-X)^n}{n}(\sin 2n\alpha\, d\alpha - \sin 2n\beta\, d\beta + \sin 2n\gamma\, d\gamma) -$$

$$- 2(\alpha\, d\alpha - \beta\, d\beta + \gamma\, d\gamma).$$

Putting $-X$ for $y$ in (3.12), we see that the coefficient of $d\log X$ is

$$-\tfrac{1}{2}\log\frac{\{(1+X)^2\cos^2\alpha + (1-X)^2\sin^2\alpha\}\{(1+X)^2\cos^2\gamma + (1-X)^2\sin^2\gamma\}}{\{(1+X)^2\cos^2\beta + (1-X)^2\sin^2\beta\}(1+X)^2}$$

$$= -\tfrac{1}{2}\log\frac{(\sin^2\gamma \cos^2\alpha + D^2)(\sin^2\alpha \cos^2\gamma + D^2)}{\sin^2\alpha \sin^2\gamma \cos^2\beta + D^2\sin^2\beta} = 0.$$

\* S, 60 (4). \hfill † S, 63 (7).

Hence, by (3.13),

$$-\tfrac{1}{2}dS(\alpha,\beta,\gamma)$$

$$= \arctan\left(\frac{1-X}{1+X}\tan\alpha\right) d\alpha -\arctan\left(\frac{1-X}{1+X}\tan\beta\right) d\beta +$$

$$+\arctan\left(\frac{1-X}{1+X}\tan\gamma\right) d\gamma$$

$$= \arctan\frac{D}{\cos\alpha\sin\gamma} d\alpha -\arctan\frac{D\tan\beta}{\sin\alpha\sin\gamma} d\beta +\arctan\frac{D}{\sin\alpha\cos\gamma} d\gamma$$

$$\tag{4.1}$$

$$= \arccos\frac{\cos\alpha\sin\gamma}{\sqrt{(\cos^2\alpha-\cos^2\beta)}} d\alpha -$$

$$-\arccos\frac{\sin\alpha\cos\beta\sin\gamma}{\sqrt{(\cos^2\alpha-\cos^2\beta)}\sqrt{(\cos^2\gamma-\cos^2\beta)}} d\beta +$$

$$+\arccos\frac{\sin\alpha\cos\gamma}{\sqrt{(\cos^2\gamma-\cos^2\beta)}} d\gamma. \tag{4.11}$$

We can now prove that

$$S(\alpha,\tfrac{1}{2}\pi,\gamma) = 2(\tfrac{1}{2}\pi-\alpha)(\tfrac{1}{2}\pi-\gamma). \tag{4.2*}$$

By (3.41), this is true when $\gamma = \tfrac{1}{2}\pi-\alpha$. It will therefore be sufficient to prove the differential relation

$$dS(\alpha,\tfrac{1}{2}\pi,\gamma) = -2(\tfrac{1}{2}\pi-\gamma) d\alpha -2(\tfrac{1}{2}\pi-\alpha) d\gamma.$$

By (4.11),

$$-\tfrac{1}{2}dS(\alpha,\tfrac{1}{2}\pi,\gamma) = (\tfrac{1}{2}\pi-\gamma) d\alpha +(\tfrac{1}{2}\pi-\alpha) d\gamma,$$

as required.

On putting $\beta = \tfrac{1}{2}\pi$ in (3.2), we obtain the interesting identity

$$\sum_1^\infty \frac{\cos^n(\alpha+\gamma)}{\cos^n(\alpha-\gamma)} \frac{\cos 2n\alpha + \cos 2n\gamma -1-\cos n\pi}{n^2} = (\tfrac{1}{2}\pi-\alpha-\gamma)^2$$

$$(0 \leqslant \alpha \leqslant \tfrac{1}{2}\pi; 0 \leqslant \gamma \leqslant \tfrac{1}{2}\pi), \tag{4.21}$$

i.e.

$$\sum_1^\infty \left(\frac{\cos x}{\cos y}\right)^n \frac{\cos nx\cos ny - \cos^2 \tfrac{1}{2}n\pi}{n^2} = \tfrac{1}{2}(\tfrac{1}{2}\pi-x)^2$$

$$(0 \leqslant x \leqslant \pi; \cos^2 x \leqslant \cos^2 y). \tag{4.22}$$

We proceed to prove the two identities†

$$S(2\alpha-\tfrac{1}{2}\pi,\tfrac{1}{2}\pi-\alpha,\tfrac{1}{6}\pi) = 4S(\alpha,\tfrac{1}{3}\pi,\tfrac{1}{6}\pi) \quad (\tfrac{1}{4}\pi \leqslant \alpha \leqslant \arccos\sqrt{\tfrac{1}{3}}), \tag{4.31}$$

$$S(\alpha,\pi-2\alpha,\alpha) = 6S(\alpha,\tfrac{1}{3}\pi,\tfrac{1}{6}\pi) \quad (0 \leqslant \alpha \leqslant \arccos\sqrt{\tfrac{1}{3}}). \tag{4.32}$$

---

* Schläfli, *Quart. J. of Math.* 2 (1858), 291. ($n = 4$, $m = 2$.)

† **S**, 64 (with $n = 4$). For the similar result

$$S(\alpha,\tfrac{1}{2}\pi-\alpha,\tfrac{1}{4}\pi) = 3S(\alpha,\tfrac{1}{4}\pi,\tfrac{1}{6}\pi) \qquad (0 \leqslant \alpha \leqslant \arccos\sqrt{\tfrac{2}{3}}),$$

see **S**, 67.

By (3.32), (3.41), and (4.2), these are true when $\alpha = \frac{1}{4}\pi$. It will therefore be sufficient to prove the corresponding differential relations. Let

$$\eta = \arccos \frac{\cos\alpha}{\sqrt{(4\cos^2\alpha - 1)}}.$$

Then, by (4.11),

$$-\tfrac{1}{2}dS(2\alpha - \tfrac{1}{2}\pi, \tfrac{1}{2}\pi - \alpha, \tfrac{1}{6}\pi) = \eta\, d(2\alpha - \tfrac{1}{2}\pi) - 2\eta\, d(\tfrac{1}{2}\pi - \alpha) = 4\eta\, d\alpha,$$

$$-\tfrac{1}{2}dS(\alpha, \tfrac{1}{3}\pi, \tfrac{1}{6}\pi) = \eta\, d\alpha,$$

$$-\tfrac{1}{2}dS(\alpha, \pi - 2\alpha, \alpha) = \eta\, d\alpha - 2\eta\, d(\pi - 2\alpha) + \eta\, d\alpha = 6\eta\, d\alpha.$$

Thus the identities are proved. (The condition $\alpha \leqslant \arccos\sqrt{\tfrac{1}{3}}$ secures the reality of $\eta$. In § 5 we shall see that this is unnecessary. The essential condition is $\alpha \leqslant \tfrac{1}{3}\pi$.)

So long as $\cos^2\alpha \cos^2\gamma \geqslant \cos^2\beta$, $X$ is unaltered when $\beta$ is replaced by its supplement. Hence, by (3.2),

$$S(\alpha, \pi - \beta, \gamma) - S(\alpha, \beta, \gamma) = (\pi - \beta)^2 - \beta^2$$
$$= \pi(\pi - 2\beta). \tag{4.4}$$

In particular, $\quad S(\tfrac{1}{6}\pi, \tfrac{2}{3}\pi, \tfrac{1}{6}\pi) - S(\tfrac{1}{6}\pi, \tfrac{1}{3}\pi, \tfrac{1}{6}\pi) = \tfrac{1}{3}\pi^2.$

But, by (4.32), $\quad S(\tfrac{1}{6}\pi, \tfrac{2}{3}\pi, \tfrac{1}{6}\pi) = 6S(\tfrac{1}{6}\pi, \tfrac{1}{3}\pi, \tfrac{1}{6}\pi).$

Therefore $\quad\quad\quad\quad S(\tfrac{1}{6}\pi, \tfrac{1}{3}\pi, \tfrac{1}{6}\pi) = \tfrac{1}{15}\pi^2. \tag{4.41}*$

By (3.41), $\quad\quad\quad\quad S(\tfrac{1}{4}\pi, \tfrac{1}{3}\pi, \tfrac{1}{6}\pi) = \tfrac{1}{48}\pi^2, \tag{4.51}$

$$S(\tfrac{1}{6}\pi, \tfrac{1}{4}\pi, \tfrac{1}{6}\pi) = \tfrac{1}{144}\pi^2, \tag{4.52}$$

$$S(\tfrac{3}{10}\pi, \tfrac{1}{3}\pi, \tfrac{1}{10}\pi) = \tfrac{1}{90}\pi^2,$$

$$S(\tfrac{1}{10}\pi, \tfrac{1}{5}\pi, \tfrac{1}{6}\pi) = \tfrac{1}{450}\pi^2, \tag{4.54}$$

$$S(\tfrac{3}{10}\pi, \tfrac{2}{5}\pi, \tfrac{1}{6}\pi) = \tfrac{19}{450}\pi^2.$$

By (4.54), putting $\alpha = \tfrac{3}{10}\pi$ in (4.31),

$$S(\tfrac{3}{10}\pi, \tfrac{1}{3}\pi, \tfrac{1}{6}\pi) = \tfrac{1}{1800}\pi^2, \tag{4.56}$$

whence, by (4.32), $\quad S(\tfrac{3}{10}\pi, \tfrac{2}{5}\pi, \tfrac{3}{10}\pi) = \tfrac{1}{300}\pi^2. \tag{4.57}$

When written out in full, these last two results become

$$\sum_1^\infty \frac{(-\sigma)^n}{n^2}(\cos\tfrac{3}{5}n\pi - \cos\tfrac{2}{3}n\pi + \cos\tfrac{1}{3}n\pi - 1) = \tfrac{13}{1800}\pi^2 \quad \left(\sigma = \frac{\sqrt5 - 1}{2}\right),$$

$$\sum_1^\infty \frac{(-\sigma)^n}{n^2}(2\cos\tfrac{3}{5}n\pi - \cos\tfrac{4}{5}n\pi - 1) = \tfrac{7}{300}\pi^2,$$

* S, 98.

i.e.

$$\sum_{1}^{\infty} \frac{\sigma^n}{n^2}(\cos\tfrac{2}{5}n\pi - \cos\tfrac{1}{3}n\pi + \cos\tfrac{2}{3}n\pi - \cos n\pi) = \tfrac{13}{1800}\pi^2, \quad (4.61)$$

$$\sum_{1}^{\infty} \frac{\sigma^n}{n^2}(2\cos\tfrac{2}{5}n\pi - \cos\tfrac{1}{5}n\pi - \cos n\pi) = \tfrac{7}{300}\pi^2. \quad (4.62)$$

Putting $\alpha = \tfrac{1}{5}\pi$, $\gamma = \tfrac{1}{10}\pi$ in (4.21),

$$\sum_{1}^{\infty} \frac{\sigma^n}{n^2}(\cos\tfrac{2}{5}n\pi + \cos\tfrac{1}{5}n\pi - 1 - \cos n\pi) = \tfrac{1}{25}\pi^2. \quad (4.63)$$

On subtracting each of (4.61), (4.62) from this, we obtain

$$S(\tfrac{1}{10}\pi, \tfrac{1}{3}\pi, \tfrac{1}{6}\pi) = \tfrac{191}{1800}\pi^2, \quad (4.64)$$

$$S(\tfrac{1}{10}\pi, \tfrac{1}{5}\pi, \tfrac{1}{10}\pi) = \tfrac{11}{300}\pi^2. \quad (4.65)$$

Again, (4.62) and (4.63) yield

$$\sum_{1}^{\infty} \frac{\sigma^n}{n^2}\cos\tfrac{1}{5}n\pi = \frac{1}{3}\sum_{1}^{\infty} \frac{\sigma^n}{n^2}(2+\cos n\pi) + \tfrac{17}{900}\pi^2$$

$$= \tfrac{2}{3}\psi(\sigma) + \tfrac{1}{3}\psi(-\sigma) + \tfrac{17}{900}\pi^2, \quad (4.66)$$

$$\sum_{1}^{\infty} \frac{\sigma^n}{n^2}\cos\tfrac{2}{5}n\pi = \frac{1}{3}\sum_{1}^{\infty} \frac{\sigma^n}{n^2}(1+2\cos n\pi) + \tfrac{19}{900}\pi^2$$

$$= \tfrac{1}{3}\psi(\sigma) + \tfrac{2}{3}\psi(-\sigma) + \tfrac{19}{900}\pi^2. \quad (4.67)$$

Evidently

$$\sum_{1}^{\infty} \frac{x^n}{n^2}\cos\tfrac{2}{3}n\pi = -\frac{1}{2}\sum_{1}^{\infty} \frac{x^n}{n^2} + \frac{3}{2}\sum_{1}^{\infty} \frac{x^{3n}}{(3n)^2}$$

$$= -\tfrac{1}{2}\psi(x) + \tfrac{1}{6}\psi(x^3), \quad (4.71)$$

$$\sum_{1}^{\infty} \frac{x^n}{n^2}\cos\tfrac{1}{3}n\pi = \sum_{1}^{\infty} \frac{(-x)^n}{n^2}\cos\tfrac{2}{3}n\pi$$

$$= -\tfrac{1}{2}\psi(-x) + \tfrac{1}{6}\psi(-x^3). \quad (4.72)$$

Hence, by (4.61) and (4.67),

$$\psi(\sigma^3) - \psi(-\sigma^3) = \psi(\sigma) - \psi(-\sigma) - \tfrac{1}{12}\pi^2.^* \quad (4.73)$$

Writing out (4.41) in full,

$$\sum_{1}^{\infty} \frac{\sigma^{2n}}{n^2}(2\cos\tfrac{1}{3}n\pi - \cos\tfrac{2}{3}n\pi - 1) = \tfrac{1}{90}\pi^2,$$

---

\* Bertrand showed that

$$\psi(\sigma) = \tfrac{1}{10}\pi^2 - (\log\sigma)^2, \qquad \psi(\sigma^2) = \tfrac{1}{15}\pi^2 - (\log\sigma)^2,$$

whence $\quad \psi(-\sigma) = \tfrac{1}{2}\psi(\sigma^2) - \psi(\sigma) = -\tfrac{1}{15}\pi^2 + \tfrac{1}{2}(\log\sigma)^2.$

whence, by (4.71) and (4.72) with $x = \sigma^2$,

$$-\psi(\sigma^6) + 2\psi(-\sigma^6) - 6\psi(-\sigma^2) = 3\psi(\sigma^2) + \tfrac{1}{15}\pi^2. \qquad (4.74)$$

Since $\cos\tfrac{4}{5}n\pi + \cos\tfrac{2}{5}n\pi = 2$ or $-\tfrac{1}{2}$ according as $n$ is or is not divisible by 5,

$$\sum_1^\infty \frac{x^n}{n^2}(\cos\tfrac{4}{5}n\pi + \cos\tfrac{2}{5}n\pi) = -\tfrac{1}{2}\psi(x) + \tfrac{1}{10}\psi(x^5).$$

Putting $\alpha = \tfrac{2}{5}\pi$, $\gamma = \tfrac{1}{5}\pi$ in (4.21),

$$\sum_1^\infty \frac{(-\sigma^2)^n}{n^2}(\cos\tfrac{4}{5}n\pi + \cos\tfrac{2}{5}n\pi - 1 - \cos n\pi) = \tfrac{1}{100}\pi^2.$$

Hence $\qquad \psi(-\sigma^{10}) - 15\psi(-\sigma^2) = 10\psi(\sigma^2) + \tfrac{1}{10}\pi^2. \qquad (4.75)$

Since obviously $\qquad \psi(-x) = \tfrac{1}{2}\psi(x^2) - \psi(x),$

we may write (4.66), (4.67), (4.73), (4.74), (4.75) in the form

$$\sum_1^\infty \frac{\sigma^n}{n^2}\cos\tfrac{1}{5}n\pi = \tfrac{1}{3}\psi(\sigma) + \tfrac{1}{6}\psi(\sigma^2) + \tfrac{17}{900}\pi^2,$$

$$\sum_1^\infty \frac{\sigma^n}{n^2}\cos\tfrac{2}{5}n\pi = -\tfrac{1}{3}\psi(\sigma) + \tfrac{1}{3}\psi(\sigma^2) + \tfrac{19}{900}\pi^2,$$

$$4\psi(\sigma^3) - \psi(\sigma^6) = 4\psi(\sigma) - \psi(\sigma^2) - \tfrac{1}{6}\pi^2,$$

$$3\psi(\sigma^4) + 3\psi(\sigma^6) - \psi(\sigma^{12}) = 3\psi(\sigma^2) - \tfrac{1}{15}\pi^2,$$

$$15\psi(\sigma^4) + 2\psi(\sigma^{10}) - \psi(\sigma^{20}) = 10\psi(\sigma^2) - \tfrac{1}{5}\pi^2.$$

The last four of these identities remain valid when every $\psi$ is replaced by $R$—see (2.3)—since the logarithmic terms cancel in virtue of the relations*

$$1 - \sigma^{2n} = (\sigma^{-n} - \sigma^n)\sigma^n, \qquad 1 - \sigma^3 = 2\sigma^2.$$

Hence

$$\sum_1^\infty \frac{\sigma^n}{n^2}\cos\tfrac{1}{5}n\pi = \tfrac{19}{300}\pi^2 - \tfrac{1}{2}(\log\sigma)^2, \qquad (4.81)$$

$$\sum_1^\infty \frac{\sigma^n}{n^2}\cos\tfrac{2}{5}n\pi = \tfrac{1}{100}\pi^2, \qquad (4.82)$$

$$4R(\sigma^3) - R(\sigma^6) = \tfrac{1}{6}\pi^2, \qquad (4.83)$$

$$3R(\sigma^4) + 3R(\sigma^6) - R(\sigma^{12}) = \tfrac{2}{15}\pi^2, \qquad (4.84)$$

$$15R(\sigma^4) + 2R(\sigma^{10}) - R(\sigma^{20}) = \tfrac{7}{15}\pi^2. \qquad (4.85)$$

\* $\sigma^{-n} - \sigma^n = 1, \quad \sqrt5, \quad 4, \quad 11, \quad 8\sqrt5, \quad 55\sqrt5,$

when $\qquad n = 1, \quad 2, \quad 3, \quad 5, \quad 6, \quad 10.$

Let us now consider the integral (2.5). The transformation

$$\alpha = \arctan \sqrt{\frac{b-x}{a-b}}$$

gives
$$\sqrt{(a-b)} \int_0^{} \frac{\arctan \sqrt{x}\, dx}{(a-x)\sqrt{(b-x)}} = -2 \int_{x=0} \arctan \sqrt{x}\, d\alpha.$$

But, by (4.1),

$$S(\alpha, \beta, \gamma) = -2 \int_{D=0} \arctan \frac{D}{\cos \alpha \sin \gamma}\, d\alpha, \qquad (4.91)$$

$D$ being given by (3.22). Comparing these results,

$$\sqrt{(a-b)} \int_0^{} \frac{\arctan \sqrt{x}\, dx}{(a-x)\sqrt{(b-x)}}$$
$$= S\!\left(\arctan \sqrt{\frac{b-x}{a-b}}, \arctan \sqrt{\frac{b+1}{a-b}}, \arctan \sqrt{\frac{1}{a}}\right)$$
$$(x \leqslant b < a). \quad (4.92)$$

By (3.2) we can now express the integral as a series, with

$$X = \frac{\sqrt{(b-x)} - \sqrt{(a-b)}\sqrt{x}}{\sqrt{(b-x)} + \sqrt{(a-b)}\sqrt{x}}. \qquad (4.93)$$

Several of the properties of $S(\alpha, \beta, \gamma)$ lead to interesting definite integrals:* (3.32) gives

$$\sqrt{(a-b)} \int_0^b \frac{\arctan \sqrt{x}\, dx}{(a-x)\sqrt{(b-x)}} = \pi\!\left(\arctan \sqrt{\frac{b+1}{a-b}} - \arctan \sqrt{\frac{1}{a}}\right) \quad (b < a);$$
$$(4.94)\dagger$$

(3.41) gives

$$\sqrt{(a-b)} \int_0^{b/(a-b+1)} \frac{\arctan \sqrt{x}\, dx}{(a-x)\sqrt{(b-x)}}$$
$$= -\!\left(\arctan \sqrt{\frac{b}{a-b+1}}\right)^2 + \left(\arctan \sqrt{\frac{b+1}{a-b}}\right)^2 - \left(\arctan \sqrt{\frac{1}{a}}\right)^2.$$
$$(4.95)$$

The case when $a = 3$, $b = 2$ is of special interest. By (4.91),

$$2S(\tfrac{1}{2}\theta, \tfrac{1}{3}\pi, \tfrac{1}{6}\pi) = \int^{\pi - \operatorname{arcsec} 3} \phi\, d\theta,$$

where
$$\sec \theta + \sec \phi + 2 = 0. \qquad (4.96)$$

---

* Coxeter, *Math. Gazette*, 13 (1926), 205.

† **L**, 42 (106) $(x = \tan^2 \phi)$. See also Bierens de Haan's *Tables* (1867), **335** (224.9).

In particular,

$$\int_0^{\pi-\operatorname{arcsec}3} \phi\, d\theta = 2S(0,\tfrac13\pi,\tfrac16\pi) = \tfrac13\pi^2.$$

Hence, by (4.64), (4.41), (4.51), (4.56) in turn,

$$\int_0 \phi\, d\theta = \tfrac{109}{900}\pi^2, \quad \tfrac15\pi^2, \quad \tfrac{7}{24}\pi^2, \quad \tfrac{299}{900}\pi^2, \tag{4.97}$$

when
$$\theta = \tfrac15\pi, \quad \tfrac13\pi, \quad \tfrac12\pi, \quad \tfrac35\pi.$$

## 5. The case $\cos\alpha\cos\gamma \leqslant \cos\beta,\ \alpha\leqslant\beta,\ \gamma\leqslant\beta$

The relation connecting Abel's $\psi$-function and Lobatschefsky's $L$-function is found as follows. Putting $y=-1$ in (3.12), and integrating with respect to $x$,

$$\frac12\sum_1^\infty \frac{(-1)^n}{n^2}\sin 2nx = -\int_0^x \log(2\cos x)\, dx \qquad (-\tfrac12\pi \leqslant x \leqslant \tfrac12\pi), \tag{5.1}$$

i.e.*
$$\frac{1}{4i}\{\psi(-e^{2ix})-\psi(-e^{-2ix})\} = L(x)-x\log 2.$$

But, by (3.11) with $\tfrac12\pi-x$ in place of $x$,
$$\tfrac12\{\psi(-e^{2ix})+\psi(-e^{-2ix})\} = x^2-\tfrac{1}{12}\pi^2.$$

Hence
$$\tfrac12 i\psi(-e^{-2ix}) = L(x)-x\log 2+\tfrac12 i(x^2-\tfrac{1}{12}\pi^2) \qquad (-\tfrac12\pi \leqslant x \leqslant \tfrac12\pi). \tag{5.2}$$

Of course
$$L(-x) = -L(x).$$

When
$$\cos\alpha\cos\gamma \leqslant \cos\beta$$

(implying $\beta \leqslant \tfrac12\pi$, since we never allow $\alpha$ or $\gamma$ to be obtuse), we define the real angle

$$\delta = \arctan\frac{D}{i\sin\alpha\sin\gamma} = \arctan\frac{\sqrt{(\cos^2\beta-\cos^2\alpha\cos^2\gamma)}}{\sin\alpha\sin\gamma}$$

$$= \arccos\frac{\sin\alpha\sin\gamma}{\sqrt{(\sin^2\alpha-\sin^2\beta+\sin^2\gamma)}}, \tag{5.3}$$

so that
$$X = e^{-2i\delta}.$$

By (3.2), replacing the cosines by pairs of exponentials,

$$S(\alpha,\beta,\gamma) = \tfrac12\{\psi(-e^{-2i(\delta+\alpha)})+\psi(-e^{-2i(\delta-\alpha)})\}-$$
$$-\tfrac12\{\psi(-e^{-2i(\delta+\beta)})+\psi(-e^{-2i(\delta-\beta)})\}+$$
$$+\tfrac12\{\psi(-e^{-2i(\delta+\gamma)})+\psi(-e^{-2i(\delta-\gamma)})\}-$$
$$-\psi(-e^{-2i\delta})-\alpha^2+\beta^2-\gamma^2.$$

* L, 54. There is a mistaken sign in his (11).

Before we can substitute for the $\psi$'s from (5.2), we must be sure that $\delta+\alpha$, $\delta+\beta$, $\delta+\gamma$ are not obtuse. This necessitates

$$\alpha, \gamma \leqslant \beta. \tag{5.31}$$

When we have substituted, the terms involving $\log 2$ and $\pi^2$ cancel among themselves, while the remaining elementary terms cancel with $-\alpha^2+\beta^2-\gamma^2$. Hence, finally,

$$iS(\alpha,\beta,\gamma) = L(\delta+\alpha)+L(\delta-\alpha)-L(\delta+\beta)-L(\delta-\beta)+ \\ +L(\delta+\gamma)+L(\delta-\gamma)-2L(\delta). \tag{5.4}$$

When $\gamma = \beta$, (5.3) shows that $\delta = \tfrac{1}{2}\pi-\beta$, whence

$$i\,S(\alpha,\beta,\beta) = L(\tfrac{1}{2}\pi-\beta+\alpha)+L(\tfrac{1}{2}\pi-\beta-\alpha)-2L(\tfrac{1}{2}\pi-\beta). \tag{5.41}$$

By Lobatschefsky's (14)* (or by integrating the identity

$$\log\sin x - \log\sin 2x = -\log 2-\log\cos x),$$

$$L(\tfrac{1}{2}\pi-x)-\tfrac{1}{2}L(\tfrac{1}{2}\pi-2x) = (\tfrac{1}{4}\pi-x)\log 2+L(x) \quad (0 \leqslant x \leqslant \tfrac{1}{2}\pi). \tag{5.5}$$

In particular, since $L(0) = 0$,

$$L(\tfrac{1}{2}\pi) = \tfrac{1}{2}\pi\log 2. \tag{5.51}$$

Hence, by (5.41) and (5.1),

$$i\,S(\alpha,\alpha,\alpha) = 2\alpha\log 2-2L(\alpha) = \sum_{1}^{\infty}\frac{(-1)^{n-1}}{n^2}\sin 2n\alpha, \tag{5.6}$$

i.e., by (5.2),

$$S(\alpha,\alpha,\alpha) = \alpha^2-\tfrac{1}{12}\pi^2-\psi(-e^{-2i\alpha}) \\ = \psi(-e^{2i\alpha})-\alpha^2+\tfrac{1}{12}\pi^2. \tag{5.61}$$

By (5.4) and (2.1), and then (5.3),

$$i\frac{\partial}{\partial\delta}S(\alpha,\beta,\gamma) = -\log\frac{\cos(\delta+\alpha)\cos(\delta-\alpha)\cos(\delta+\gamma)\cos(\delta-\gamma)}{\cos(\delta+\beta)\cos(\delta-\beta)\cos^2\delta} \\ = -\log\frac{(\cos^2\delta-\sin^2\alpha)(\cos^2\delta-\sin^2\gamma)}{(\cos^2\delta-\sin^2\beta)\cos^2\delta} = 0.$$

Hence

$$i\,dS(\alpha,\beta,\gamma) = \log\frac{\cos(\delta-\alpha)}{\cos(\delta+\alpha)}\,d\alpha+\log\frac{\cos(\delta+\beta)}{\cos(\delta-\beta)}\,d\beta+\log\frac{\cos(\delta-\gamma)}{\cos(\delta+\gamma)}\,d\gamma. \tag{5.7}$$

By using (5.7) instead of (4.11), we can extend the range of validity of (4.31) and (4.32) up to $\alpha = \tfrac{1}{3}\pi$. (The condition $\alpha \leqslant \tfrac{1}{3}\pi$ comes from (5.31).)

Of greater interest is the similar identity

$$S(\alpha,\tfrac{1}{2}\pi-\alpha,\tfrac{1}{4}\pi) = 3S(\alpha,\tfrac{1}{4}\pi,\tfrac{1}{6}\pi) \quad (0 \leqslant \alpha \leqslant \tfrac{1}{4}\pi). \tag{5.8}$$

* L, 54.

When $\alpha \leqslant \arccos\sqrt{\tfrac{2}{3}}$, this is proved by means of (4.11). For our present purposes we take $\alpha \geqslant \arccos\sqrt{\tfrac{2}{3}}$. On both sides of (5.8),

$$\delta = \arctan\sqrt{(2-\cot^2\alpha)}. \tag{5.81}$$

Now (5.8) is true when $\alpha = \arccos\sqrt{\tfrac{2}{3}}$, since then both sides vanish. By (5.7), therefore, it will be sufficient to prove that

$$\log\frac{\cos(\delta-\alpha)}{\cos(\delta+\alpha)} - \log\frac{\cos(\delta+\tfrac{1}{2}\pi-\alpha)}{\cos(\delta-\tfrac{1}{2}\pi+\alpha)} = 3\log\frac{\cos(\delta-\alpha)}{\cos(\delta+\alpha)}.$$

By (5.81),

$$-\log\frac{\cos(\delta+\tfrac{1}{2}\pi-\alpha)}{\cos(\delta-\tfrac{1}{2}\pi+\alpha)} = \log\frac{\sin(\alpha+\delta)}{\sin(\alpha-\delta)} = \log\frac{\tan\alpha+\tan\delta}{\tan\alpha-\tan\delta}$$

$$= \log\frac{\tan\alpha+\sqrt{(2-\cot^2\alpha)}}{\tan\alpha-\sqrt{(2-\cot^2\alpha)}} = \log\left(\frac{\cot\alpha+\sqrt{(2-\cot^2\alpha)}}{\cot\alpha-\sqrt{(2-\cot^2\alpha)}}\right)^2$$

$$= 2\log\frac{\cot\alpha+\tan\delta}{\cot\alpha-\tan\delta} = 2\log\frac{\cos(\delta-\alpha)}{\cos(\delta+\alpha)},$$

as required.

Putting $x = \tfrac{1}{4}\pi-\delta$ in (5.5),

$$L(\delta+\tfrac{1}{4}\pi)+L(\delta-\tfrac{1}{4}\pi) = \delta\log 2+\tfrac{1}{2}L(2\delta) \qquad (-\tfrac{1}{4}\pi \leqslant \delta \leqslant \tfrac{1}{4}\pi).$$

By (5.8) and (5.4), therefore,

$$L(\tfrac{1}{2}\pi-\alpha+\delta)-L(\tfrac{1}{2}\pi-\alpha-\delta)+2L(\alpha+\delta)-2L(\alpha-\delta)+$$
$$+3L(\tfrac{1}{6}\pi+\delta)-3L(\tfrac{1}{6}\pi-\delta)-4L(\delta)-2L(2\delta) = 4\delta\log 2$$
$$(0 \leqslant \delta \leqslant \tfrac{1}{4}\pi;\ \alpha = \operatorname{arccot}\sqrt{(2-\tan^2\delta)}). \tag{5.9}$$

When $\alpha = \delta = \tfrac{1}{4}\pi$, this becomes

$$3L(\tfrac{5}{12}\pi)+3L(\tfrac{1}{12}\pi)-4L(\tfrac{1}{4}\pi) = \tfrac{1}{2}\pi\log 2.$$

But, by (5.5) (with $x = \tfrac{1}{12}\pi$),

$$L(\tfrac{5}{12}\pi)-L(\tfrac{1}{12}\pi)-\tfrac{1}{2}L(\tfrac{1}{3}\pi) = \tfrac{1}{6}\pi\log 2.$$

Hence 
$$12L(\tfrac{1}{12}\pi) = 8L(\tfrac{1}{4}\pi)-3L(\tfrac{1}{3}\pi),$$

i.e. 
$$L(\tfrac{1}{12}\pi) = \tfrac{2}{3}L(\tfrac{1}{4}\pi)-\tfrac{1}{4}L(\tfrac{1}{3}\pi). \tag{5.91}$$

When $\delta = \tfrac{1}{10}\pi$, so that $\alpha = \tfrac{1}{5}\pi$, (5.9) becomes

$$3L(\tfrac{4}{15}\pi)-3L(\tfrac{1}{15}\pi)+2L(\tfrac{3}{10}\pi)-6L(\tfrac{1}{10}\pi)+L(\tfrac{2}{5}\pi)-3L(\tfrac{1}{5}\pi) = \tfrac{2}{5}\pi\log 2.$$

But, by (5.5) (with $x = \tfrac{1}{10}\pi, \tfrac{1}{5}\pi$),

$$L(\tfrac{2}{5}\pi) = \tfrac{1}{2}L(\tfrac{3}{10}\pi)+L(\tfrac{1}{10}\pi)+\tfrac{3}{20}\pi\log 2,$$
$$L(\tfrac{1}{5}\pi) = L(\tfrac{3}{10}\pi)-\tfrac{1}{2}L(\tfrac{1}{10}\pi)-\tfrac{1}{20}\pi\log 2.$$

Hence
$$3L(\tfrac{4}{15}\pi)-3L(\tfrac{1}{15}\pi) = \tfrac{1}{2}\{L(\tfrac{3}{10}\pi)+7L(\tfrac{1}{10}\pi)+\tfrac{1}{5}\pi\log 2\}$$
$$= 3L(\tfrac{2}{5}\pi)-L(\tfrac{1}{5}\pi)-\tfrac{2}{5}\pi\log 2,$$

i.e. 
$$L(\tfrac{4}{15}\pi)-L(\tfrac{1}{15}\pi) = L(\tfrac{2}{5}\pi)-\tfrac{1}{3}L(\tfrac{1}{5}\pi)-\tfrac{2}{15}\pi\log 2. \tag{5.92}$$

## 6. Spherical or elliptic space

Schläfli showed that a double-rectangular tetrahedron of angles $\alpha$, $\beta$, $\gamma$ exists in spherical space if $\sin^2\alpha \sin^2\gamma > \cos^2\beta$, and in Euclidean space if $\sin\alpha \sin\gamma = \cos\beta$. In the former case he called the volume

$$\tfrac{1}{8}\pi^2 f(\alpha,\beta,\gamma).$$

Since an infinitesimal spherical figure has the same angles as a Euclidean figure, this function must vanish when*

$$\sin\alpha \sin\gamma = \cos\beta.$$

Its differential† is given by

$$\tfrac{1}{4}\pi^2 \, df(\alpha,\beta,\gamma) = \arccos \frac{\sin\alpha\cos\gamma}{\sqrt{(\sin^2\alpha-\cos^2\beta)}} \, d\alpha \;+$$

$$+ \arccos \frac{\cos\alpha\cos\beta\cos\gamma}{\sqrt{(\sin^2\alpha-\cos^2\beta)}\sqrt{(\sin^2\gamma-\cos^2\beta)}} \, d\beta \;+$$

$$+ \arccos \frac{\cos\alpha\sin\gamma}{\sqrt{(\sin^2\gamma-\cos^2\beta)}} \, d\gamma$$

$$(0 \leqslant \alpha \leqslant \pi;\; 0 \leqslant \beta \leqslant \pi;\; 0 \leqslant \gamma \leqslant \pi;\; \sin^2\alpha\sin^2\gamma \geqslant \cos^2\beta).$$

On comparing these formulae with (3.5) and (4.11), we see that

$$S(\alpha,\beta,\gamma) = \tfrac{1}{2}\pi^2 f(\tfrac{1}{2}\pi-\alpha, \beta, \tfrac{1}{2}\pi-\gamma). \tag{6.1}$$

Thus, subject to the condition

$$\cos^2\alpha \cos^2\gamma > \cos^2\beta,$$

$S(\alpha,\beta,\gamma)$ *is equal to four times the volume of a double-rectangular tetrahedron of angles*

$$\tfrac{1}{2}\pi-\alpha, \qquad \beta, \qquad \tfrac{1}{2}\pi-\gamma. \tag{6.2}$$

One consequence of this result is that the order of the group‡

$$[k_1, k_2, k_3]$$

(whose fundamental region is a double-rectangular tetrahedron) is

$$\frac{8\pi^2}{S(\tfrac{1}{2}\pi-\pi/k_1, \pi/k_2, \tfrac{1}{2}\pi-\pi/k_3)}.$$

The orders of the actual groups

$$[3,3,3], \; [3,3,4], \; [3,4,3], \; [3,3,5]$$

are given by (4.41), (4.51), (4.52), (4.56), respectively.

Schläfli regards the tetrahedron as lying on a hypersphere of unit

---

* In Schläfli's notation (**S**, 54, 56), $\Delta(1234) = 0$, or $\Delta_4(\alpha,\beta,\gamma) = 0$.
† **S**, 97.
‡ J. A. Todd, *Proc. Cambridge Phil. Soc.* 27 (1931), 214.

radius, but it can just as well be regarded as lying in elliptic space with unit space-constant. The volume is still

$$\tfrac{1}{4}S(\alpha,\beta,\gamma).$$

## 7. Hyperbolic space

In the notation of § 1, the angles of the double-rectangular tetrahedron (6.2) are

$$(1\,2) = \tfrac{1}{2}\pi-\alpha, \qquad (2\,3) = \beta, \qquad (3\,4) = \tfrac{1}{2}\pi-\gamma,$$
$$(1\,3) = (1\,4) = (2\,4) = \tfrac{1}{2}\pi. \tag{7.1}$$

For this tetrahedron to exist in hyperbolic space, we require

$$\cos\alpha\cos\gamma < \cos\beta \quad (0 < \alpha < \tfrac{1}{2}\pi;\ 0 < \beta < \tfrac{1}{2}\pi;\ 0 < \gamma < \tfrac{1}{2}\pi).$$

Since the angular excesses at the four vertices are

$$(2\,3)+(3\,4)+(2\,4)-\pi = \beta-\gamma,$$
$$(3\,4)+(1\,3)+(1\,4)-\pi = \tfrac{1}{2}\pi-\gamma,$$
$$(1\,2)+(1\,4)+(2\,4)-\pi = \tfrac{1}{2}\pi-\alpha,$$
$$(1\,2)+(2\,3)+(1\,3)-\pi = \beta-\alpha,$$

all the vertices will be actual if

$$\alpha, \gamma < \beta.$$

If either of these inequalities is replaced by equality, the tetrahedron extends to infinity (although, as we shall see, its volume remains finite). In fact, if $\alpha = \beta > \gamma$ or $\alpha < \beta = \gamma$, one vertex lies on the absolute; and if $\alpha = \beta = \gamma$, two vertices lie on the absolute.

If $\alpha > \beta$ or $\gamma > \beta$, the tetrahedron has one or two *ideal* vertices, and we cannot hope to obtain an intelligible expression for its volume. But, so long as (5.31) holds, we should expect the volume to be

$$\tfrac{1}{4}S(\alpha,\beta,\gamma),$$

possibly multiplied by $i$. In order to prove that this is in fact the case, we proceed to identify (5.4) with Lobatschefsky's formula*

$$4P = L(\tfrac{1}{2}\pi-\mu+\delta)-L(\tfrac{1}{2}\pi-\mu-\delta)-$$
$$-L(\tfrac{1}{2}\pi-\xi+\delta)+L(\tfrac{1}{2}\pi-\xi-\delta)+$$
$$+L(\tfrac{1}{2}\pi-B+\delta)-L(\tfrac{1}{2}\pi-B-\delta)-2L(\delta).$$

In his Fig. 19,† let us number the faces of the tetrahedron in the order $ABC$, $ACD$, $ABD$, $BCD$. Then (7.1) enables us immediately to identify his $\mu$, $B$ with our $\tfrac{1}{2}\pi-\alpha$, $\tfrac{1}{2}\pi-\gamma$. It remains to be proved

* L, 97 (160). † L, 92.

that $\frac{1}{2}\pi-\xi$ is the dihedral angle at the edge $AD$, and that his $\delta$ is the same as ours.

By drawing a sphere with centre $D$, we obtain a right-angled spherical triangle, whose hypotenuse is $\gamma'$; one of its acute angles is $B$, and the other is equal to the dihedral angle in question, which is therefore
$$\operatorname{arccot}(\cos\gamma'\tan B).$$

The required equation
$$\tan\xi = \cos\gamma'\tan B$$

follows from Lobatschefsky's (134) and (159).*

In order to identify $\delta$ in his (156) and our (5.3), we have to prove that
$$\sqrt{(\sin^2\xi-\sin^2\mu\sin^2B)}/\cos\mu\cos B = \cos h'\tan B. \qquad (7.2)$$

By Lobatschefsky's (26)† applied to the triangle $DCB$,
$$\sin\alpha' = \sin h'\cos\mu.$$

Also‡
$$\cos\alpha' = \sin\xi/\sin B.$$

Therefore
$$\sin^2\xi/\sin^2B = 1-\sin^2h'\cos^2\mu = \sin^2\mu+\cos^2h'\cos^2\mu,$$

whence (7.2) follows at once.

Thus, finally,
$$\tfrac{1}{2}\pi-\mu = \alpha, \qquad \tfrac{1}{2}\pi-\xi = \beta, \qquad \tfrac{1}{2}\pi-B = \gamma$$
and
$$P = \tfrac{1}{4}iS(\alpha,\beta,\gamma). \qquad (7.3)$$

There are two different conventions as to the interpretation of length in hyperbolic space. Sommerville,§ following Lobatschefsky, regards length, and therefore volume, as real; but Coolidge‖ (in order that the formulae of elliptic geometry may be carried over without change) makes them pure-imaginary. With the former convention, taking the space-constant to be 1, the volume of our tetrahedron is given by (7.3); with the latter, taking the space-constant to be $i$, it is
$$\tfrac{1}{4}S(\alpha,\beta,\gamma),$$

precisely as in the elliptic case.

The formula (5.41) gives four times the (real) volume of an *infinite*

---

\* L, 93, 96. † L, 55.

‡ L, 98 (near the bottom of the page).

§ D. M. Y. Sommerville, *The Elements of Non-Euclidean Geometry* (London, 1914).

‖ J. L. Coolidge, *The Elements of Non-Euclidean Geometry* (Oxford, 1909).

tetrahedron $DBCA$, with $D$ on the absolute;* $\frac{1}{2}\pi-\alpha$, $\frac{1}{2}\pi-\beta$ are the dihedral angles at the edges $AC$, $BB'$ respectively. Our formula agrees with Lobatschefsky's (89), since

$$\tfrac{1}{2}\pi-\alpha = a', \qquad \tfrac{1}{2}\pi-\beta = B.$$

($CC'$ and $CB$ are respectively parallel and perpendicular to $BB'$, and $a'$ is the *parallel-angle* $BCC'$.)

## 8. Regular polyhedra inscribed in the hyperbolic absolute

In the hyperbolic plane the (positive) area of a triangle of angles $\alpha$, $\beta$, $\gamma$ is $\pi-\alpha-\beta-\gamma$. A regular $n$-gon of angle $2\alpha$ can be divided into $2n$ triangles of angles $\frac{1}{2}\pi$, $\pi/n$, $\alpha$, whence its area is

$$2n(\tfrac{1}{2}\pi-\pi/n-\alpha) = (n-2)\pi-2n\alpha.$$

More generally, the area of a regular polygon† $\{n/d\}$ of angle $2\alpha$ is

$$2n(\tfrac{1}{2}\pi-d\pi/n-\alpha) = (n-2d)\pi-2n\alpha.$$

When the polygon is inscribed in the absolute, we have $\alpha = 0$, and the area becomes $(n-2d)\pi.$

In hyperbolic space, a polyhedron $\{k_1, k_2\}$ of dihedral angle $2\alpha$, with $E$ edges,‡ can be divided into $4E$ double-rectangular tetrahedra of angles $\pi/k_1$, $\pi/k_2$, $\alpha$, whence its (real) volume is

$$EiS(\tfrac{1}{2}\pi-\pi/k_1, \pi/k_2, \tfrac{1}{2}\pi-\alpha).$$

When the polyhedron is inscribed in the absolute, we have

$$\pi/k_2+\alpha = \tfrac{1}{2}\pi,$$

and, by (5.41), the volume becomes§

$$EiS(\tfrac{1}{2}\pi-\pi/k_1, \pi/k_2, \pi/k_2)$$
$$= E[L(\pi-\pi/k_1-\pi/k_2)+L(\pi/k_1-\pi/k_2)-2L(\tfrac{1}{2}\pi-\pi/k_2)]. \quad (8.1)$$

By (4.31), (4.32), and (5.6), the volume of a regular tetrahedron

---

* See Lobatschefsky's Fig. 16 (L, 82).

† e.g. a pentagram $\{\frac{5}{2}\}$. For the notation see Coxeter, *Proc. Cambridge Phil. Soc.* 27 (1931), 201. The area of a star polygon is defined as the sum of the areas of the triangles that join its centre to its sides. Consequently certain portions are counted several times over.

‡ When $k_1$, $k_2$ are integers, the number of edges is given by

$$1/E = 1/k_1+1/k_2-\tfrac{1}{2}.$$

The Kepler-Poinsot polyhedra all have 30 edges.

§ When $k_1 > k_2$, we naturally replace the term $+L(\pi/k_1-\pi/k_2)$ by $-L(\pi/k_2-\pi/k_1)$.

$\{3, 3\}$ inscribed in the absolute (i.e. the greatest volume that a tetrahedron can possibly have) is

$$\tfrac{3}{2}i\, S(\tfrac{1}{3}\pi, \tfrac{1}{6}\pi, \tfrac{1}{6}\pi) = i\, S(\tfrac{1}{3}\pi, \tfrac{1}{3}\pi, \tfrac{1}{3}\pi)$$
$$= \tfrac{1}{2}\pi \log 2 - 3L(\tfrac{1}{6}\pi) = \tfrac{2}{3}\pi \log 2 - 2L(\tfrac{1}{3}\pi)^*$$
$$= \frac{3\sqrt{3}}{4}\left(1 - \frac{1}{2^2} + \frac{1}{4^2} - \frac{1}{5^2} + \frac{1}{7^2} - \frac{1}{8^2} + \cdots\right)$$
$$= \frac{\sqrt{3}}{2}\left(1 + \frac{1}{2^2} - \frac{1}{4^2} - \frac{1}{5^2} + \frac{1}{7^2} + \frac{1}{8^2} - \cdots\right).$$

Geometrically, the regular tetrahedron (of dihedral angle $\tfrac{1}{3}\pi$) can be divided into six double-rectangular tetrahedra of angles $\tfrac{1}{3}\pi$, $\tfrac{1}{6}\pi$, $\tfrac{1}{3}\pi$ by joining one vertex to all the medians of the opposite face; it can also be divided into four double-rectangular tetrahedra of angles $\tfrac{1}{6}\pi$, $\tfrac{1}{3}\pi$, $\tfrac{1}{6}\pi$ by joining each of two opposite edges to the mid-point of the other.

By (5.8), the volume of an *octahedron* $\{3, 4\}$ inscribed in the absolute is

$$4i\, S(\tfrac{1}{4}\pi, \tfrac{1}{4}\pi, \tfrac{1}{4}\pi) = 2\pi \log 2 - 8L(\tfrac{1}{4}\pi) = 4\left(1 - \frac{1}{3^2} + \frac{1}{5^2} - \frac{1}{7^2} + \cdots\right).$$

Geometrically, the octahedron (of dihedral angle $\tfrac{1}{2}\pi$) can be divided into sixteen double-rectangular tetrahedra of angles $\tfrac{1}{4}\pi$, $\tfrac{1}{4}\pi$, $\tfrac{1}{4}\pi$ by drawing all the planes of symmetry through one pair of opposite vertices and also the plane that reflects those vertices into one another.

By a double application of (5.5), the volume of a *hexahedron*† $\{4, 3\}$ inscribed in the absolute is

$$\tfrac{10}{3}\pi \log 2 - 10L(\tfrac{1}{3}\pi) = 5i\, S(\tfrac{1}{3}\pi, \tfrac{1}{3}\pi, \tfrac{1}{3}\pi),$$

i.e. just five times that of the regular tetrahedron. The hexahedron can in fact be built up by taking one regular tetrahedron and placing another on each face.

The expression for the volume of the *great icosahedron* $\{3, \tfrac{5}{2}\}$ can be simplified by means of (5.92). It is not known whether any similar reduction is possible for the icosahedron itself.

---

\* The identity $2L(\tfrac{1}{3}\pi) - 3L(\tfrac{1}{6}\pi) = \tfrac{1}{6}\pi \log 2$ is merely a special case of (5.5).

† The name *cube* seems inappropriate when the dihedral angle is $\tfrac{1}{3}\pi$ instead of $\tfrac{1}{2}\pi$.

**2**

# INTEGRAL CAYLEY NUMBERS

Reprinted, with the editors' permission,
from the *Duke Mathematical Journal*,
Vol. 13, No. 4, December, 1946.

The first part of this paper contains a simple proof of the famous "eight square theorem" which expresses the product of any two sums of eight squares as a single sum of eight squares (§2). This proof, like that of Dickson [8; 158–159], employs the system of Cayley numbers, which is somewhat analogous to the field of complex numbers and to the quasi-field of quaternions.

Since the integral complex numbers and integral quaternions were extensively studied by Gauss and Hurwitz (see §3), it is strange that the analogous set of Cayley numbers has been comparatively neglected. There is a short paper by Kirmse [14], but that is marred by a rather serious error, as we shall see in §4. Bruck showed me a simple way to remedy this defect. §§5–12 are concerned with a verification that Kirmse's set of Cayley numbers, as corrected by Bruck, is indeed a set of *integral* elements, according to a precise definition due to Dickson. A geometrical representation is found to be helpful (§6), and it appears that the integral Cayley numbers correspond to the closest packing of spheres in eight dimensions, just as the integral quaternions correspond to the closest packing in four dimensions.

I wish to thank H. G. Forder (of University College, Auckland, N. Z.) for arousing my interest in Cayley numbers, and R. H. Bruck (of the University of Wisconsin) for pointing the way out of the impasse into which Kirmse's mistake had led me.

1. **Cayley numbers: associative and anti-associative triads.** In the notation of Cartan and Schouten [1; 944] the Cayley numbers are sets of eight real numbers

$$[a_0 , a_1 , a_2 , a_3 , a_4 , a_5 , a_6 , a_7]$$

$$= a_0 + a_1 e_1 + a_2 e_2 + a_3 e_3 + a_4 e_4 + a_5 e_5 + a_6 e_6 + a_7 e_7 ,$$

which are added like vectors and multiplied according to the rules

$$e_r^2 = -1, \qquad e_{r+1}e_{r+3} = e_{r+2}e_{r+6} = e_{r+4}e_{r+5} = e_r ,$$

$$e_{r+3}e_{r+1} = e_{r+6}e_{r+2} = e_{r+5}e_{r+4} = -e_r , \qquad e_{r+7} = e_r .$$

These rules may be written in the concise form

(1.1) $\quad e_r^2 = e_1 e_2 e_4 = e_2 e_3 e_5 = e_3 e_4 e_6 = e_4 e_5 e_7 = e_5 e_6 e_1 = e_6 e_7 e_2 = e_7 e_1 e_3 = -1,$

provided we interpret "$e_1 e_2 e_4 = -1$" to mean

$$e_2 e_4 = -e_4 e_2 = e_1 , \qquad e_4 e_1 = -e_1 e_4 = e_2 , \qquad e_1 e_2 = -e_2 e_1 = e_4 ,$$

like the famous relations $i^2 = j^2 = k^2 = ijk = -1$ of Hamilton [12; 339].[*]

[*] For references, see pages 38-39.

The seven *associative triads*

$$e_1e_2e_4 \,, \qquad e_2e_3e_5 \,, \qquad e_3e_4e_6 \,, \qquad e_4e_5e_7 \,, \qquad e_5e_6e_1 \,, \qquad e_6e_7e_2 \,, \qquad e_7e_1e_3$$

correspond to the lines of the finite projective geometry of seven lines and seven points, represented diagrammatically in Fig. 1. Thus any operation of the simple 168-group generated by the permutations

$$(1\ \ 2)(3\ \ 6) \qquad \text{and} \qquad (1\ \ 2\ \ 3\ \ 4\ \ 5\ \ 6\ \ 7)$$

will leave the multiplication table unchanged, provided we make suitable changes of sign. (The relevance of this group seems to have been first noticed by Mathieu [16; 354].) Moreover, we may change the sign of all the $e$'s except those of an associative triad, without making any permutation. Such changes of sign form, with the identity, an Abelian group of order 8 and type $(1, 1, 1)$. Hence the symmetry group of the multiplication table is of order $8 \cdot 168 = 1344$, and the above is one of

$$2^7\ 7!/1344 = 480$$

possible notations; so it is not surprising that every author uses a different notation.

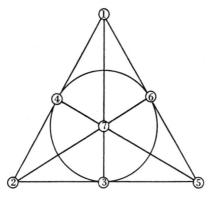

Fig. 1

Although the associative law holds for the triads in (1.1) and for any triad with a repeated element, such as

$$e_1^2 e_2 = -e_2 = e_1 \cdot e_1 e_2 \qquad \text{or} \qquad e_1 e_2 \cdot e_1 = e_2 = e_1 \cdot e_2 e_1 \,,$$

it does not hold universally. In fact, the remaining 28 triads are *anti-associative*:

$$e_1 e_2 \cdot e_3 = -e_6 = -e_1 \cdot e_2 e_3 \,, \qquad e_1 e_2 \cdot e_5 = \quad e_7 = -e_1 \cdot e_2 e_5 \,,$$

$$e_1 e_2 \cdot e_6 = \quad e_3 = -e_1 \cdot e_2 e_6 \,, \qquad e_1 e_3 \cdot e_5 = -e_4 = -e_1 \cdot e_3 e_5 \,,$$

and so on, using the cyclic permutation (1 2 3 4 5 6 7). (The distinction between the two kinds of triad was pointed out by Cayley [3].)

**2. The alternating-associative law and the Eight Square Theorem.** For any Cayley number $a = a_0 + a_1 e_1 + \cdots + a_7 e_7$, we define the *conjugate*

$$\bar{a} = a_0 - a_1 e_1 - \cdots - a_7 e_7 ,$$

the *norm*

$$\mathrm{N}a = \bar{a}a = a\bar{a} = a_0^2 + a_1^2 + \cdots + a_7^2 ,$$

and, if $a \neq 0$, the *inverse*

$$a^{-1} = (\mathrm{N}a)^{-1}\bar{a},$$

which has the property $a^{-1}a = aa^{-1} = 1$. It is easily verified (as in the case of quaternions) that $\overline{ab} = \bar{b}\bar{a}$, but it is not so easy to deduce that

$$(2.1) \qquad\qquad \mathrm{N}a \cdot \mathrm{N}b = \mathrm{N}(ab).$$

As a first step, we prove that the associative law holds for $\bar{a}ab$. (See Kirmse [14; 67].) When written out in full, the expressions $\bar{a}a \cdot b$ and $\bar{a} \cdot ab$ agree term by term, except possibly for terms such as

$$-a_p e_p\, a_q e_q \cdot b_r e_r$$

where $e_p e_q e_r$ is an anti-associative triad. But they actually agree here too; for,

$$e_p e_q \cdot e_r = -e_q e_p \cdot e_r = e_q \cdot e_p e_r ,$$

and therefore $-a_p e_p a_q e_q \cdot b_r e_r = -a_q e_q \cdot a_p e_p b_r e_r$ . (The same argument shows that $a^2 b = a \cdot ab$.) Similarly

$$ab \cdot \bar{b} = a \cdot b\bar{b}.$$

From the identity $\bar{a}a \cdot b = \bar{a} \cdot ab$, multiplication by $(\mathrm{N}a)^{-1}$ gives

$$b = a^{-1} \cdot ab.$$

Similarly $ab \cdot b^{-1} = a$. Hence

$$(ab)^{-1} = b^{-1} \cdot b(ab)^{-1} = b^{-1} \cdot (a^{-1} \cdot ab)(ab)^{-1} = b^{-1}a^{-1}.$$

(See Moufang [18; 418].)

We are now ready to prove (2.1). If $a$ or $b$ vanishes, it is obvious. If not, we have

$$\mathrm{N}(ab) = \overline{ab}\, ab = \bar{b}\bar{a} \cdot ab = (\mathrm{N}b \cdot b^{-1} \cdot \mathrm{N}a \cdot a^{-1})ab$$

$$= \mathrm{N}b \cdot \mathrm{N}a \cdot b^{-1}a^{-1} \cdot ab = \mathrm{N}a \cdot \mathrm{N}b \cdot (ab)^{-1}ab = \mathrm{N}a \cdot \mathrm{N}b.$$

When written out in full, (2.1) is the Eight Square Theorem:

$$(a_0^2 + a_1^2 + \cdots + a_7^2)(b_0^2 + b_1^2 + \cdots + b_7^2)$$
$$= (a_0b_0 - a_1b_1 - a_2b_2 - a_3b_3 - a_4b_4 - a_5b_5 - a_6b_6 - a_7b_7)^2$$
$$+ \sum(a_0b_1 + a_1b_0 + a_2b_4 + a_3b_7 - a_4b_2 + a_5b_6 - a_6b_5 - a_7b_3)^2,$$

where the $\sum$ implies summation of seven squares given by cyclic permutation of the suffix numbers 1, 2, 3, 4, 5, 6, 7, leaving 0 unchanged. This was first given (in a less systematic notation) by Degen in 1818 [8; 164].

3. **The arithmetic of an algebra.** The complex number $a = a_0 + a_1i$, the quaternion $a = a_0 + a_1i + a_2j + a_3k$, and the Cayley number

$$a = a_0 + a_1e_1 + a_2e_2 + a_3e_3 + a_4e_4 + a_5e_5 + a_6e_6 + a_7e_7 ,$$

all satisfy a "rank equation" with real coefficients:

$$x^2 - 2a_0x + Na = 0$$

$(2a_0 = a + \bar{a}, \qquad Na = \bar{a}a)$.

A set of elements selected from an algebra is called a set of *integral* elements [10; 141–142] if it satisfies the following four conditions:

(i) For each element, the coefficients of the rank equation are integers,
(ii) The set is closed under subtraction and multiplication,
(iii) The set contains 1,
(iv) The set is not a subset of a larger set satisfying (i), (ii), (iii).

According to this definition, the integral elements of the algebra of complex numbers are the *Gaussian integers* $a_0 + a_1i$, where $a_0$ and $a_1$ are ordinary integers. These are represented in the Argand diagram by the ordinary lattice points, which are the vertices of the regular tessellation of squares, $\{4, 4\}$. The *units*, i.e., elements of unit norm, are $\pm1$ and $\pm i$, and the corresponding points

$$(\pm1, 0), \qquad (0, \pm1)$$

form a square: the *vertex figure* of the tessellation, as indicated by the second 4 in the Schläfli symbol $\{4, 4\}$.

Analogously, the *integral quaternions* are $a_0 + a_1i + a_2j + a_3k$, where the four $a$'s are either integers or all halves of odd integers. These have been investigated by Hurwitz [13]. The 24 units

$$\pm1, \qquad \pm i, \qquad \pm j, \qquad \pm k, \qquad \tfrac{1}{2}(\pm1\pm i\pm j\pm k)$$

are the elements of the "binary tetrahedral group" [6; 372]. If we represent the quaternion $a_0 + a_1i + a_2j + a_3k$ by the point $(a_0, a_1, a_2, a_3)$ in Euclidean 4-space, we find that the 24 units are represented by the vertices

$$(\pm1, 0, 0, 0), \qquad (0, \pm1, 0, 0), \qquad (0, 0, \pm1, 0), \qquad (0, 0, 0, \pm1),$$
$$(\pm\tfrac{1}{2}, \pm\tfrac{1}{2}, \pm\tfrac{1}{2}, \pm\tfrac{1}{2})$$

of the 24-cell {3, 4, 3}, one of the regular polytopes discovered by Schläfli [19; 51] about 1850. Hence the integral quaternions are represented by the vertices of the regular honeycomb {3, 3, 4, 3}, whose vertex figure is {3, 4, 3}.

A typical cell of the polytope {3, 4, 3} is the octahedron {3, 4} whose vertices represent the six quaternions $1, i, \frac{1}{2}(1 + i \pm j \pm k)$. A typical cell of the honeycomb {3, 3, 4, 3} is the *cross polytope* {3, 3, 4} whose eight vertices represent these same quaternions along with 0 and $1 + i$.

In terms of the combinations

$$l_1 = \tfrac{1}{2}(1 + k), \qquad l_2 = \tfrac{1}{2}(1 - k), \qquad l_3 = \tfrac{1}{2}(i + j), \qquad l_4 = \tfrac{1}{2}(i - j),$$

of norm $\frac{1}{2}$, we have

$$1 = l_1 + l_2, \qquad i = l_3 + l_4, \qquad j = l_3 - l_4, \qquad k = l_1 - l_2,$$

$$\tfrac{1}{2}(1 + i + j + k) = l_1 + l_3.$$

By addition and subtraction we can derive all expressions of the form $\pm l_r \pm l_s$. Hence these (with $r \neq s$) are the 24 units, and *the integral quaternions consist of all expressions*

$$x_1 l_1 + x_2 l_2 + x_3 l_3 + x_4 l_4$$

*where the $x$'s are integers and $\sum x$ is even.* In other words, the vertices of {3, 3, 4, 3} (of edge $2^{\frac{1}{2}}$) admit as coordinates all sets of four integers having an even sum [4; 347]. The eight quaternions represented by a typical cell may now be expressed in the simple form

$$l_1 \pm l_s \qquad\qquad (s = 1, 2, 3, 4).$$

Hurwitz's integral quaternions evidently satisfy Dickson's conditions (i)-(iii). Since they are closed under subtraction, we see that the vertices of {3, 3, 4, 3} form a *lattice*. Condition (iv) may be verified geometrically as follows. A point situated as far as possible from the nearest vertex of {3, 3, 4, 3} must be at the center of a cell. Now, the circum-radius of {3, 3, 4} (of unit edge) is $2^{-\frac{1}{2}}$; e.g., this is the distance between the points representing 0 and $l_1$. Since $2^{-\frac{1}{2}} < 1$, we cannot add further points to the lattice without reducing the minimum distance between pairs of points.

The same argument in two dimensions (where the cell is a *square* of circum-radius $2^{-\frac{1}{2}}$) explains why the integral complex numbers $a_0 + a_1 i$ admit only integral values for $a_0$ and $a_1$. On the other hand, the quaternions $a_0 + a_1 i + a_2 j + a_3 k$ with integral $a$'s are represented by the four-dimensional cubic lattice {4, 3, 3, 4}; and this *can* be augmented, since the circum-radius of the hyper-cube {4, 3, 3} is exactly 1.

4. **Kirmse's mistake.** Integral Cayley numbers have been discussed by Kirmse, whose notation is related to that of Cartan and Schouten (1.1) as follows:

$$i_1 = e_2, \quad i_2 = e_1, \quad i_3 = -e_4, \quad i_4 = e_3, \quad i_5 = -e_5, \quad i_6 = e_6, \quad i_7 = e_7.$$

Thus Kirmse [14; 64] considers Cayley numbers $a_0 + a_1 i_1 + \cdots + a_7 i_7$ , where

(4.1) $\quad i_r^2 = i_1 i_2 i_3 = i_1 i_5 i_4 = i_3 i_4 i_6 = i_3 i_5 i_7 = i_2 i_6 i_5 = i_1 i_6 i_7 = i_2 i_4 i_7 = -1.$

From these he selects an eight-dimensional *module*, i.e., a set of rational Cayley numbers which is closed under subtraction and contains eight linearly independent members. As usual, a module is called an *integral domain* if it is closed under multiplication. A simple instance is the module $J_0$ consisting of all Cayley numbers

$$\gamma_0 + \gamma_1 i_1 + \cdots + \gamma_7 i_7$$

where the $\gamma$'s are integers [14; 68]. He then defines a *maximal* (umfassendste) integral domain over $J_0$ as an extension of $J_0$ which cannot be further extended without ceasing to be an integral domain. He states that there are *eight* such domains, one of which he calls $J_1$ and describes in detail [14; 70]. Actually there are only *seven*, which presumably are the remaining seven of his eight. We easily verify that $J_1$ itself is not closed under multiplication. In fact, the product

$$\tfrac{1}{2}(1 + i_1 + i_2 + i_3) \cdot \tfrac{1}{2}(1 + i_1 + i_4 + i_5) = \tfrac{1}{2}(i_1 + i_2 + i_4 + i_7)$$

is not one of the 240 "units" [14; 76]. When I pointed this out to Bruck, he sent me a revised description of such a domain.

Bruck's domain $J$ can be derived from $J_1$ by transposing two of the $i$'s. We shall verify in §5 that it is closed under multiplication, and in §11 that it is maximal. Since the 168-group (§1) is doubly transitive on the seven $i$'s, *any* transposition will serve to rectify $J_1$ in the desired manner. But there are only seven such domains, since the $\binom{7}{2} = 21$ possible transpositions fall into 7 sets of 3, each set having the same effect. In each of the seven domains, one of the $i$'s plays a special role, viz., that one which is not affected by any of the three transpositions.

Comparing Kirmse's multiplication table (4.1) with Cayley's

(4.2) $\quad i_r^2 = i_1 i_2 i_3 = i_1 i_4 i_5 = i_3 i_4 i_7 = i_3 i_6 i_5 = i_2 i_5 i_7 = i_1 i_7 i_6 = i_2 i_4 i_6 = -1$

[2], we see that either can be derived from the other by interchanging $i_6$ with $i_7$ and reversing the sign of $i_5$ . Accordingly, Kirmse's $J_1$ could be used as it stands if we replaced his multiplication table by Cayley's. In that case the special unit would be $i_1$ ; but we shall find it more convenient to specialize $i_4$ .

## 5. The integral domain $J$ and its 240 units.

The relation between the notations of Dickson, Cayley, and Cartan-Schouten may be expressed as follows:

$$i = i_1 = e_2 , \qquad j = i_2 = e_1 , \qquad k = i_3 = -e_4 , \qquad e = i_4 = e_3 ,$$

$$ie = i_5 = e_5 , \qquad je = i_6 = e_7 , \qquad ke = i_7 = e_6 .$$

Thus $a_0 + a_1i_1 + \cdots + a_7i_7 = q + Qe$, where $q$ and $Q$ are quaternions:

$$q = a_0 + a_1i + a_2j + a_3k, \qquad Q = a_4 + a_5i + a_6j + a_7k.$$

In other words [8; 158], Cayley numbers are derived from quaternions by adjoining a new unit $e$, which enters into multiplication according to the rule

$$(q + Qe)(r + Re) = qr - \overline{R}Q + (Rq + Q\overline{r})e,$$

where the quaternions $\overline{R}$ and $\overline{r}$ are the conjugates of $R$ and $r$.

Now, the three Cayley numbers

$$i, \quad j, \quad \text{and} \quad h = \tfrac{1}{2}(i + j + k + e)$$

generate (by multiplication) $k = ij$ and $ih, jh, kh$. We proceed to verify that the module $J$ based on the eight Cayley numbers

(5.1) $$1, \quad i, \quad j, \quad k, \quad h, \quad ih, \quad jh, \quad kh$$

is closed under multiplication. In fact,

$$i^2 = j^2 = k^2 = h^2 = -1,$$

$$jk = -kj = -ih \cdot h = i, \quad ki = -ik = -jh \cdot h = j, \quad ij = -ji = -kh \cdot h = k,$$

$$i \cdot ih = j \cdot jh = k \cdot kh = -h,$$

$$ih \cdot i = -h \cdot ih = h - i, \quad jh \cdot j = -h \cdot jh = h - j, \quad kh \cdot k = -h \cdot kh = h - k,$$

$$(ih)^2 = hi = -1 - ih, \quad (jh)^2 = hj = -1 - jh, \quad (kh)^2 = hk = -1 - kh,$$

$$jh \cdot k = -j - ih, \qquad kh \cdot i = -k - jh, \qquad ih \cdot j = -i - kh,$$

$$kh \cdot j = -k + ih, \qquad ih \cdot k = -i + jh, \qquad jh \cdot i = -j + kh,$$

$$jh \cdot ih = k - h - ih, \quad kh \cdot jh = i - h - jh, \quad ih \cdot kh = j - h - kh,$$

$$kh \cdot ih = -j + h - ih, \quad ih \cdot jh = -k + h - jh, \quad jh \cdot kh = -i + h - kh,$$

$$k \cdot jh = -j \cdot kh = 1 + j - k + ih, \quad i \cdot kh = -k \cdot ih = 1 + k - i + jh,$$

$$j \cdot ih = -i \cdot jh = 1 + i - j + kh.$$

The accuracy of this table may be efficiently checked by applying the associative law to three elements of which two are equal; e.g., $(k \cdot jh)k = k(jh \cdot k)$. It is interesting to observe also that

$$(ih)^3 = (jh)^3 = (kh)^3 = 1,$$

and that the conjugates of $h$ and $ih$ are $-h$ and $-ih - 1$.

This integral domain $J$ includes the elements

$$e = 2h - i - j - k, \quad ie = 2ih + 1 + j - k,$$

$$je = 2jh + 1 + k - i, \quad ke = 2kh + 1 + i - j,$$

$$\tfrac{1}{2}(-1 - j + k + ie) = ih, \qquad \tfrac{1}{2}(i + e + je - ke) = h + jh - kh - i,$$

$$\tfrac{1}{2}(-1 - k + i + je) = jh, \qquad \tfrac{1}{2}(j + e + ke - ie) = h + kh - ih - j,$$

$$\tfrac{1}{2}(-1 - i + j + ke) = kh, \qquad \tfrac{1}{2}(k + e + ie - je) = h + ih - jh - k,$$

$$\tfrac{1}{2}(-1 + ie + je + ke)$$

$$= 1 + ih + jh + kh, \tfrac{1}{2}(i + j + k + e) = h,$$

$$\tfrac{1}{2}(-1 + i + e + ie) = h + ih - k, \tfrac{1}{2}(-j - k + je - ke) = jh - kh - i,$$

$$\tfrac{1}{2}(-1 + j + e + je) = h + jh - i, \tfrac{1}{2}(-k - i + ke - ie) = kh - ih - j,$$

$$\tfrac{1}{2}(-1 + k + e + ke) = h + kh - j, \tfrac{1}{2}(-i - j + ie - je) = ih - jh - k,$$

all of which are of unit norm. By adding or subtracting 1, $i$, $j$, $k$, $e$, $ie$, $je$ or $ke$, we find altogether 240 units:

| $\pm 1,$ | $\pm i,$ | $\pm j,$ | $\pm k,$ | $\pm e,$ | $\pm ie,$ | $\pm je,$ | $\pm ke,$ |
|---|---|---|---|---|---|---|---|

$$\tfrac{1}{2}(\pm 1 \pm j \pm k \pm ie), \qquad\qquad \tfrac{1}{2}(\pm i \pm e \pm je \pm ke),$$

$$\tfrac{1}{2}(\pm 1 \pm k \pm i \pm je), \qquad\qquad \tfrac{1}{2}(\pm j \pm e \pm ke \pm ie),$$

$$\tfrac{1}{2}(\pm 1 \pm i \pm j \pm ke), \qquad\qquad \tfrac{1}{2}(\pm k \pm e \pm ie \pm je),$$

$$\tfrac{1}{2}(\pm 1 \pm ie \pm je \pm ke), \qquad\qquad \tfrac{1}{2}(\pm i \pm j \pm k \pm e),$$

$$\tfrac{1}{2}(\pm 1 \pm i \pm e \pm ie), \qquad\qquad \tfrac{1}{2}(\pm j \pm k \pm je \pm ke),$$

$$\tfrac{1}{2}(\pm 1 \pm j \pm e \pm je), \qquad\qquad \tfrac{1}{2}(\pm k \pm i \pm ke \pm ie),$$

$$\tfrac{1}{2}(\pm 1 \pm k \pm e \pm ke), \qquad\qquad \tfrac{1}{2}(\pm i \pm j \pm ie \pm je).$$

(See Kirmse [14; 76].) The last seven rows of this table have the following properties: two elements in the same row have no common terms, but a ny two elements not in the same row have just two common terms, and the four remaining terms of two such elements form another element of the set. Since all the basic units (5.1) are included, it follows that only three types of Cayley number

$$(5.2) \qquad a = a_0 + a_1 i + a_2 j + a_3 k + a_4 e + a_5 ie + a_6 je + a_7 ke$$

can occur in $J$: the constituents $a_0$, $a_1$, $\cdots$, $a_7$ may be integers, or four or eight of them may be halves of odd integers. Moreover, if four of them are halves of odd integers, these must be distributed as in the above table. In particular, this shows that the table is complete: $J$ contains *only* 240 units.

**6. A geometrical representation.** For further investigation of the integral domain $J$, we shall find it convenient to replace the basis (5.1) by

$$(6.1) \qquad 1, \quad j, \quad e, \quad ke, \quad -h, \quad ih, \quad jh, \quad eh,$$

where, as before, $h = \frac{1}{2}(i + j + k + e)$. These eight units will serve equally well, since

$$kh = -ih - jh - eh - 2,$$
$$i = j + ke - 2kh - 1 \qquad \text{and}$$
$$k = 2h - i - j - e.$$

Their special virtue lies in the possibility of associating them with the nodes of a certain tree (Fig. 2) in such a way that two of them, say $a$ and $b$, satisfy

$$N(a + b) = 1 \quad \text{or} \quad 2$$

according as the corresponding nodes are adjacent or non-adjacent.

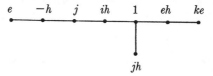

Fig. 2

Let us see what this means geometrically, when we represent each Cayley number by the point in Euclidean 8-space whose Cartesian coordinates are its eight constituents (i.e., the coefficients of $1$, $i$, $j$, $k$, $e$, $ie$, $je$, $ke$). The domain $J$ is now represented by a certain lattice, which we shall identify with the vertices of a uniform honeycomb called $5_{21}$.

Since the origin represents the Cayley number $0$, we can equally well say that we are representing the Cayley number $a$ by a vector $\mathbf{a}$ of magnitude $(Na)^{\frac{1}{2}}$. Then the unit vectors representing (5.1) or (6.1) generate the lattice. Two unit vectors $\mathbf{a}$ and $\mathbf{b}$ make an angle whose cosine is the scalar part of $\bar{a}b$, i.e.,

$$\cos(\mathbf{ab}) = a_0b_0 + a_1b_1 + \cdots + a_7b_7 = \frac{1}{2}(\bar{a}b + \bar{b}a).$$

Since $N(a + b) = (\bar{a} + \bar{b})(a + b) = Na + \bar{a}b + \bar{b}a + Nb = \bar{a}b + \bar{b}a + 2$, we have $(\mathbf{ab}) = \frac{2}{3}\pi$ or $\frac{1}{2}\pi$ according as $N(a + b) = 1$ or $2$. In the latter case it is natural to speak of $a$ and $b$ as *orthogonal* Cayley numbers.

More generally, two Cayley numbers $a$ and $b$ (of arbitrary norms) are said to be orthogonal whenever $\bar{a}b + \bar{b}a = 0$. Then

$$N(x_1a + x_2b) = N(x_1a) + N(x_2b)$$

for any ordinary numbers $x_1$ and $x_2$.

For use later on, let us see whether we can find a unit of $J$ orthogonal to all of (6.1) except $e$. (This would enable us to add an extra branch and node to Fig. 2 on the left.) Being orthogonal to 1, $j$ and $ke$, it must have the form

$$\tfrac{1}{2}(\pm\ k\ \pm\ e\ \pm\ ie\ \pm\ je).$$

Being orthogonal to $h$, $ih$, $jh$, $eh$, it must have opposite signs for $k$ and $e$, opposite signs for $k$ and $ie$, the same sign for $k$ and $je$, opposite signs for $ie$ and $je$. Hence it is either

(6.2) $$\tfrac{1}{2}(k\ -\ e\ -\ ie\ +\ je)\ =\ 2k\ -\ h\ -\ ih\ +\ jh$$

or the negative of this.

### 7. Reflections in eight-dimensional Euclidean space.

Let $\Lambda_a$ or $\Delta_a$ denote the operation of multiplying every Cayley number on the left or right (respectively) by a Cayley number $a$ of unit norm. These are congruent transformations of the Euclidean 8-space, since the squared distance between the points representing two Cayley numbers $x$ and $y$ is $N(y\ -\ x)$, which is the same as $N(ay\ -\ ax)$ or $N(ya\ -\ xa)$ in virtue of (2.1). (The corresponding transformations of elliptic 7-space were investigated by Cartan and Schouten [1].)

Let $\Phi_a$ denote the reflection in the hyperplane perpendicular to the unit vector **a**. Since $x$ and $-\bar{x}$ differ only in the sign of their scalar parts, $\Phi_1$ changes every $x$ into $-\bar{x}$. Since $\mathbf{a} = 1^{\Delta_a}$ (meaning the vector 1 operated on by $\Delta_a$), the principle of transformation gives

$$\Phi_a\ =\ \Delta_a^{-1}\Phi_1\Delta_a\ =\ \Delta_{a^{-1}}\Phi_1\Delta_a\ =\ \Delta_{\bar{a}}\Phi_1\Delta_a\ ,$$

which changes $x$ into $-\overline{x\bar{a}}\ a\ =\ -(a\bar{x})a$. Similarly $\mathbf{a} = 1^{\Lambda_a}$, and $\Phi_a\ =\ \Lambda_{\bar{a}}\Phi_1\Lambda_a\ ,$ which changes $x$ into $-a\ \overline{\bar{a}x}\ =\ -a(\bar{x}a)$. Thus the associative law holds for any product of the form $aba$ (with $Na\ =\ 1$, and consequently also without this restriction; see [17; 216]), and we see that $\Phi_a$ is the transformation

(7.1) $$x\ \rightarrow\ -a\bar{x}a.$$

The following alternative proof is suggested by Witt's treatment of quaternions [20; 308]. The reflection $\Phi_a$ transforms any vector **x** into $\mathbf{x}\ -\ 2(\mathbf{x}\cdot\mathbf{a})\mathbf{a}$. It therefore transforms the corresponding Cayley number $x$ into

$$x\ -\ (x\bar{a}\ +\ a\bar{x})a\ =\ -(a\bar{x})a \qquad\qquad (Na\ =\ 1).$$

Since $(\mathbf{x}\cdot\mathbf{a})\mathbf{a}\ =\ \mathbf{a}(\mathbf{a}\cdot\mathbf{x})$, we may equally well write this as

$$x\ -\ a(\bar{a}x\ +\ \bar{x}a)\ =\ -a(\bar{x}a).$$

### 8. The polytope $4_{21}$.

The domain $J$ is evidently invariant under the transformation (7.1) when $a$ is any one of the 240 units; in particular, when $a$ is one of the basic units (6.1). We are thus led to consider reflections in the hyperplanes through the origin perpendicular to the corresponding eight vectors. Of the

28 dihedral angles between pairs of these eight hyperplanes, seven are $\frac{1}{3}\pi$, arranged as in Fig. 2, while the remaining 21 are right angles. The eight reflections generate a group known as $[3^{4,2,1}]$. (The indices are the numbers of branches of the tree that emanate from the central node 1 in various directions.)

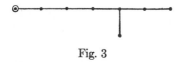

Fig. 3

Fig. 3 (or briefly, $4_{21}$) is the symbol for the eight-dimensional polytope which has one vertex on the line of intersection of seven of the eight hyperplanes (viz., all except the one whose node is ringed) while its remaining vertices are the transforms of this one by the group $[3^{4,2,1}]$. The position of the first vertex is determined by the Cayley number (6.2), orthogonal to all of (6.1) except $e$. Since this is one of the 240 units, we conclude that all the vertices of $4_{21}$ represent units of $J$. But $4_{21}$ is known to have 240 vertices [7; §11.8]. Hence *the* 240 *vertices of* $4_{21}$ *represent the* 240 *units of* $J$.

9. **The honeycomb** $5_{21}$. We can go further, and identify the lattice of $J$ with the vertices of the eight-dimensional honeycomb $5_{21}$, whose vertex figure is $4_{21}$. The eight hyperplanes indicated by the nodes in Fig. 2 or 3 form an angular region at the origin (like a trihedron in ordinary space). This is a fundamental region for the finite group $[3^{4,2,1}]$. The fundamental region for the analogous infinite group $[3^{5,2,1}]$ is the simplex cut off from this angular region by a hyperplane perpendicular to the direction of (6.2) but *not* through the origin. The corresponding node of the tree is ringed in Fig. 4. The point of intersection of the first eight hyperplanes is, of course, the origin. This is one vertex of the honeycomb $5_{21}$, and the remaining vertices are derived from it by applying $[3^{5,2,1}]$. One of them is the point representing $2k - h - ih + jh$, provided we take the ninth hyperplane to lie midway between this and the origin. This point is one of 240, derived from it by means of the subgroup $[3^{4,2,1}]$. These are the vertices of $4_{21}$ (representing the 240 units) and terminate the 240 edges of $5_{21}$ that radiate from the origin. Each of these edges is perpendicularly bisected by a hyperplane of symmetry, the transform of the "ninth hyperplane" by some element of $[3^{4,2,1}]$. There is also, perpendicular to each edge, a hyperplane of symmetry through the origin. The product of the reflections in these two parallel hyperplanes is just the translation along the edge. Thus all such translations are elements of the group $[3^{5,2,1}]$. The eight edges terminated by the points representing (6.1) yield eight translations which generate the lattice of $J$. Hence every element of $J$ is represented by a vertex of $5_{21}$. Conversely, every vertex of $5_{21}$ represents an element of $J$; for, the lattice includes all the 240 neighbors of the origin, and so also all the 240 neighbors of any other vertex.

The consequent coordinates for the vertices of $5_{21}$ were first obtained by Du Val [5; 185].

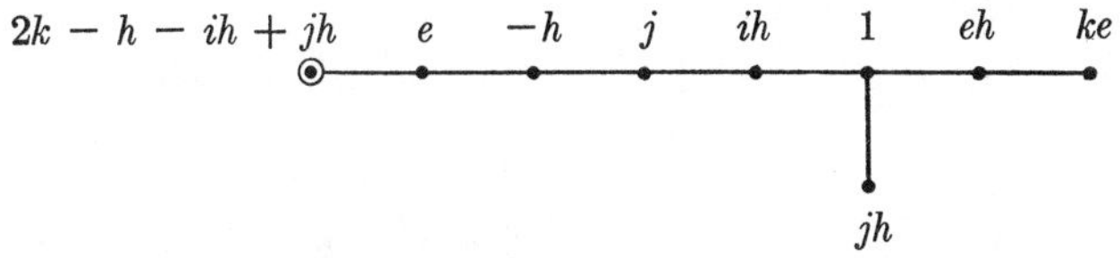

Fig. 4

The polytope $4_{21}$ was discovered by Gosset in 1897. It has cells of two kinds, whose symbols are derived from Fig. 3 by omitting one of the terminal nodes (with its branch): a seven-dimensional regular simplex $\alpha_7$, by omitting the node marked $jh$ in Fig. 2, and a seven-dimensional "cross polytope" (or octahedron-analogue) $\beta_7$, by omitting the node marked $ke$. Actually [11; 48] there are 17280 $\alpha_7$'s and 2160 $\beta_7$'s, so that the order of the symmetry group $[3^{4,2,1}]$ is

$$17280 \cdot 8! = 192 \cdot 10!.$$

Analogously, the honeycomb $5_{21}$ has cells $\alpha_8$ and $\beta_8$, whose symbols are derived by omitting (in turn) these same nodes from Fig. 4. Each vertex is surrounded by 17280 $\alpha_8$'s and 2160 $\beta_8$'s.

**10. An alternative notation.** In terms of the combinations

$$l_1 = \tfrac{1}{2}(1 + e), \qquad l_3 = \tfrac{1}{2}(i + ie), \qquad l_5 = \tfrac{1}{2}(j + je), \qquad l_7 = \tfrac{1}{2}(k + ke),$$

$$l_2 = \tfrac{1}{2}(1 - e), \qquad l_4 = \tfrac{1}{2}(i - ie), \qquad l_6 = \tfrac{1}{2}(j - je), \qquad l_8 = \tfrac{1}{2}(k - ke),$$

of norm $\tfrac{1}{2}$, we have

$$1 = l_1 + l_2, \qquad i = l_3 + l_4, \qquad j = l_5 + l_6, \qquad k = l_7 + l_8,$$

$$h = \tfrac{1}{2}(l_1 - l_2 + l_3 + l_4 + l_5 + l_6 + l_7 + l_8),$$

$$ih = \tfrac{1}{2}(-l_1 - l_2 + l_3 - l_4 - l_5 - l_6 + l_7 + l_8),$$

$$jh = \tfrac{1}{2}(-l_1 - l_2 + l_3 + l_4 + l_5 - l_6 - l_7 - l_8),$$

$$kh = \tfrac{1}{2}(-l_1 - l_2 - l_3 - l_4 + l_5 + l_6 + l_7 - l_8).$$

By addition and subtraction we can derive all expressions of the form $\pm l_r \pm l_s$, and also $\tfrac{1}{2}(\pm l_1 \pm l_2 \pm l_3 \pm l_4 \pm l_5 \pm l_6 \pm l_7 \pm l_8)$ with any odd number of minus signs. Hence these (with $r \neq s$) are the $112 + 128$ units, and *J consists of all expressions*

(10.1) $$x_1 l_1 + \cdots + x_8 l_8,$$

*where the x's are either eight integers with an even sum or eight halves of odd integers with an odd sum.* In other words, the vertices of $5_{21}$ (of edge $2^{\frac{1}{2}}$) admit as coordinates all sets of eight integers with an even sum, or eight halves of odd integers with an odd sum [4; 385]. These are rectangular Cartesian coordinates; for, since the $l$'s are mutually orthogonal,

$$N(x_1l_1 + \cdots + x_8l_8) = N(x_1l_1) + \cdots + N(x_8l_8) = \tfrac{1}{2}(x_1^2 + \cdots + x_8^2).$$

A typical cell $\alpha_8$ represents the nine Cayley numbers

$$\tfrac{1}{2}(l_1 - l_2 - \cdots - l_8), \qquad l_1 - l_s \qquad (s = 1, 2, \cdots, 8);$$

and a typical cell $\beta_8$ represents the sixteen Cayley numbers

$$l_1 \pm l_s .$$

(Note that this $\alpha_8$ and $\beta_8$ have a common $\alpha_7$ .)

Defining $E = l_1 = \tfrac{1}{2}(1 + e)$, so that

$$l_1 = E, \qquad l_3 = iE, \qquad l_5 = jE, \qquad l_7 = kE,$$
$$l_2 = \overline{E}, \qquad l_4 = i\overline{E}, \qquad l_6 = j\overline{E}, \qquad l_8 = k\overline{E},$$

we may express (10.1) as

$$(x_1 + x_3i + x_5j + x_7k)E + (x_2 + x_4i + x_6j + x_8k)\overline{E}.$$

Every product of two $l$'s is half one of the units $\pm l_r \pm l_s$ ; in fact, we have

$$E\overline{E} \;\;= \overline{E}E = -(iE)^2 = -(i\overline{E})^2 = \tfrac{1}{2},$$
$$E^2 \;\;= -\overline{E}^2 = iE \cdot i\overline{E} = -i\overline{E} \cdot iE = \tfrac{1}{2}e,$$
$$E \cdot iE = iE \cdot \overline{E} = i\overline{E} \cdot E = \overline{E} \cdot i\overline{E} = jE \cdot k\overline{E} = -k\overline{E} \cdot jE = j\overline{E} \cdot kE$$
$$= -kE \cdot j\overline{E} = \tfrac{1}{2}i,$$
$$iE \cdot E = \overline{E} \cdot iE = -E \cdot i\overline{E} = -i\overline{E} \cdot \overline{E} = -jE \cdot kE = kE \cdot jE = j\overline{E} \cdot k\overline{E}$$
$$= -k\overline{E} \cdot j\overline{E} = \tfrac{1}{2}ie,$$

and other rules derived from these by cyclic permutation of $i, j, k$.

**11. The interstices of the lattice.** A point situated as far as possible from the nearest vertex of $5_{21}$ must be at the center of a cell. Since the circum-radii of $\alpha_8$ and $\beta_8$ (of unit edge) are $\tfrac{2}{3}$ and $2^{-\frac{1}{2}}$, the latter being the greater, such a point is actually at the center of a $\beta_8$ , and its distance from the nearest lattice point (or rather, from the sixteen nearest lattice points) is $2^{-\frac{1}{2}}$. In fact, this is the distance from the center $l_1$ of the cell $l_1 \pm l_s$ to any of its vertices, such as 0. Hence *every point in the 8-space is distant $2^{-\frac{1}{2}}$ or less from some vertex of $5_{21}$* .

**12. Dickson's criterion.** We are now ready to verify that $J$ is a set of *integral* elements according to Dickson's definition (see §3). We consider the four conditions in turn.

(i) For each type of element (5.2) we observe that $2a_0$ and $Na$ are integers.

(ii) Being a module, $J$ is closed under subtraction; and we saw in §5 that it is closed under multiplication as well.

(iii) The element 1 occurs in the basis (5.1).

(iv) By §11, since $2^{-\frac{1}{2}} < 1$, we cannot add further points to the lattice of $J$ without reducing the minimum distance between pairs of points.

**13. Integral Cayley numbers of norm 1, 2, 3 or 4.** As we saw in §8, the 240 integral Cayley numbers of norm 1 are represented by the 240 vertices of $4_{21}$ (Fig. 3), or let us simply say that these units *are* the 240 vertices of $4_{21}$. In the notation of §10, they consist of

$$112 \text{ like } \pm l_1 \pm l_2 \text{ and } 128 \text{ like } \tfrac{1}{2}(-l_1 + l_2 + l_3 + \cdots + l_8)$$

(with an odd number of minus signs). Similarly, the integral Cayley numbers of norm 2 are

$$16 \text{ like } \pm 2l_1, \qquad 1120 \text{ like } \pm l_1 \pm l_2 \pm l_3 \pm l_4,$$

and $\qquad 1024 \text{ like } \tfrac{1}{2}(3l_1 + l_2 + l_3 + \cdots + l_8)$

(with an even number of minus signs); those of norm 3 are

$$1344 \text{ like } \pm 2l_1 \pm l_2 \pm l_3, \ 1792 \text{ like } \pm l_1 \pm l_2 \pm l_3 \pm l_4 \pm l_5 \pm l_6,$$

and $\qquad 3584 \text{ like } \tfrac{1}{2}(3l_1 + 3l_2 + l_3 + l_4 + \cdots - l_8)$

(with an odd number of minus signs); and those of norm 4 are the doubles of the 240 units and also

$$8960 \text{ like } \pm 2l_1 \pm l_2 \pm l_3 \pm l_4 \pm l_5,$$

$$128 \text{ like } l_1 + l_2 + \cdots + l_8$$

(with an even number of minus signs),

$$7168 \text{ like } \tfrac{1}{2}(3l_1 + 3l_2 + 3l_3 + l_4 + \cdots + l_8)$$

(with an even number of minus signs), and

$$1024 \text{ like } \tfrac{1}{2}(-5l_1 + l_2 + l_3 + \cdots + l_8)$$

(with an odd number of minus signs).

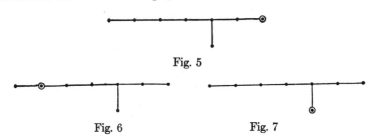

Fig. 5

Fig. 6       Fig. 7

A typical cell $\alpha_7$ of $4_{21}$ has the 8 vertices

$$\tfrac{1}{2}(l_1 - l_2 - \cdots - l_8), \qquad l_1 - l_s \qquad (s = 2, \cdots, 8);$$

and a typical cell $\beta_7$ has the 14 vertices

$$l_1 \pm l_s \qquad\qquad (s = 2, \cdots, 8).$$

(All the other cells could be derived from these by applying the group $[3^{4,2,1}]$.) Since the center of this $\beta_7$ is $l_1$, the 2160 integral Cayley numbers of norm 2 [14; 76] are the centers of the 2160 $\beta_7$'s of a $4_{21}$ of edge 2, or the 2160 vertices of a $2_{41}$ (Fig. 5). Moreover, since the mean of the two Cayley numbers $l_1 + l_2$ and $l_1 + l_3$ is $\tfrac{1}{2}(2l_1 + l_2 + l_3)$, the 6720 integral Cayley numbers of norm 3 are the mid-points of the 6720 edges of that same $4_{21}$, or the 6720 vertices of the "truncated $4_{21}$" whose symbol is shown in Fig. 6. Again, since the center of the above $\alpha_7$ is

$$\frac{3}{16}(5l_1 - l_2 - l_3 - \cdots - l_8),$$

the 17280 integral Cayley numbers of norm 4 (other than the doubles of those of norm 1) are the centers of the 17280 $\alpha_7$'s of a $4_{21}$ of edge 8/3, or the 17280 vertices of $1_{42}$ (Fig. 7). (If we went further in this direction we would obtain symbols having more than one ringed node.)

A still more convenient notation, for some purposes, is obtained by defining

$$y_0 = \frac{1}{3} \sum_1^8 x_r , \qquad y_s = x_s - \frac{1}{6} \sum_1^8 x_r \qquad (s = 1, \cdots, 8)$$

and

$$l_0 = \tfrac{1}{2}(1 + i + j + k) = \tfrac{1}{2} \sum_1^8 l_r ,$$

so that

$$\sum_0^8 y_r = 0, \qquad x_s = y_s + \tfrac{1}{2}y_0 ,$$

and

$$\sum_1^8 x_r l_r = \sum_1^8 y_r l_r + \tfrac{1}{2}y_0 \sum_1^8 l_r = \sum_0^8 y_r l_r.$$

Since $y_s = x_s - \tfrac{1}{2} \sum x_r + \tfrac{1}{3} \sum x_r$, where $x_s - \tfrac{1}{2} \sum x_r$ is an integer for every Cayley number (10.1), the nine $y$'s are either integers, or integers $+\tfrac{1}{3}$, or integers $-\tfrac{1}{3}$. Conversely, such $y$'s (with sum zero) lead to values of $x_s = y_s + \tfrac{1}{2}y_0$ which are either integers with an even sum $3y_0$ or halves of odd integers with an odd sum $3y_0$. Hence $J$ *consists of all expressions*

$$y_0 l_0 + y_1 l_1 + \cdots + y_8 l_8 ,$$

*where the nine $y$'s have sum zero and are either integers or integers plus $\frac{1}{3}$ or integers minus $\frac{1}{3}$.*

Moreover, since

$$\sum_{1}^{8} x_r^2 = \sum_{1}^{8} (y_r + \tfrac{1}{2}y_0)^2 = \sum_{1}^{8} y_r^2 + y_0 \sum_{1}^{8} y_r + 2y_0^2 = \sum_{0}^{8} y_r^2 ,$$

we have

$$\mathrm{N}(y_0 l_0 + y_1 l_1 + \cdots + y_8 l_8) = \tfrac{1}{2}(y_0^2 + y_1^2 + \cdots + y_8^2),$$

and the above sets of $y$'s are Cartesian coordinates for the vertices of $5_{21}$ (of edge $2^{\frac{1}{2}}$) in the 8-space $y_0 + y_1 + \cdots + y_8 = 0$ (See [4; 393]).

Thus the 240 integral Cayley numbers of norm 1 consist of

$$72 \text{ like } l_0 - l_1 \qquad \text{and} \qquad 168 \text{ like } \pm\left(l_0 + l_1 + l_2 - \tfrac{1}{3} \sum_{0}^{8} l_r\right).$$

Since $\sum l_r = 3l_0$, these may be expressed as

$$l_r - l_s \qquad \text{and} \qquad \pm(l_r + l_s + l_t - l_0),$$

where $r$, $s$, $t$ take any three distinct values among $0, 1, \cdots, 8$. Similarly [4; 397], the 2160 integral Cayley numbers of norm 2 are

$$l_p + l_q - l_r - l_s , \ \pm(2l_r + l_s - l_0) \text{ and } \pm(l_p + l_q + l_r + l_s - l_t - l_0).$$

$(756 + 144 + 1260 = 2160.)$

Perhaps the greatest advantage of this notation is that it enables us to replace the nine Cayley numbers named in Fig. 4 by

$$l_8 - l_7 , \ l_7 - l_6 , \ l_6 - l_5 , \ l_5 - l_4 , \ l_4 - l_3 , \ l_3 - l_2 , \ l_2 - l_1 , \ l_1 - l_0 ,$$
$$1.$$

**14. Postscript.** After I had written the above, Olga Taussky Todd drew my attention to a recent paper by Mahler [15], which refers to Dickson [9]. I thus became aware that Dickson himself had discovered a system of integral Cayley numbers as long ago as 1923. As we saw in §4, the seven possible systems may be described as specializing (in turn) the seven units $i$, $j$, $k$, $e$, $ie$, $je$, $ke$, or $i_1$, $i_2$, $i_3$, $i_4$, $i_5$, $i_6$, $i_7$. In defining $J$, I chose to specialize $e$, thus preserving the cyclic symmetry of $i$, $j$, $k$. Dickson obscured this symmetry by specializing $i$ (or $j$ or $k$) instead of $e$. In fact, he asserted [9; 324] that there are only three systems which include all the eight basic units and satisfy the conditions for a set of integral elements. He missed the remaining four of the seven systems by adding the very artificial requirement [9; 321] that the integral Cayley numbers should include the integral quaternions $\frac{1}{2}(\pm 1 \pm i \pm j \pm k)$.

Possibly Mahler's work would have been simplified by specializing $e$ instead of $i$. He says the following theorem [15; 127] is "not easy to prove:"

For every Cayley number $X$ there is an integral Cayley number $G$ such that $N(X - G) \leq \frac{1}{2}$.

Dickson had succeeded only in proving the statement with $5/4$ in place of $\frac{1}{2}$. We now recognize this refinement as an immediate consequence of §11.

Finally, Mahler proved that every left (or right) ideal in the ring of all integral Cayley numbers is a principal ideal, generated by a multiple (in the ordinary sense) of an integral Cayley number of norm 1 or 2 or 4; but not every integral Cayley number of norm 2 or 4 will serve. It would be interesting to find out what subsets of the 2160 of norm 2 and of the 17280 of norm 4 are involved here, and what polytopes are formed by the corresponding sets of points in eight-dimensional space.

### REFERENCES

1. ELIE CARTAN AND J. A. SCHOUTEN, *On Riemannian geometries admitting an absolute parallelism*, Koninklijke Akademie van Wetenschappen te Amsterdam, Proceedings of the Section of Sciences, vol. 29(1926), pp. 933–946.

2. ARTHUR CAYLEY, *On Jacobi's elliptic functions, in reply to the Rev. B. Bronwin; and on quaternions*, The Collected Mathematical Papers, vol. 1(1889), p. 127.

3. ARTHUR CAYLEY, *Note on a system of imaginaries*, The Collected Mathematical Papers, vol. 1(1889), p. 301.

4. H. S. M. COXETER, *The polytopes with regular-prismatic vertex figures*, Part 1, Philosophical Transactions of the Royal Society of London (A), vol. 229(1930), pp. 329–425.

5. H. S. M. COXETER, *The polytopes with regular-prismatic vertex figures*, Part 2, Proceedings of the London Mathematical Society (2), vol. 34(1932), pp. 126–189.

6. H. S. M. COXETER, *The binary polyhedral groups, and other generalizations of the quaternion group*, this Journal, vol. 7(1940), pp. 367–379.

7. H. S. M. COXETER, *Regular Polytopes*, New York, 1963.

8. L. E. DICKSON, *On quaternions and their generalization and the history of the eight square theorem*, Annals of Mathematics (2), vol. 20(1919), pp. 155–171.

9. L. E. DICKSON, *A new simple theory of hypercomplex integers*, Journal de Mathématiques Pures et Appliquées (9), vol. 2(1923), pp. 281–326.

10. L. E. DICKSON, *Algebras and their Arithmetics*, Chicago, 1923.

11. THOROLD GOSSET, *On the regular and semi-regular figures in space of n dimensions*, Messenger of Mathematics, vol. 29(1900), pp. 43–48.

12. W. R. HAMILTON, *Note respecting the researches of John T. Graves, Esq.*, Transactions of the Royal Irish Academy, vol. 21(1848), pp. 338–341.

13. ADOLF HURWITZ, *Über die Zahlentheorie der Quaternionen*, Mathematische Werke, vol. 2 (1933), pp. 303–330.

14. JOHANNES KIRMSE, *Über die Darstellbarkeit natürlicher ganzer Zahlen als Summen von acht Quadraten und über ein mit diesem Problem zusammenhängendes nichtkommutatives und nichtassoziatives Zahlensystem*, Bericht über die Verhandlungen der Sächsischen Akademie der Wissenschaften zu Leipzig, vol. 76(1924), pp. 63–82.

15. KURT MAHLER, *On ideals in the Cayley-Dickson algebra*, Proceedings of the Royal Irish Academy (A), vol. 48(1943), pp. 123–133.

16. EMILE MATHIEU, *Sur une formule relative à la théorie des nombres*, Journal für die reine und angewandte Mathematik, vol. 60(1862), pp. 351–356.

17. RUTH MOUFANG, *Alternativkörper und der Satz vom vollständigen Vierseit ($D_9$)*, Abhandlungen aus dem Mathematischen Seminar der Hamburgischen Universität, vol. 9 (1933), pp. 207–222.

18. Ruth Moufang, *Zur Struktur von Alternativkörpern*, Mathematische Annalen, vol. 110 (1934), pp. 416–430.

19. Ludwig Schläfli, *Theorie der vielfachen Kontinuität*, Denkschriften der Schweizerischen naturforschenden Gesellschaft, vol. 38(1901), pp. 1–237.

20. Ernst Witt, *Spiegelungsgruppen und Aufzählung halbeinfacher Liescher Ringe*, Abhandlungen aus dem Mathematischen Seminar der Hansischen Universität, vol. 14(1941), pp. 289–322.

# 3

# WYTHOFF'S CONSTRUCTION FOR UNIFORM POLYTOPES

Reprinted, with the editors' permission,
from the *Proceedings of the London Mathematical Society,*
Ser. 2, Vol. 38 (1935).

1. A polyhedron is said to be uniform if its faces are regular and if it has a group of symmetries which is transitive on the vertices. Thus the uniform polyhedra are the regular and Archimedean solids, prisms and antiprisms[*]. A polytope is said to be uniform if its bounding figures are uniform and it has a group of symmetries which is transitive on the vertices. Various families of uniform polytopes have been discovered by Schläfli, Gosset, Mrs. Boole Stott, Schoute, and Elte[†].

Wythoff[‡] gave a new construction for Mrs. Stott's polytopes. The object of this paper is to apply that construction exhaustively, obtaining all the Schläfli-Gosset-Stott-Schoute-Elte polytopes and many others. It may be true that all *finite* polytopes are Wythoffian; but this would certainly be difficult to prove, since non-Wythoffian degenerate polytopes ("nets") are known[§].

2. If a discrete set of primes is symmetrical by reflexion in every one of them, it effects a partition of space into congruent *fundamental regions*[||]. If the reflexions leave a direction invariant, they transform into itself any prime normal to this direction, and so they can be regarded as operating in a space of fewer dimensions. We shall always suppose this reduction to have been carried out. If the reflexions leave a point invariant, they

---

[*] Kepler, 1, 116. (For references, see pages 52–53.)

[†] Schläfli, 1, 386, 389, 390. Gosset, 1, 45–48. Stott, 1. Schoute, 1, 35–67; 2, 73–104; 3. Elte, 1, 104, 117, 118, 123.

[‡] Wythoff, 1.

[§] Coxeter, 2, 184 (§ 20 . 3).

[||] *Cf.* Klein, 1, 20,

transform into itself any sphere* drawn around this point, and the angular fundamental region is conveniently replaced by its intersection with the sphere. It can be proved† that this spherical region is a simplex (all of whose dihedral angles are submultiples of $\pi$).

If the reflexions leave no point invariant, the fundamental region may be finite or infinite. If finite, it is a (Euclidean) simplex or simplicial prism‡; if infinite, it can be regarded as joining a finite simplex or simplicial prism to a simplex at infinity§.

3. Spherical and Euclidean simplexes whose dihedral angles are submultiples of $\pi$ have been completely enumerated‖, and are conveniently represented by diagrams consisting of linked dots¶. Each dot denotes a bounding prime of the simplex; a pair of dots not directly linked denotes a pair of perpendicular primes. A link marked $k$ denotes a dihedral angle $\pi/k$ $(k>2)$; but it is convenient to omit the mark when (as so often happens) $k=3$. Such a diagram is what topologists call a *graph***.

In the spherical case, the graph is a *forest* (with very simple trees). When the forest contains more than one tree††, the vertices of the simplex fall into sets which lie in absolutely perpendicular spaces, each set forming a simplex represented by one of the trees.

In the Euclidean case, the graph is either a tree or an isolated circuit (of unmarked links)‡‡. The new graph obtained by juxtaposing several such trees or circuits can be regarded as representing a prism whose constituents are the corresponding simplexes (left undetermined as to size).

The most general case that we shall consider is when the connected pieces of the graph fall into two categories: trees representing spherical simplexes, and trees or circuits representing Euclidean simplexes. The pieces of the first category together represent a spherical simplex which we regard as lying on a sphere of infinite radius. The region represented by the whole graph is obtained by joining this simplex at infinity to the prism represented by the pieces of the second category.

---

* We use the word *sphere* for the analogue in any number of dimensions.
† Coxeter, **5**, 596 (Theorem 4).
‡ Called *prismotope* in Schoute, **3**.
§ Coxeter, **5**, 599 (Theorem 7).
‖ Coxeter, **2**, 144.
¶ Coxeter, **3**, 133.
** Whitney, **1**. (We use the word *dot* for *vertex*, to avoid confusion with vertices of polytopes.)
†† Coxeter, **2**, 137 (§ 15.3).
‡‡ Coxeter, **2**, 138 (§ 15.4).

4. With every discrete group generated by reflexions we can associate the graph that represents its fundamental region. The dots of the graph correspond to a set of generators of the group, and the abstract definition can be read off in the manner described by Coxeter, **2**, 164. If the graph is disconnected, the group is the direct product of the groups represented by the several trees or circuits. For the orders of these groups, see Coxeter, **5**, 618, 619, or **6**.

5. Take first the case when the fundamental region is a simplex. We use capital letters to denote its vertices, and corresponding small letters for the opposite bounding primes. Consider any vertex $A$, and its transforms under all the operations of the group, *i.e.* the $A$-vertices of all the simplexes that arise by repeated reflexion in the primes*. These points are the vertices of a uniform polytope which can be symbolized by drawing a little ring around the $a$-dot of the graph. The graph with both the $a$-dot and the $b$-dot ringed symbolizes the polytope whose typical vertex is the point of intersection of the edge $AB$ of the simplex with the bisector of the dihedral angle $ab$. When the $a$-, $b$-, and $c$-dots are ringed, the typical vertex is the point of concurrence of the three lines in which the plane $ABC$ is met by the bisectors of the angles $bc$, $ca$, $ab$†, and so on. Finally, the graph with all dots ringed symbolizes the polytope whose vertices are the in-centres of all the simplexes‡.

If the graph is disconnected, the polytope is the prism whose constituents are symbolized by those connected pieces which contain ringed dots. There may be unringed pieces, but they will have no effect. It is, therefore, natural to make an unringed graph symbolize a mere point. In particular, the null graph symbolizes a point.

Consider the sub-graph derived by removing certain dots along with all the links that are incident in those dots. The corresponding polytope is an element of the polytope symbolized by the whole graph, and every element can be obtained in this manner§. Unringed pieces of the sub-graph have no effect on the shape of the element; but, in order to calculate the *number* of elements of a particular type, we must make the unringed pieces as extensive as possible (and discard all sub-graphs that could be derived from others by reducing unringed pieces). The number is then $\Gamma/\gamma$, where $\Gamma$ and $\gamma$ are the orders of the groups represented by the whole

---

* *Cf.* Klein, **1**, 23.
† For a diagram, see Robinson, **1**, 71.
‡ *Cf.* Möbius, **1**, 661, 677, 691.
§ *Cf.* Schoute, **1**, 16; **2**, 15, 80.

graph and sub-graph regardless of rings. In particular, the *vertices* are given by removing all the ringed dots (and their links)*; *e.g.*, the polytope

has $\Gamma/1 = 696729600$ vertices.

In order that $\pi_r$ and $\pi_s$ $(r > s)$ may be *incident* elements (of $r$ and $s$ dimensions), all ringed pieces of the sub-graph symbolizing $\pi_s$ must occur in the sub-graph symbolizing $\pi_r$. The number of such $\pi_r$'s incident in each $\pi_s$ is then $\gamma/\gamma'$, where $\gamma$ is the order of the group represented by the sub-graph symbolizing $\pi_s$ (regardless of rings), while $\gamma'$ is the order of the group represented by the common part of the two sub-graphs. Thus the number is 1 whenever the sub-graph symbolizing $\pi_r$ contains the whole of that symbolizing $\pi_s$; *e.g.*, in the particular polytope mentioned above, any vertex belongs to one edge of each of eight distinct types, symbolized by the separate ringed dots.

6. When the fundamental region is a prism, we must stipulate that every connected piece of the graph has at least one ringed dot. The rule of § 5 then gives a definite point in each constituent of the prism, and the typical vertex of the derived polytope is constructed from such points in an obvious manner. It is always possible to adjust the relative size of the constituents so as to make the polytope equilateral, and so uniform. In order to obtain the elements of the polytope (which is a degenerate prism†), we now have to remove at least one dot from *every* connected piece.

This construction can be extended to the case when the graph involves pieces of both categories, but then it yields merely semi-degenerate‡ prisms.

7. In order to enumerate the polytopes that can be derived by such a construction, it will be sufficient to consider *connected* graphs, in which case the fundamental regions must occur among the simplexes considered in Coxeter, **3**. All other polytopes of this kind can then be derived as prisms.

---

* *Cf.* Robinson, **1**, 73.

† Coxeter, **1**, 353; **2**, 187.

‡ Coxeter, **1**, 354. The semi-degenerate prism $[\delta_{n-1}, \alpha_1]$ is described by Gosset, **1**, 45.

7.1. If the simplex* is $(k_1, k_2, ..., k_{m-1})$, so that the graph is a simple chain of $m-1$ links, we shall number the dots $0, 1, ..., m-1$. When the 0-dot is ringed, the polytope is $\{k_1, k_2, ..., k_{m-1}\}$†. When the $n$-dot is ringed, it is $t_n\{k_1, k_2, ..., k_{m-1}\}$‡. Hence we naturally let

$$t_{f, g, ..., l}\{k_1, k_2, ..., k_{m-1}\} \quad (0 \leqslant f < ... < l < m)$$

denote the polytope symbolized when the $f$-, $g$-, ..., $l$-dots are all ringed§. Robinson‖ has shown that the operator $t_{f, ..., l}$ is equivalent to Mrs. Stott's $ce_f ... e_l$, with the convention that $ce_0$ is identity. In fact, $e_i$ adds a ring to the $i$-dot, while $c$ (or $c_0$) removes the ring from the 0-dot.

7.2. If the simplex is $\begin{pmatrix} 3^n \\ 3^p \\ 3^q \end{pmatrix}$, so that the graph consists of three simple chains, of $n, p, q$ links, all emanating from one special dot¶, we shall call the special dot 0 and number the remaining dots $1, 2, ...$ along each chain. The polytope derived by ringing certain dots will be denoted by a symbol of the form

$$(n_{r, ..., t}\, p_{u, ..., w}\, q_{x, ..., z}) \quad \text{or} \quad (n_{r, ..., t}\, p_{u, ..., w}\, q_{x, ..., z})_0$$

$$(n \geqslant r > ... > t > 0, \quad p \geqslant u > ... > w > 0, \quad q \geqslant x > ... > z > 0),$$

where the suffixes indicate which dots are ringed**.

In particular,

$$(n_n\, p\, q) = n_{pq}\,††,$$

$$(n_{n-l}\, p\, q) = t_l n_{pq} \quad (l < n),$$

$$(n\, p\, q)_0 = 0_{npq},$$

$$(n\, p_1\, q_1) = t_{n+1} n_{pq},$$

$$(n\, p_{p'+1}\, q_{p'+1}) = t_{p'\, p'}\, n_{pq} \quad (p' < p,\ p' < q)‡‡.$$

---

\* Coxeter, **3**, 133.

† Todd, **1**.

‡ Coxeter, **1**, 354.   $t_1\{3, 3, 5\}$ is the *octicosahedric* of Gosset, **1**, 46.

§ This agrees with the notation $t_{0,1}$, $t_{1,2}$ of Coxeter, **4**. See also Du Val, **1**, 48 (foot-note).

‖ Robinson, **1**, 72.

¶ Coxeter, **3**, 133.

\*\* Thus the polytope used as an example in § 5 is $(4_{1,3,2,1}\, 2_{2,1}\, 1_1)_l$.

†† The polytopes $n_{21}$ ($n = 0, 1, 2, 3, 4, 5$) are described by Gosset, **1**, 45, 47, 48. $1_{22}$, $2_{31}$, $1_{32}$, $2_{41}$ are Elte's $V_{72}$, $V_{126}$, $V_{576}$, $V_{2160}$, respectively. Du Val, **1**, 73, mistakenly describes $n_{pq}$ as $(n_1 p\, q)$.

‡‡ Coxeter, **1**, 412 (" the case $p' = q'$"). The symbols $t_{01}$, $t_{02}$ of Du Val, **1**, 37–39, should have contained commas: $t_{0,1}\, 5_{21} = (5_{5,4}\, 2\, 1)$, $t_{0,2}\, 5_{21} = (5_{5,3}\, 2\, 1)$.

The rest of the polytopes $(2_1\,2\,1)$, ..., $(5_{5,4,3,2,1}\,2_{2,1}\,1_1)_0$,

$$(3_1\,3\,1), \quad ..., \quad (3_{3,2,1}\,3_{3,2,1}\,1_1)_0,$$

$$(2_1\,2\,2), \quad ..., \quad (2_{2,1}\,2_{2,1}\,2_{2,1})_0$$

are now published for the first time*.

It is sometimes convenient to allow the final zero suffix to be brought inside the brackets and appended to one of $n, p, q$ (*i.e.* to associate the 0-dot with a particular one of the three chains); *e.g.*†

$$(n_0\,p\,q) = (n\,p_0\,q) = (n\,p\,q_0) = (n\,p\,q)_0.$$

With this convention,

$$(n_{n-f,\,...,\,n-l}\,1\,1) = t_{f,\,...,\,l}\,\beta_{n+3} \qquad (l \leqslant n)$$

and

$$(n_{n-f,\,...,\,n-l}\,1_1\,1_1) = t_{f,\,...,\,l,\,n+1}\,\beta_{n+3}.$$

Since‡

$$h\gamma_{n+3} = 1_{n1} = (n\,1_1\,1)$$

and

$$t_{f,\,...,\,l}\,\gamma_{n+3} = t_{n+2-l,\,...,\,n+2-f}\,\beta_{n+3} = (n_{l-2,\,...,\,f-2}\,1\,1) \quad (f \geqslant 2),$$

it is natural to define

$$h_{f,\,...,\,l}\,\gamma_{n+3} = (n_{l-2,\,...,\,f-2}\,1_1\,1) \quad (1 < f < \,...\, < l < n+3).$$

Then, in Schoute's notation§,

$$h_{f,\,...,\,l}\,\gamma_m = e_f \,...\, e_l H M_m.$$

We are now numbering the dots in the graph for $\begin{pmatrix} 3^{m-3} \\ 3 \\ 3 \end{pmatrix}$ as follows:

**7.3.** The graph for $\begin{pmatrix} 3^{m-4},\,4 \\ 3 \\ 3 \end{pmatrix}$ differs from this in having the final link marked 4‖. By ringing the dots, we obtain derivatives of $\delta_m$ analogous to

\* Except those mentioned in the preceding foot-note.

† *Cf.* Coxeter, **1**, 372 (7.36), 411 (12.31).

‡ Coxeter, **1**, 372 (7.43).

§ Schoute, **2**, 87.   Robinson, **1**, 73–74 (§ 3).

‖ Coxeter, **3**, 135 (i).

the above derivatives of $\beta_m$ and $\gamma_m$. In particular\*, $h\delta_m$ is

and by ringing the $f$-, ..., $l$- dots we obtain Schoute's "*hmpd.* nets"†

$$h_{f,\,...,\,l}\delta_m = e_{f-1}...e_{l-1}NH_{m-1} \quad (1 < f < ... < l < m).$$

7.4. The simplex‡ $\begin{pmatrix} 3, & & 3 \\ & 3^{m-5}, & \\ 3, & & 3 \end{pmatrix}$ leads first to alternative symbols

for some of the above polytopes, in accordance with the rule that

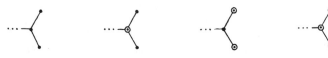

can be replaced by

respectively. But

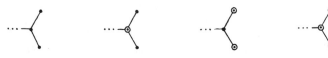

(with $m$ dots altogether) appears to be a new discovery (for $m > 5$). We shall call it§ $q\delta_m$, and ring the $f$-, ..., $l$- dots to derive

$$q_{f,\,...,\,l}\delta_m \quad (1 < f < ... < l < m-2).$$

(Most of these polytopes will have two alternative symbols, on account of the symmetry of the unringed graph; *e.g.*, $q_3\delta_6 = q_2\delta_6$.)

In terms of Mrs. Stott's *partial operations*‖, we find that

$$h_{f,\,...,\,l} = \tfrac{1}{2}e_0 ce_f...e_l \quad \text{and} \quad q_{f,\,...,\,l} = \tfrac{1}{2}e_{m-1}h_{f,\,...,\,l}.$$

The vertices of $q\delta_m$ (edge $\sqrt{2}$) can be selected from the totality of points having $m-1$ integral coordinates by the following rule: if $r_n$ of the coordinates are congruent to $n \pmod 4$, so that $r_0+r_1+r_2+r_3 = m-1$, then

$$r_1 \equiv r_2 \equiv r_3 \pmod 2.$$

---

\* Coxeter, **1**, 365. $h\delta_4$ is the *simple tetroctahedric check* of Gosset, **1**, 46.

† Schoute, **2**, 98.

‡ Coxeter, **3**, 135 (ii).

§ $h$ for *half* ; $q$ for *quarter*.

‖ Stott, **1**, 15.

It is therefore proper to let $q\delta_4$ denote

(Andreini's* Fig. 15).

7.5. The simplex† $\begin{pmatrix} 3 \\ 3 \\ 3 \\ 3 \end{pmatrix}$ gives nothing new. In fact, with an obvious

extension of the $(n\,p\,q)$ notation,

$$(1_1 1 1 1) \quad = h\delta_5 \quad = \{3, 3, 4, 3\}$$

$$(1_1 1_1 1 1) \quad = q\delta_5 \quad = t_1\delta_5$$

$$(1_1 1_1 1_1 1) = h_3\delta_5 = t_2\{3, 3, 4, 3\}$$

$$(1_1 1_1 1_1 1_1) = t_{1,3}\delta_5 = t_1\{3, 4, 3, 3\}$$

$$(1 1 1 1)_0 \quad = t_2\delta_5 \quad = \{3, 4, 3, 3\}$$

$$(1_1 1 1 1)_0 \quad = h_2\delta_5 \quad = t_{0,1}\{3, 3, 4, 3\}$$

$$(1_1 1_1 1 1)_0 \quad = q_2\delta_5 \quad = t_{1,2}\delta_5$$

$$(1_1 1_1 1_1 1)_0 \quad = h_{2,3}\delta_5 = t_{1,2}\{3, 3, 4, 3\}$$

$$(1_1 1_1 1_1 1_1)_0 = t_{1,2,3}\delta_5 = t_{0,1}\{3, 4, 3, 3\}.$$

7.6. Finally, a circuit‡ of $m$ dots and $m$ links with one dot ringed symbolizes§ $a_{m-1}h$; and by ringing further dots we derive Schoute's *simplex nets*‖. The symbolism is perfect, since the numbers of links between consecutive ringed dots form Schoute's *partition cycle*¶. In particular, a polygon with every vertex ringed is the symbol for the space-filling investigated by Hinton**.

8. As an example, let us enumerate the elements of the six-dimensional polytope $(2_2\,2\,1_1)$ or

(for which $\Gamma = 51840$).

(In the figures below, for greater clarity, the omitted links have been allowed to appear as broken lines.)

---

* Andreini, **1**, 107. Fig. 13 will be mentioned later. Fig. 16 is not uniform. Figs. 12, 14, 17, *et seq.*, are respectively $h\delta_4$, $t_{1,2}\delta_4$, $t_{0,1}\delta_4$, $t_1\delta_4$, $h_3\delta_4$, $t_{0,2}\delta_4$, $t_{0,1,2}\delta_4$, $t_{0,1,3}\delta_4$, $h_{2,3}\delta_4$, $h_2\delta_4$, $t_{0,1,2,3}\delta_4$.

  † Coxeter, **3**, 135 (iii).

  ‡ *Ibid.*, (iv).

  § Coxeter, **1**, 366.

  ‖ Schoute, **1**, 35–67.

  ¶ Schoute, **1**, 49.

  ** Hinton, **1**. (He considers explicitly only the cases $m = 3, 4, 5$. For the general case, see Schoute, **1**, 59.)

ELEMENTS OF $(2_2 2 1_1)$.

|  |  | |  |  |
|---|---|---|---|---|
| $h\gamma_5$ | $\Gamma/1920 = 27$ | | $a_0$ | $\Gamma/120 = 432$ |
| $[\alpha_4, \alpha_1]$ | $\Gamma/120.2 = 216$ | | $a_4$ | $\Gamma/120 = 432$ |
| $[\alpha_2, \alpha_1]$ | $\Gamma/6.2.6 = 720$ | | $a_1$ | $\Gamma/2.2.6 = 2160$ |
| $t_{0,3}\alpha_4$ | $\Gamma/120.2 = 216$ | | $a_3$ | $\Gamma/24.2 = 1080$ |
| $h_4\gamma_5$ | $\Gamma/1920 = 27$ | | $\beta_4$ | $\Gamma/192 = 270$ |
| $\alpha_5$ | $\Gamma/720 = 72$ | | $\beta_2$ | $\Gamma/2.2.6 = 2160$ |
| $\alpha_4$ | $\Gamma/120 = 432$ | | $[\alpha_2, \alpha_1]$ | $\Gamma/6.2.2 = 2160$ |
| $\alpha_3$ | $\Gamma/24.2 = 1080$ | | $[\alpha_3, \alpha_1]$ | $\Gamma/24.2 = 1080$ |
| $\alpha_2$ | $\Gamma/6.6 = 1440$ | | $\alpha_2$ | $\Gamma/6.2 = 4320$ |
| $\alpha_1$ | $\Gamma/2.24 = 1080$ | | $\alpha_3$ | $\Gamma/24 = 2160$ |

As a check, we can apply the extension of Euler's theorem, obtaining

$$432 - (1080 + 2160) + (1440 + 2160 + 4320)$$

$$- (720 + 1080 + 1080 + 2160 + 2160) + (216 + 432 + 432 + 270 + 1080)$$

$$- (27 + 216 + 27 + 72) = 0.$$

9. When only one dot is ringed, the vertex figure* of the polytope can be obtained by the following rule. If all the links incident in the ringed dot are unmarked, remove it and them, and ring the dots to which they lead. (*E.g.*, the vertex figure of $a_4 h$ or is or $t_{0,3} a_4$, formerly called $ea_4$.) On the other hand, if any of those links are marked (in which case the graph is necessarily a tree), the rule as just stated gives a forest, symbolizing a prism with one constituent for each removed link; whenever the link is marked $k$, the corresponding constituent must be given edge $2 \cos \pi/k$; *e.g.*, the vertex figure of the cuboctahedron

is the rectangle or $[a_1, \beta_1]$.

The calculations of Coxeter, 1, 381-383, were based on the assumption that the polytopes $n_{21}$ exist. We can now justify this assumption without the use of coordinates.

10. The vertex figure is again simple when *every* dot is ringed. If the polytope is finite, its vertex figure is a simplex having the same vertices as the spherical polar of the fundamental region. If the polytope is degenerate, its vertex figure is the reciprocal of the fundamental region with respect to a concentric† sphere of suitable radius.

11. Since the product of an even number of reflexions is a *positive* operation, it is always possible to regard the fundamental regions as being alternately black and white. We have seen that the graph with all dots ringed symbolizes a polytope with one vertex in every region. Another polytope can be constructed by taking a vertex in every region of one colour‡. But it is seldom possible to choose the position of the typical vertex in its region so as to make the polytope equilateral, and so uniform. (When this is possible, the vertex will not, in general, be the in-centre of the region.)

Apart from three semi-degenerates§, the actual cases are as follows.

---

\* Coxeter, 1, 336.

† *I.e.*, concentric with the *inscribed* sphere that the region must have in this case.

‡ Klein, 1, 23. Möbius, 1, 656, 669, 688.

§ The *tetroctahedric semi-check* of Gosset, 1, 47, corresponds to the graph that consists of a triangle and an isolated dot.

| Fundamental region | Graph | Polytope |
|---|---|---|
| $(2^{m-1})$* | • • • ... (m dots) | $h\gamma_m$ |
| $\gamma_{m-1}$ | •—∞—• •—∞—• ... (m−1 pieces) | $h\delta_m$ |
| $(k)$ | •—$k$—• | $\{k\}$ |
| $(2, k)$ | • •—$k$—• | $\{k\}$-antiprism |
| $(3, 3)$ | •—•—• | $\{3, 5\}$ |
| $(3, 4)$ | •—•—$4$—• | The snub cube $s\left\{{3\atop4}\right\}$ |
| $(3, 5)$ | •—•—$5$—• | The snub dodecahedron $s\left\{{3\atop5}\right\}$ |
| $(3, 6)$ | •—•—$6$—• | $s\left\{{3\atop6}\right\}$ (Kepler's† $L$) |
| $(4, 4)$ | •—$4$—•—$4$—• | $s\left\{{4\atop4}\right\}$ (Kepler's $N$) |
| $\alpha_2$ | △ | $\{3, 6\}$ |
| $\left({3\atop{3\atop3}}\right)$ | (Y-shaped graph) | $s\{3, 4, 3\}$ (Gosset's *tetricosahedric*)‡ |
| $[\alpha_2, \alpha_1\sqrt{\tfrac{2}{3}}]$ | △ •—∞—• | Gosset's *complex tetroctahedric check*‡ (Andreini's§ Fig. 13) |
| $\left({3\atop{3\atop{3\atop3}}}\right)$ | (X-shaped graph) | A new degenerate polytope $s\{3, 4, 3, 3\}$ |

\* Meaning $(2, 2, \ldots , 2)$.

† Kepler, **1**, 117 (XIX).

‡ Gosset, **1**, 46.

§ Andreini, **1**, 106.

We might symbolize such a *snub**\** polytope by ringing all the dots in the graph, and then removing the dots (leaving just the rings and links). For the dots represent reflections which are not symmetries of the polytope.

12. $s\{3, 4, 3\}$ is bounded by 24 icosahedra and $96+24$ tetrahedra. Its 96 vertices can be selected from the 120 vertices of $\{3, 3, 5\}$ by removing 24 that belong to an inscribed $\{3, 4, 3\}$. The icosahedra are the actual vertex figures of $\{3, 3, 5\}$ at the 24 removed vertices, and the tetrahedra all occur among the 600 bounding tetrahedra of $\{3, 3, 5\}$.

$s\{3, 4, 3, 3\}$ has, at each vertex, five $\alpha_4$'s, one $\beta_4$, and four $s\{3, 4, 3\}$'s.

Mrs. Stott has pointed out that $s\{3, 4, 3\}$ can be constructed by taking one vertex in every edge of $\{3, 4, 3\}$, not at the mid-point (which would give $t_1\{3, 4, 3\}$), but at one of the two points of "golden section" (ratio $\tau : 1$†). $s\{3, 4, 3, 3\}$ is similarly derivable from $\{3, 4, 3, 3\}$. This process can be applied to any regular polytope for which $k_2$ is even, the remaining cases being

$$s\beta_3 = \{3, 5\} \text{ (ratio } \tau : 1), \quad s\delta_3 = s\begin{Bmatrix} 4 \\ 4 \end{Bmatrix} \text{ (ratio } \sqrt{3} : 1),$$

$$s\{3, 6\} = \{3, 6\} \text{ (ratio } 2 : 1).$$

### References.

A. Andreini, 1.   "Sulle reti di poliedri regolari e semiregolari", *Mem. di mat. e di fis. della soc. ital. delle sci.* (3), 14 (1907), 75–110.

H. S. M. Coxeter, 1.   "The polytopes with regular-prismatic vertex figures", *Phil. Trans. Royal Soc.* (A), 229 (1930), 329–425.

————, 2.   (The same; Part 2), *Proc. London Math. Soc.* (2), 34 (1932), 126–189).

————, 3.   "Groups whose fundamental regions are simplexes", *Journal London Math. Soc.*, 6 (1931), 132–136.

————, 4.   "The densities of the regular polytopes", *Proc. Camb. Phil. Soc.*, 27 (1931), 211.

————, 5.   "Discrete groups generated by reflections", *Annals of Math.*, 35 (1934), 588–621.

————, 6.   "The complete enumeration of finite groups of the form $R_i^2 = (R_i R_j)^{k_{ij}} = 1$", *Journal London Math. Soc.*, 10 (1935).

P. Du Val, 1.   "On the directrices of a set of points in a plane", *Proc. London Math. Soc.* (2), 35 (1933), 23–74.

E. L. Elte, 1.   *The semiregular polytopes of the hyperspaces* (Groningen, 1912).

T. Gosset, 1.   "On the regular and semi-regular figures in space of $n$ dimensions", *Messenger of Math.*, 29 (1900), 43–48.

---

\* The word "snub" is a free translation of Kepler's *simus*.

† Here $\tau = \frac{1}{2}(\sqrt{5}+1)$. Convenient coordinates for the vertices of $s\{3, 4, 3\}$ (edge 2) are the even permutations of $\pm(\tau^2, \tau, 1, 0)$.

C. H. Hinton, 1.  *The fourth dimension* (London, 1906), 135, 225.

J. Kepler, 1.  *Harmonia Mundi*, lib. II [Opera Omnia, 5 (Frankfurt, 1864), 114–127.]

F. Klein, 1.  *Vorlesungen über das Ikosaeder und die Auflösung der Gleichungen vom fünften Grade* (Leipzig, 1884).

A. F. Möbius, 1.  "Symmetrische Figuren", *Gesammelte Werke*, 2 (1886), 561–708.

G. de B. Robinson, 1.  " On the fundamental region of a group, and the family of configurations which arise therefrom ", *Journal London Math. Soc.*, 6 (1931), 70–75.

L. Schläfli, 1.  " Réduction d'une intégrale multiple qui comprend l'arc du cercle et l'aire du triangle sphérique comme cas particuliers ", *Journal de Math.*, 20 (1855), 359–394.

P. H. Schoute, 1.  "Analytical treatment of the polytopes regularly derived from the regular polytopes ", *Ver. der K. Akad. van Wet., Amsterdam* (1), 11.3 (1910).

————, 2.  *Ibid.*, 11.5.

————, 3.  " The characteristic numbers of the prismotope ", *Proc. Royal Acad. of Sci. Amsterdam*, 14 (1911), 424–428.

A. Boole Stott, 1.  "Geometrical deduction of semi-regular from regular polytopes and space fillings ", *Ver. der K. Akad. van Wet., Amsterdam* (1), 11.1 (1910).

J. A. Todd, 1.  "The groups of symmetries of the regular polytopes", *Proc. Camb. Phil. Soc.*, 27 (1931), 213–217.

H. Whitney, 1.  "Non-separable and planar graphs ", *Trans. American Math. Soc.*, 34 (1932), 339–340.

W. A. Wythoff, 1.  "A relation between the polytopes of the $C_{600}$-family ", *Proc. Royal Acad. of Sci., Amsterdam*, 20 (1918), 966–970.

# 4

# THE CLASSIFICATION
# OF ZONOHEDRA
# BY MEANS OF PROJECTIVE
# DIAGRAMS

Reprinted, with the editors' permission,
from *Journal de Mathématiques pures et appliquées*,
tome 41 (1962).

**1. Introduction.** — In Euclidean 3-space we may define, for any lattice (of the kind used in crystallography and the geometry of numbers), a *Voronoi polyhedron*, consisting of all the points which are as near to one lattice point as to any other.   An affine variant of such a polyhedron is the *parallelohedron*, which may be described as a convex fundamental region for a group of translations.   In 1885, Fedorov enumerated the five types of parallelohedra in the course of his crystallographic investigations.   In 1933, Delaunay discovered a simpler approach to the same enumeration.   The present paper advocates a third procedure which is claimed to be still simpler. Meanwhile, a fourth method has been discovered by Moser [16b].[*] Like Fedorov, we consider first the more general *zonohedron* : a solid bounded by centrally symmetrical polygons.   After a preliminary account of the analogous problem in two dimensions, we develop, in the real projective plane, a representative configuration consisting of a finite number of lines with all their points of intersection.   We find, in paragraph **9**, that the parallelohedra are distinguished from other zonohedra by a simple property of the configuration, namely, that every line contains either two or three of the points.   This means that the configuration for a parallelohedron cannot be more complicated than a complete quadrangle. Finally, in paragraphs **11-14** we illustrate a possible extension to four dimensions by two particular examples which have a connection with the geometry of numbers.

[*]For bibliography, see pages 73–74.

**2.** CENTRALLY SYMMETRICAL POLYGONS. — In the affine plane, a polygon is said to be *centrally symmetrical* if it is transformed into itself by a half-turn (or " central inversion ", or " reflection in a point "). Clearly, a centrally symmetrical polygon has an even number of vertices, occurring in opposite pairs whose joins are all bisected by the centre. Any two opposite sides are equal and parallel. When there are only four sides (and four vertices) the polygon is a parallelogram.

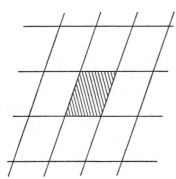 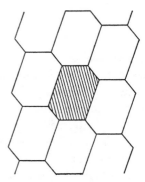

Fig. 1.

By examining figure 1 we see that any centrally symmetrical polygon which has four or six sides can be repeated by translations to form a tessellation filling the whole plane (without overlapping and without interstices). Conversely, if a convex polygon can be repeated by translations to fill the plane, the number of its sides cannot exceed six. To prove this, we observe that a suitable subgroup of the group of translations will serve as the fundamental group for a torus ([9], pp. 14, 16) on which we can draw a finite map locally identical with the given infinite tessellation. If such a map has V vertices, E edges, and F $p$-gonal faces, we easily verify that

(2.1) $$3V \leq 2E = pF.$$

Since the Euler-Poincaré characteristic is zero, so that

$$E - V = F,$$

we obtain

$$\tfrac{1}{2}\,p\mathrm{F} \;=\; \mathrm{E} \;\le\; 3\,(\mathrm{E} - \mathrm{V}) \;=\; 3\mathrm{F},$$

whence

$$p \le 6.$$

A different proof, yielding the number $2(2^n - 1)$ in $n$ dimensions, was given by Voronoi (*see* [3], p. 225).

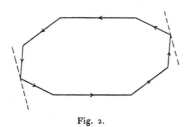

Fig. 2.

Figure 2 shows vectors along the successive sides of a centrally symmetrical $2m$-gon.  Figure 3 shows the same $2m$ vectors all issuing from one point so as to form a *star*, which consists of $m$ line-segments with a common midpoint O.

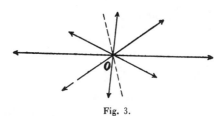

Fig. 3.

A line through O in an arbitrary direction (such as the broken line in figure 3) lies in a pair of opposite angles of the star and represents a pair of parallel *lines of support* of the polygon (*fig.* 2). In fact, the vertex angles of the polygon are the supplements of the angles in the star.  Thus the $m$ pairs of opposite vertices are represented by the $m$ pairs of opposite angular regions into which the $m$ lines of the star decompose the plane.   If we embed the affine plane in a projective plane by adding a line at infinity, we can say that the $m$ lines of the star decompose the line at infinity (which is a real projective line) into $m$ segments which represent the $m$ pairs of opposite vertices of the polygon.

Evidently the star, like the polygon, has central symmetry. Giving the plane a Euclidean metric, we can say further that any symmetry-operation of the polygon belongs also to the star, and *vice versa*.   In particular, if the polygon is a rectangle or a rhombus (so that its symmetry group is the four-group $\mathbb{D}_2$), the star consists of the diagonals of a rhombus or a rectangle, respectively; and if the polygon is a regular $2m$-gon (so that its group is the dihedral group $\mathbb{D}_{2m}$ of order $4m$), the star joins opposite vertices of another regular $2m$-gon.

An important example of a centrally symmetrical polygon in the Euclidean plane arises as the *Dirichlet region* of a lattice : the polygon whose interior consists of all the points of the Euclidean plane that are nearer to one lattice point than to any other ([11], p. 216). Since the lattice itself is centrally symmetrical, and each side of the region is part of the locus of points equidistant from two lattice points, a given $2m$-gon can be used as a Dirichlet region if and only if it satisfies the following three conditions :

(i) it is centrally symmetrical;

(ii) $m = 2$ or $3$;

(iii) the lines joining the midpoints of pairs of opposite sides are perpendicular to those sides; that is to say, the vertices all lie on a circle.

Thus the only parallelograms admissible as Dirichlet regions are rectangles (including squares).

**5. Zonohedra.** — In affine 3-space, a *zonohedron* ([12], p. 257) is a convex polyhedron each of whose faces is centrally symmetrical. Every edge determines a *zone* of faces in which every two adjacent faces meet in an edge parallel to the given edge.   If edges occur in $n$

different directions, there are $n$ zones. Since two zones containing a common face meet again in an antipodal face, the whole zonohedron is centrally symmetrical [5]. Given the zonohedron, a practical method for computing $n$ is to count the edges that have to be traversed in taking a minimal journey from any vertex to the opposite vertex.

In a manner analogous to the transition from figures 2 to 3, we can represent a zonohedron that has $n$ zones by a 3-dimensional star consisting of $n$ line-segments with a common midpoint O. Each line-segment is parallel to a zone, and twice as long as the edges that belong to the zone. Representing any two parallel planes of support of the zonohedron by a parallel plane through O, we see that the 3-dimensional star contains a 2-dimensional star for each pair of opposite faces of the zonohedron. Thus any given zonohedron can be represented by a star.

Conversely, given the star, we can reconstruct the zonohedron. To see this, denote the vectors in the star by

$$\pm e_1, \quad \pm e_2, \quad \ldots, \quad \pm e_n,$$

with an arbitrary choice of sign for each pair. The corresponding zonohedron consists of the points whose position-vectors are

$$\theta_1 e_1 + \theta_2 e_2 + \ldots + \theta_n e_n,$$

where $0 \leq \theta_k \leq 1$ $(k = 1, 2, \ldots, n)$. Instead of $0$ to $1$, the range of $\theta_k$ could equally well have been chosen from $-\frac{1}{2}$ to $\frac{1}{2}$, thus we can interchange the symbols $e_k$ and $-e_k$ without altering the zonohedron. (Such a change amounts to shifting the origin in the solid by the vector $e_k$.)

**4.** THE FIRST PROJECTIVE DIAGRAM. — If we embed the affine space in a projective space by adding a plane at infinity, we may consider, instead of the lines and planes of the star, their sections by the plane at infinity (which is a real projective plane). In other words, we can abstractly identify the lines and planes of the star with points and lines in a real projective plane. The resulting configuration consists of $n$ points joined in every possible way by lines : one line for each pair of opposite faces of the zonohedron.   If the faces are $2m$-gons, the line contains $m$ of the points.

Although any given zonohedron determines a star and a consequent unique configuration, a given set of $n$ points in the plane at infinity determines not just one zonohedron but a whole family of them having edges in given directions but of various lengths.   Thus a given configuration in the real projective plane, considered abstractly, represents a family of "isomorphic" zonohedra ([8], p. 106).

The inequality (2.1) remains valid for a convex polyhedron having faces of various kinds, provided we take $p$ to be the *average* number of vertices of a face. Since now $E - V = F - 2$, we conclude that $p \leq 6 (F - 2)/F < 6$. Since the faces are $2m$-gons ($m = 2, 3, \ldots$), the number $p$ is the average of a set of numbers which can only take the values $4, 6, \ldots$. Since $p < 6$, the value 4 must occur at least once. Hence the faces of a zonohedron must include at least one pair of parallelograms ($m = 2$). In the diagram, this means that the joins of $n$ points in the real projective plane must include at least one line containing only two of the points. Thus the theory of polyhedra provides a proof for Sylvester's famous conjecture ([17], p. 452). Since the joins of $n$ points actually include at least $\dfrac{3n}{7}$ lines containing only two of the points ([14], p. 218), every zonohedron with $n$ zones has at least $\dfrac{6n}{7}$ parallelogram faces : a lower bound which is actually attained (with $n = 7$) by figure 115 of Fedorov ([12], p. 276; see also [8], p. 32, Plate II, No. 5). This solid is derived from the rhombic dodecahedron by truncating the vertices at which four faces meet ([1]), p. 19, *fig.* 6).

If the $n$ points are of general position (no three collinear) they have $\dbinom{n}{2}$ joins, and all the faces of the zonohedron are parallelograms ([8], p. 27); hence, if all the faces of a convex polyhedron are parallelograms, their number is of the form $n(n-1)$.   In the simplest instance $n = 3$ : the configuration is a triangle and the zonohedron is a parallelepiped, which has six faces.

If $m$ of the $n$ points are collinear, their line counts $\binom{m}{2}$ times among the $\binom{n}{2}$ joins. In terms of the zonohedron, this remark yields Fedorov's formula

$$(n-1)n = 1.2f_2 + 2.3f_3 + \ldots + (m-1)mf_m + \ldots,$$

where $f_m$ is the number of pairs of opposite $2m$-gons ([12], p. 277; Fedorov's $p$ is our $n$). For instance, a $2m$-gonal prism has $f_2 = m$, $f_m = 1$, and $n = m + 1$.

In this kind of projective diagram, each point represents a zone (or a set of parallel edges) and each line joining $m$ of the points represents a pair of opposite faces, namely $2m$-gons. Moreover, by considering a pair of parallel planes of support, and the parallel plane

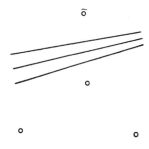

Fig. 4.

in the star, we see that each pair of opposite vertices of the zonohedron is represented in the diagram by an infinite class of lines, related to one another by continuous variations without ever passing through one of the points (*fig. 4*). Every line that avoids all the points belongs to one such class, and the number of pairs of opposite vertices appears as the number of classes. (In *Regular Polytopes* [8], p. 28, this was expressed as "the number of ways in which $n$ points in the real projective plane can be separated by a line". Of course, a line does not decompose the projective plane into two regions. However, even as few as two points, say A and B, may be said to be differently separated by a line $c$ according to which one of the two segments AB is cut by $c$.)

**5.** The second projective diagram. — This rather awkward division of lines into classes is transformed into a familiar concept by applying the principle of *duality* (both projective and topological; see *fig.* 5). In future, therefore, we shall represent each zonohedron by a new kind of projective diagram consisting of $n$ lines with all their points of intersection. Each line represents a zone, and each point represents a pair of opposite faces. If $m$ lines pass through one of the

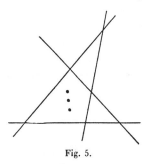

Fig. 5.

points, the corresponding faces are $2m$-gons. Each segment of a line represents a pair of opposite edges, and each *region* (bounded by such segments) represents a pair of opposite vertices. The number of sides of the region is equal to the number of edges at a vertex, and the number of points on a line is half the number of faces in a zone.

For instance, three non-concurrent lines decompose the plane into four triangular regions which represent the four pairs of opposite vertices of a parallelepiped.

**6.** Polyhedra whose faces are parallelograms. — White [20] has investigated the arrangements of $n$ lines, no three concurrent, for $n \leqq 7$. Each of these arrangements represents a zonohedron whose faces consist of $n(n-1)$ parallelograms. Since each line is decomposed into $n-1$ segments by the remaining $n-1$ lines, the zonohedron has $2(n-1)$ edges in each of its $n$ zones. The diagram has $\binom{n}{2}+1$ regions, representing the $n(n-1)+2$ vertices.

White ([20], p. 60-61) observes that there is essentially only one such arrangement of $n$ lines when $n = 3$, 4 or 5. Hence the only convex polyhedra bounded by 6, 12 or 20 parallelograms are combinatorially isomorphic to the cube, rhombic dodecahedron and rhombic icosahedron ([12], p. 261; [8], p. 29). Since the complete quadrangle forms 7 regions, namely 4 triangles and 3 quadrangles, the rhombic dodecahedron has 14 vertices : 6 where 3 faces meet, and 8 where 4 faces meet. Since the complete pentagram (*fig.* 6) has 11 regions, namely 5 triangles, 5 quadrangles and 1 pentagon, the rhombic icosahedron has 22 vertices : 10 where 3 faces meet, 10 where 4 faces meet, and 2 where 5 faces meet. It follows from

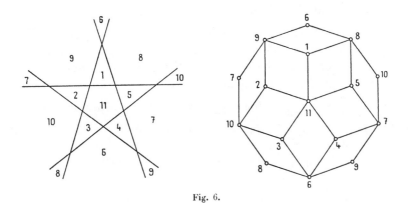

Fig. 6.

White's further investigations that there are 4 combinatorially distinct rhombic triacontahedra (one of which is the classical triaconta-hedron [15]) and at least as many rhombic 42-hedra.

The rhombic icosahedron is the case $n = 5$ of C. H. H. Franklin's *polar* zonohedron ([8], p. 29) for which the diagram consists of $n$ lines bounding a convex $n$-gon. When $n$ is odd, such a complete $n$-line can be constructed in the Euclidean plane by prolonging the sides of a regular $n$-gon.

7. Zonohedra whose faces are regular. — Another kind of zono-hedron, in Euclidean space, is the uniform polyhedron $t\begin{Bmatrix} p \\ q \end{Bmatrix}$ whose vertices are the incentres of the characteristic spherical triangles

belonging to a finite group $[p, q]$ generated by reflections ([7], p. 394; [8], p. 82).

When $p = 2$, two of the reflecting planes or *mirrors*, inclined at $\frac{\pi}{q}$, are perpendicular to the third, the network of characteristic triangles is formed by the equator and $q$ complete meridians, and the zonohedron $t\begin{Bmatrix} 2 \\ q \end{Bmatrix}$ is a $2q$-gonal *prism*. In other cases the network is formed by great circles lying in the planes of symmetry of the Platonic solid $\{p, q\}$, which has (say) V vertices, E edges, and F $p$-gonal faces. It follows that the zonohedron $t\begin{Bmatrix} p \\ q \end{Bmatrix}$ has $4$E vertices, $6$E edges, and $E + F + V$ faces : E squares, F $2p$-gons and V $2q$-gons. Each great circle yields a zone; in fact, the lines of the projective diagram may be taken to be the sections of the planes of symmetry by an arbitrary plane, or by the plane at infinity.

For instance, $t\begin{Bmatrix} 3 \\ 3 \end{Bmatrix}$ is the *truncated octahedron* whose faces consist of six squares and eight hexagons. In this case the projective diagram is a complete quadrangle (*fig.* 7) whose six sides are sections of the six planes of symmetry of the regular tetrahedron $\{3, 3\}$.

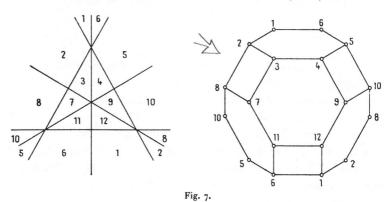

Fig. 7.

More generally, for any zonohedron we can construct a metrical version of the diagram by drawing, for each zone, the perpendicular plane through the centre, and considering the section by the plane at infinity.   In other words, the diagram is just the crystallographers' *gnomonic projection*.

**8. CONTRACTION.** — A zonohedron with $n$ zones has edges in $n$ different directions.   The edges in any one direction all have the same length, and this length can be varied without changing the essential nature of the zonohedron.   More drastically, the edges in one direction can be shrunk till their lengh is zero.   Then we are left with a new zonohedron having only $n-1$ zones.   In the diagram, this procedure corresponds to removing one of the $n$ lines.   For instance, the removal of the horizontal line in the left half of figure 7 represents the shrinking of all the vertical edges of the truncated

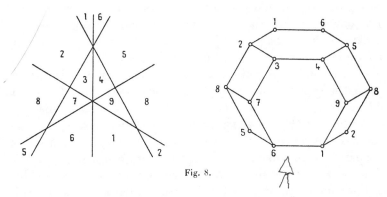

Fig. 8.

octahedron on the right.   In this manner we reduce a zone of two pairs of opposite hexagons and one pair of opposite parallelograms to two pairs of opposite parallelograms and one pair of opposite edges. The result is an *elongated dodecahedron* (*fig. 8*), whose faces consist of eight parallelograms and four hexagons.   The elongated dodecahedron has zones of two distinct types, whose reduction yields respectively a rhombic dodecahedron and a hexagonal prism.   (This connection between the rhombic dodecahedron and the elongated dodecahedron shows how the latter acquired its name.)

For a zonohedron whose faces are parallelograms, contraction simply means removing a zone.   Thus a contracted triacontahedron is a rhombic icosahedron, and a contracted rhombic icosahedron is a new kind of rhombic dodecahedron, apparently first recognized by Bilinski[4]. Its faces are equals rhombs, the same shape as those of the triacontahedron.    It has the symmetry of a rectangular parallele-piped.    Although in affine geometry it is indistinguishable from the ordinary rhombic dodecahedron, its occurence in Euclidean space shows that the number of " isozonohedra " (having rhombic faces, all alike) is not four (namely, the cube or any other rhombohedron, the ordinary rhombic dodecahedron, the rhombic icosahedron, and the triacontahedron) but five.

**9. PARALLELOHEDRA.** — Consider a convex solid that can be repeated by a group of translations to fill the whole affine space, without overlapping and without interstices.    In other words, consi-der a convex fundamental region for a three-dimensional group of translations. It can be proved [16a] that such a solid is a centrally-symmetrical polyhedron (see also [15 a]).

The space-filling is naturally called a honeycomb.    If any two cells that have more than an edge in common *share a whole face*, the cell is called a *parallelohedron* ([12], p. 285).    It has been conjectured that the proviso about sharing a whole face is unnecessary, i. e., that no new shapes are introduced when we allow neighbouring cells to share part of a face, like bricks in a wall ([19], p. 155).

Since any two opposite faces of a parallelohedron are related by a translation and also by a central inversion, each face is a centrally-symmetrical polygon.    Hence *every parallelohedron is a zonohedron*.

In the honeycomb of parallelohedra, each zone of any cell deter-mines an infinite layer of cells in which every cell shares the faces of just (this) one zone with adjacent cells.    After stretching the layer sufficiently by prolonging all the edges in the zonal direction, we can take the section by a suitable plane to obtain a tessellation of the kind shown in figure 1.    The vertices of the tessellation are sections of the prolonged edges.    Since the tessellation consists of either parallel-ograms or hexagons, *every zone of a parallelohedron contains either four or six faces*.

Conversely, if every zone of a zonohedron contains at most six faces, the zonohedron is a parallelohedron.    For, the existence of the corresponding tessellation ensures the possibility of repeating such a zonohedron by translations to make a complete layer, and by using such layers in various directions we can eventually fill the whole space.

Thus the problem of enumerating the combinatorially distinct parallelohedra reduces to the problem of enumerating the projective configurations consisting of $n$ lines and all their points of intersection, with the extra condition that *each line contains either two or three of the points.*

Since each line must contain at least two of the points, $n$ cannot be less than three, and the only suitable diagram with $n = 3$ is a triangle.

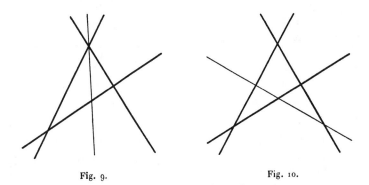

Fig. 9.                     Fig. 10.

If there is a fourth line, it either passes through a vertex of the triangle, as in figure 9, or cuts all three sides in distinct points, as in figure 10. These are the only possibilities with $n = 4$.

The only way to add a fifth line to figure 9 (without introducing a fourth point on some line) is to draw it through one of the three points where only two lines already meet (*fig.* 11). Likewise, the only suitable fifth line to add to the complete quadrilateral (*fig.* 10) is a diagonal (yielding figure 11 again), and the only admissible sixth line is another diagonal (*fig.* 12). We have now exhausted the possibilities with $n = 5$ or 6. Figure 12 (where $n = 6$) may be regarded as a complete quadrangle with its three diagonal points. This configuration is " saturated " in the sense that no further lines can be added; for instance, the line joining two of the three diagonal points would meet two further sides of the quadrangle, and would thus contain four points.

The enumeration of parallelohedra is now complete. The results are summarized in the first three columns of the accompanying Table. Each parallelohedron is typified by its most symmetrical Euclidean

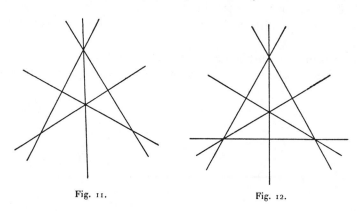

Fig. 11.                         Fig. 12.

variety. Instead of "cube" we might have written "parallelepiped", but then, for consistency, the rhombic dodecahedron would no longer be " rhombic ".

TABLE OF THE FIVE PARALLELOHEDRA.

| Diagram. | $n.$ | Parallelohedron. | Number of types of Voronoi polyhedra. |
|---|---|---|---|
| Triangle............... | 3 | Cube................... | 3 |
| Triangle with a cevian..... | 4 | Hexagonal prism......... | 3 |
| Complete quadrilateral..... | 4 | Rhombic dodecahedron... | 5 |
| Complete quadrilateral with a diagonal............ | 5 | Elongated dodecahedron... | 5 |
| Complete quadrangle...... | 6 | Truncated octahedron .... | 8 |

**10.** VORONOI POLYHEDRA. — In paragraph **2** we mentioned the conditions for a $2m$-gon to be the Dirichlet region of a lattice. We are now ready for the extension to three dimensions. Dirichlet himself does not seem to have troubled to take this step, but Voronoi worked in Euclidean space of an arbitrary number of dimensions. Accordingly we define, for any lattice, a *Voronoi polyhedron*, whose interior consists of all the points that are nearer to one lattice point than to any other.

By applying the translations that belong to the lattice, we see that every Voronoi polyhedron is a parallelohedron. Since each face is part of the locus of points equidistant from two lattice points, a given parallelohedron can be used as a Voronoi polyhedron if and only if the lines joining the centres of pairs of opposite faces are perpendicular to those faces. Delaunay ([10], p. 135) distinguishes 24 types of lattice according to the shape of the Voronoi polyhedron. The distribution of these 24 is shown in the last column of our Table. For instance, the three possibilities with $n = 3$ are the cube itself, the square prism, and the general rectangular parallelepiped. Again, although the hexagonal prism must be a "right" prism, its base may be either a regular hexagon, or one having a circumcircle and just two perpendicular lines of symmetry, or the general 2-dimensional Dirichlet region.

**11.** SPACES OF MORE THAN THREE DIMENSIONS. — The extension of the above ideas to affine and Euclidean spaces of higher dimension is complicated by the existence, in such spaces, of parallelohedra which are not zonohedra. It would not be very difficult to enumerate the zonohedra in affine 4-space; but the result would make no substantial contribution towards solving the more important problem of the enumeration of parallelohedra (which include the Voronoi polytopes).

In affine $l$-space, we naturally define a *zonohedron* to be a convex polytope whose $k$-dimensional elements are centrally symmetrical $(k = 1, 2, \ldots, l)$. Actually it suffices to specify this property for the single value $k = 2$; for it is trivial when $k = 1$, and the argument at the beginning of paragraph 5 can be used for $k = 3, 4, \ldots, l$. Thus we could have defined a zonohedron to be *a convex polytope whose 2-dimensional faces are centrally symmetrical*.

Naturally, also, we define a *parallelohedron* to be a convex polytope which can be repeated by translations to fill the whole affine $l$-space in such a way that, in the honeycomb so formed, any two cells which have more than an $(l-2)$-dimensional element in common share a whole $(l-1)$-dimensional element (" face ").

**12.** A PARALLELOHEDRON WHICH IS NOT A ZONOHEDRON. — A simple example of a 4-dimensional parallelohedron whose 2-dimensional faces are triangles (so that it cannot be a zonohedron) is the regular 24-cell, Schläfli's $\{3, 4, 3\}$ (*see* [8], p. 148-158), which is the Voronoi polytope for the lattice $\{3, 3, 4, 3\}$ consisting of all the points in Euclidean 4-space whose Cartesian coordinates are integers having an even sum. Minkowski [16] recognized this to be the lattice of centres in the densest lattice-packing of spheres in Euclidean 4-space. In other words, the densest lattice-packing is attained by the inspheres of the cells of the regular honeycomb $\{3, 4, 3, 3\}$.

The analogous lattice in 3-space is the face-centred cubic lattice, whose Voronoi polyhedron is the rhombic dodecahedron.

**13.** POLYTOPES WHOSE CELLS ARE PARALLELEPIPEDS. — For zonohedra in four dimensions, as in three, we can construct " stars " and projective diagrams. The " second " diagram is a configuration in real projective 3-space, consisting of $n$ planes (one for each family of parallel edges) and all their intersections. White [21] has investigated these configurations for $n \leq 7$ in the " general " case when no more than three planes meet in any point. Since the number of points is $\binom{n}{3}$, such a configuration represents a 4-dimensional zonohedron whose cells consist of $\frac{1}{3}n(n-1)(n-2)$ parallelepipeds. Since each line of intersection of 2 planes is decomposed into $n-2$ segments by the remaining $n-2$ planes, there are altogether $\frac{1}{2}n(n-1)(n-2)$ segments, and the polytope has $n(n-1)(n-2)$ faces (parallelograms). Since each plane is decomposed into $\binom{n-1}{2}+1$ regions by the remaining $n-1$ planes, the polytope has $n^2-3n+4$ edges in each of its $n$ zones. Since the whole configuration has $\binom{n}{3}+\binom{n}{1}$ solid regions, the polytope has $\frac{1}{3}n(n^2-3n+8)$ vertices. White observes that there is essentially only one such arrangement of $n$ planes when $n = 4$, 5 or 6. Hence there is (disregarding combinatorial isomorphism) only one polytope bounded by 8, 20 or 40 parallelepipeds. The first case is simply a parallelotope (e. g., the 4-dimensional cube $\gamma_4$) ([8], p. 123). The second is typified by the reciprocal of the *expanded simplex* $e\alpha_4$ or $e_3 S(5)$ or $t_{0,3}\alpha_4$, whose cells consist of 10 tetrahedra and 20 triangular prisms ([6], p. 336).

**14.** A PARALLELOHEDRON WHICH IS A ZONOHEDRON. — The " stretching " procedure that we used in paragraph **9** shows that those 4-dimensional zonohedra which are parallelohedra are represented by diagrams in which every line of the configuration contains either two or three of the points. The most complicated example of such a configuration is the *complete pentagon* formed by five points of general position and the 10 planes that join them in threes. These 10 planes decompose the 3-space into 60 tetrahedral regions, representing the 120 vertices of the corresponding 4-dimensional zonohedron.

In terms of redundant projective coordinates $x_1, \ldots, x_5$ with sum zero, we may take the 10 planes to be

(14.1) $$x_i = x_j \qquad (i < j),$$

so that the five points are

$$(-4, 1, 1, 1, 1), \quad \ldots, \quad (1, 1, 1, 1, -4).$$

Then the interior points of the 60 regions correspond to the 60 possible relations of order among the five coordinates, opposite orders being considered equivalent (because we could multiply all five coordinates by the same negative number). In other words, one region is given by

$$x_1 < x_2 < x_3 < x_4 < x_5 \qquad \text{or} \qquad x_1 > x_2 > x_3 > x_4 > x_5,$$

and the other 59 regions are derived from this one by permuting the $5x$'s.

In a manner resembling the procedure in paragraph **7**, we may identify the 10 planes (14.1) in projective 3-space with 10 hyperplanes through a point in Euclidean 4-space, namely, the 10 hyperplanes of symmetry of the regular simplex $\{3, 3, 3\}$. The five vertices of this simplex have the same coordinates as the five points considered above, except that now these are regarded as Cartesian coordinates in a Euclidean 5-space, and the simplex lies in the 4-dimensional subspace

$$x_1 + \ldots + x_5 = 0.$$

The 60 regions in the projective 3-space yield 120 " angular " (or " conical ") regions corresponding to all the 5! relations of order among the five coordinates. Reflections in the 10 hyperplanes correspond to transpositions of pairs of coordinates. Five of these ten reflections, suitably chosen, generate the complete symmetry group of the regular simplex. This is a well-known representation of the symmetric group of degree 5.

As a point within the angular region

$$x_1 < x_2 < x_3 < x_4 < x_5,$$

equidistant from the four bounding hyperplanes

$$x_1 = x_2, \qquad x_2 = x_3, \qquad x_3 = x_4, \qquad x_4 = x_5,$$

we may take

$$(-2, -1, 0, 1, 2).$$

Applying the 5! permutations to this point, we obtain the 120 vertices of a polytope having 24 edges in each of 10 directions, namely the directions perpendicular to the 10 hyperplanes. This is the most symmetrical Euclidean variety of the 4-dimensional zonohedron-parallelohedron represented by the complete pentagon : the 4-dimensional analogue of the truncated octahedron.

This polytope, and the honeycomb of its translated replicas, were discussed at considerable length by Hinton ([13], p. 135, 225). They were extended to $n$ dimensions by Schoute ([18], p. 59) who denoted the 4-dimensional polytope by

$$e_1 e_2 e_3 \, S(5).$$

In the notation of Wythoff's construction ([6], p. 334), Hinton's polytope and honeycomb are respectively

Deleting the nodes in turn, we verify that the polytope is a cell of the honeycomb, and that the solid cells of the polytope are of two kinds ([7], p. 394), 10 truncated octahedra and 20 hexagonal prisms :

The vertices of the honeycomb consist of all the points in the 4-space $\Sigma x_i = 0$ whose five coordinates are integers mutually *incongruent* modulo 5. The centres of the cells are all the points in the same 4-space whose five coordinates are mutually *congruent* modulo 5. This lattice, whose Voronoi polytope is Hinton's polytope, is believed by Bambah [2 a] to be the lattice of centres of spheres in the thinnest lattice-covering of Euclidean 4-space. In other words, according to Bambah's conjecture, the thinnest lattice-covering is attained by the circumspheres of the cells of Hinton's honeycomb.

The analogous lattice in 3-space is the body-centred cubic lattice, whose Voronoi polyhedron is the truncated octahedron. In this case [2] the thinnest covering is indeed attained by the circumspheres of the cells of the honeycomb.

BIBLIOGRAPHY.

[1] T. Bakos, *Octahedra inscribed in a cube* (*Math. Gazette*, t. 43, 1959, p. 17-20).

[2] R. P. Bambah, *On lattice coverings by spheres* (*Proc. Nat. Inst. Sc.*, India, t. 20, 1954, p. 25-52).

[2 a] R. P. Bambah, *Lattice coverings with four-dimensional spheres* (*Proc. Camb. Phil. Soc.*, t. 50, 1954, p. 203-208).

[3] R. P. Bambah and H. Davenport, *The covering of n-dimensional space by spheres* (*J. London Math. Soc.*, t. 27, 1952, p. 224-229).

[4] S. Bilinski, *Über die Rhombenisoeder* (*Glasnik*, t. 15, 1960, p. 251-263).

[5] J. J. Burckhardt, *Über konvexe Körper mit Mittelpunkt* (*Vierteljahrsschrift der Naturforschenden Gesellschaft in Zürich*, t. 85, 1940, p. 149-154).

[6] H. S. M. Coxeter, *Wythoff's construction for uniform polytopes* [*Proc. London Math. Soc.*, (2), t. 38, 1935, p. 327-339].

[7] H. S. M. Coxeter, *Regular and semi-regular polytopes* (*Math. Z.*, t. 46, 1940, p. 380-407).

[8] H. S. M. Coxeter, *Regular Polytopes*, New York, 1963.

[9] H. S. M. Coxeter, *Map-coloring problems* (*Scripta Mathematica*, t. 23, 1957, p. 11-25).

[10] B. Delaunay, *Neue Darstellung der geometrischen Kristallographie* (*Z. Krist.*, t. 84, 1933, p. 109-149).

[11] G. L. Dirichlet, *Über die Reduction der positiven quadratischen Formen mit drei unbestimmten ganzen Zahlen* (*J. reine angew. Math.*, t. 40, 1850, p. 209-227).

[12] E. S. Fedorov (Е. С. Фёдоров), Начала Учения о Фигурах, Leningrad, 1953.

[13] C. H. Hinton, *The Fourth Dimension*, London, 1906.

[14] L. M. Kelly and W. O. J. Moser, *On the number of ordinary lines determined by n points* (*Canad. J. Math.*, t. 10, 1958, p. 210-219).

[15] G. Kowalewski, *Der Keplersche Körper und andere Bauspiele*, Leipzig, 1938.

[15 a] A. M. Macbeath, *On convex fundamental regions for a lattice* (*Canad. J. Math.*, t. 13, 1961, p. 177-178).

[16] H. Minkowski, *Diskontinuitätsbereich für arithmétische. Äquivalenz* (*Ges. Math. Abh.*, t. 2, 1911, p. 43-100; see especially p. 75).

[16 a] H. Minkowski, *Allgemeine Lehrsätze über die konvexen Polyeder* (*Ges. Math. Abh.*, t. 2, 1911, p. 103-121; see especially p. 120).

[16 b] W. O. J. Moser, *An enumeration of the five parallelohedra* (*Canad. Math. Bull.*, t. 4, 1961, p. 45-47).

[17] Th. Motzkin, *The lines and planes connecting the points of a finite set* (*Trans. Amer. Math. Soc.*, t. 70, 1951, p. 451-464).

[18] P. H. Schoute, *Analytical treatment of the polytopes regularly derived from the regular polytopes* (*Verh. K. Akad. Wetensch. te Amsterdam*, t. 11, 1911, No 3).

[19] H. Steinhaus, *Mathematical Snapshots*, New York, 1952.

[20] H. S. White, *The plane figures of seven real lines* (*Bull. Amer. Math. Soc.*, t. 38, 1932, p. 59-65).

[21] H. S. White, *The convex cells formed by seven planes* [*Proc. Nat. Acad. Sci.* (*U. S. A.*), t. 25, 1939, p. 147-153).

*Reference 6 is Chapter 3 of this book.

# 5

# REGULAR SKEW POLYHEDRA IN THREE AND FOUR DIMENSIONS, AND THEIR TOPOLOGICAL ANALOGUES

Reprinted, with the editors' permission, from the *Proceedings of the London Mathematical Society*, Ser. 2, Vol. 43, 1937.

The five regular ("Platonic") polyhedra,

$$\{3, 3\}, \quad \{3, 4\}, \quad \{4, 3\}, \quad \{3, 5\} \quad \{5, 3\}^*,$$

have been known and admired for thousands of years. As analogues of these, Kepler drew attention to the three regular plane-fillings,

$$\{4, 4\}, \quad \{3, 6\}, \quad \{6, 3\},$$

which may be regarded as regular polyhedra with infinitely many faces. He also introduced two of the four *star* polyhedra,

$$\{5, \tfrac{5}{2}\}, \quad \{\tfrac{5}{2}, 5\}, \quad \{3, \tfrac{5}{2}\}, \quad \{\tfrac{5}{2}, 3\}\dagger;$$

the remaining two were added by Poinsot, but it was Cauchy‡ who first proved that there are *only* four. These have the disadvantage of possessing "false vertices", *i.e.* points other than vertices, where three or more faces meet.

One day in 1926, J. F. Petrie told me with much excitement that he had discovered two new regular polyhedra; infinite, but free from false vertices. When my incredulity had begun to subside, he described them to me: one consisting of squares, six at each vertex (Fig. i), and one

---

* The meaning of these symbols, introduced by Schläfli, is clear from the following example : the *cube* is $\{4, 3\}$, because it is bounded by *squares, three* at each vertex.

† This extension of the above notation is also due to Schläfli. The four polyhedra are, respectively, the *great dodecahedron*, the *small stellated dodecahedron*, the *great icosahedron*. the *great stellated dodecahedron*.

‡ Cauchy, 1. *Cf.* Coxeter, 1, 203. **(For references, see page 105.)**

consisting of hexagons, four at each vertex (Fig. ii). It was useless to protest that there is *no room* for more than four squares round a vertex. The trick is, to let the faces go up and down in a kind of zig-zag formation, so that the faces that adjoin a given "horizontal" face lie alternately "above" and "below" it. When I understood this, I pointed out a third possibility: hexagons, six at each vertex (Fig. iii).

Fig. i: {4, 6|4}.

Fig. ii: {6, 4|4}.

Fig. iii: {6, 6|3}.

It then occurred to us that, although these new polyhedra are infinite, we might find analogous *finite* polyhedra by going into four-dimensional space. Petrie cited one consisting of $n^2$ squares, four at each vertex. We call such figures *skew polyhedra*, by analogy with " skew polygons ". They must not be confused with *polytopes* (which have solid faces). Actually, every regular skew polyhedron is intimately associated with a definite *uniform** polytope. *E.g.*, Fig. i consists of half the squares of $\{4, 3, 4\}$ or $\delta_4$, the net of cubes; Fig. ii consists of all the hexagons of $t_{1,2}\delta_4$, the net of truncated octahedra; and Fig. iii† consists of all the hexagons of $q\delta_4$, the net of tetrahedra and truncated tetrahedra‡.

The reader naturally asks, What do we mean by saying that these new polyhedra are *regular*? To explain this, we have to go back to first principles.

## 1. *Definitions.*

We define a *polyhedron* as a connected set of ordinary plane polygons‖, such that every side of every polygon belongs also to just one other polygon of the set. We stipulate that any two of the polygons shall have in common either a side (and two vertices), or a single vertex, or nothing at all (thus ruling out the Kepler-Poinsot polyhedra). We define a *symmetry* (or " symmetry operation ") of any figure as a congruent transformation of the figure into itself (*i.e.*, a combination of translations, rotations, and reflections). A completely irregular figure has no symmetry save identity.

A polyhedron is said to be *regular* if it possesses two particular symmetries: one which cyclically permutes the vertices of any face $c$, and one which cyclically permutes the faces that meet at a vertex C, C being a vertex of $c$. It follows that these two symmetries, say $R$ and $S$, generate a group which is transitive on the vertices, on the edges, and on the faces. Moreover, the faces are regular polygons (in the most elementary sense). This clearly agrees with our preconceived notion of regularity.

Perhaps this will seem clearer if we give the analogous definition for a regular *polygon*. A polygon (which may be skew) is said to be regular if it possesses a symmetry which cyclically permutes the vertices (and therefore also the sides) of the polygon. Regular polygons can be classified

---

\* Coxeter, **4**.

† For this picture some apology is required. The lines which divide each hexagon into six triangles are to be ignored, since they arise merely from the manner in which this particular model was built up.

‡ Andreini, **1**, 106 (Fig. 14), 107 (Fig. 15).

‖ In § 5, we relax this definition, and consider a connected set of *topological* polygons.

according to the nature of this symmetry. The cases that arise in ordinary space are as follows:

| *Symmetry.* | *Polygon.* |
|---|---|
| Reflection | Digon, $\{2\}$, with two vertices and two coincident sides. |
| Rotation (through $2\pi/n$) | Ordinary $n$-gon, which we denote by $\{n\}$. |
| Rotary reflection (involving rotation through $\pi/n$) | Finite skew $2n$-gon. |
| Translation | Apeirogon, $\{\infty\}$ (*i.e.*, an infinite straight line broken into equal segments). |
| Glide | Plane zig-zag (dividing the plane into two equal parts). |
| Screw | Helical polygon. |

Since the "square" of a rotary reflection is a pure rotation, the vertices of a regular *finite* skew polygon lie alternately on two circles which reflect into one another in a plane parallel to the planes of the circles (and therefore half-way between them). In fact, the sides of such a polygon are the lateral edges of an *antiprism* (*e.g.*, Fig. iv). Hence the number of sides must be *even*. (If there are only four sides, these are edges of a tetragonal bisphenoid, which can be regarded as a *digonal* antiprism.)

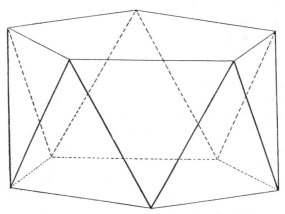

Fig. iv: The skew decagon in a pentagonal antiprism.

The line joining two alternate vertices of an ordinary regular $n$-gon (of unit side) is of length $2\cos(\pi/n)$, and is called the *vertex figure* of the $\{n\}$.

For a regular polyhedron (of unit edge), consider those vertices which are joined by edges to any one vertex. These can be regarded as the vertices of a finite polygon whose sides are vertex figures of faces of the polyhedron. This polygon is called the *vertex figure* of the polyhedron. It is a *regular* polygon, since its sides are cyclically permuted by the special symmetry $S$.

If a regular polyhedron is four-dimensional and finite, its vertices are equidistant from their centroid, which we call the *centre* of the polyhedron. Hence those vertices which are joined by edges to any one vertex lie on the intersection of two hyper-spheres, *i.e.* on an ordinary sphere. Thus in this case, as well as when the polyhedron (finite or infinite) is three-dimensional, the vertex figure lies in a three-space. By taking as vertex figure a plane polygon, we obtain an ordinary regular solid or plane-filling; but the vertex figure may just as well be a skew polygon, whose vertices lie alternately in two parallel planes.

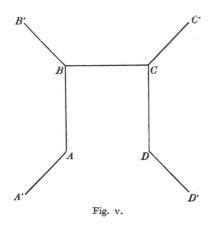

Fig. v.

Let us first suppose the polyhedron to lie in ordinary space. It divides space into two parts, one "inside" and one "outside" the polyhedron. When the vertex figure is a skew polygon, the inside and outside are alike (*i.e.* congruent) in the neighbourhood of one vertex, and are therefore alike altogether. (In fact, $R$ and $S$ each interchange the inside and outside.) Hence the polyhedron must be infinite. On account of the zig-zag nature of the skew polygon, the faces that surround a given "horizontal" face are alternately "above" and "below" it. Hence $R$, like $S$, is a rotary reflection.

The situation when the polyhedron is finite in four dimensions is very similar. For, after projecting the polyhedron on to its circumscribed hyper-sphere, we have again a surface lying in a manifold of three dimensions, and dividing that manifold into two equal parts. $R$ and $S$ are still rotary reflections, when regarded as operating in the spherical three-space.

Let A′ABB′ ..., B′BCC′ ..., C′CDD′ ... (Fig. v) be three faces of any regular polyhedron. We may suppose that $R$ cyclically permutes the edges B′B, BC, CC′, ... which belong to the second of these faces, while $S$ cyclically permutes the edges CD, CC′, CB, ... which meet at C. Then the product $RS$ reverses the edge BC, while the "quotient" $RS^{-1}$ cyclically permutes the edges AB, BC, CD, .... Hence ABCD... is a regular polygon (but may be skew). The lines AA′, BB′, CC′, DD′, ... are clearly concurrent, say at O. (When the faces are squares, so that these edges are parallel, we say that O is at infinity.) If the polyhedron is in three dimensions, the planes ABB′, DCC′ are equally inclined to the plane B′BCC′, on the same side of it; therefore ABCD ... is a plane polygon. If the polyhedron is finite in four dimensions, the symmetry $RS^{-1}$, being the product of two rotary reflections, is a *positive* transformation. Since it leaves the centre invariant, this must be either a simple rotation or a double rotation. Since it also leaves the point O invariant, it can only be a simple rotation. Hence ABCD... is again a plane polygon.

In Figs. i, ii, iii, this polygon appears as a hole (triangular in Fig. iii, square in Figs. i and ii), and it is convenient to adopt the term *hole* in other cases, too. The hole may be described as a path along edges, such that at the end of each edge we leave two faces on (say) the left. In other words, at any vertex of the path, the edges to be selected are not adjacent but alternate. When (as above) the hole is a plane polygon, we may determine it from its vertex figure, which is the join of two alternate vertices of the vertex figure of the polyhedron. In the case of the ordinary polyhedron $\{l, m\}$, whose vertex figure is an $\{m\}$ of side $2\cos(\pi/l)$, it follows that the "hole"* is an $\{n\}$, where

$$2\cos(\pi/n) = 2\cos(\pi/l) \cdot 2\cos(\pi/m),$$

*i.e.*

(1.1) $$\cos(\pi/n) = 2\cos(\pi/l)\cos(\pi/m).$$

We shall use the symbol $\{l, m\,|\,n\}$ to denote any regular polyhedron which is uniquely determined by

$$\begin{cases} l, \text{ the number of vertices or sides of a face,} \\ m, \text{ the number of edges or faces at a vertex,} \\ n, \text{ the number of vertices or sides of a hole.} \end{cases}$$

---

* Here the term " hole " is not entirely appropriate. But triangular " pits " are clearly visible in the star polyhedron $\{5, \frac{5}{2}\}$ (Fig. xiii).

We shall see that every regular polyhedron in three dimensions, or finite in four, is so determined. From this point of view,

$$\{n, 3\} \quad \text{is} \quad \{n, 3 \,|\, n\},$$
$$\{3, n\} \quad \text{is} \quad \{3, n \,|\, n\},$$
$$\{4, 4\} \quad \text{is} \quad \{4, 4 \,|\, \infty\}.$$

## 2. *The trigonometrical criterion for* $\{l, m \,|\, n\}$.

Consider first the general infinite regular polyhedron in three dimensions. Let $\lambda$, $\lambda'$ denote the planes that perpendicularly bisect two adjacent edges, BC, CC', belonging to the face B' BCC' ... (Fig. v), and let $\mu$, $\mu'$ denote the planes that bisect the dihedral angles at those same edges. These four planes bound a tetrahedron, whose angles we proceed to calculate.

The planes $\lambda$, $\mu$, and likewise $\lambda'$, $\mu'$, are obviously perpendicular. The planes $\lambda$, $\lambda'$ are both perpendicular to the plane BCC', and meet it in the perpendicular bisectors of two adjacent sides of the $\{l\}$ B' BCC' ... ; so the angle between them is $2\pi/l$. The bisecting planes of the dihedral angles at the edges through C all pass through the axis of the rotary reflection $S$; since there are $m$ such edges, the angle between two consecutive bisecting planes, such as $\mu$, $\mu'$, is $2\pi/m$. The planes $\lambda$, $\mu'$ are both perpendicular to the plane BCD, and meet it in the bisectors of the side BC and angle BCD of the $\{n\}$ ABCD... ; so the angle between them is $\pi/n$. Collecting results, we see that the six dihedral angles of the tetrahedron $\lambda\lambda'\,\mu\mu'$ are

$$(2.1) \quad \begin{cases} (\lambda\mu) = (\lambda'\,\mu') = \tfrac{1}{2}\pi, \\ \qquad\quad (\lambda\lambda') = 2\pi/l, \\ \qquad\quad (\mu\mu') = 2\pi/m, \\ (\lambda\mu') = (\lambda'\,\mu) = \pi/n. \end{cases}$$

We have now reduced the problem of enumerating the possible regular polyhedra to that of enumerating the possible tetrahedra of a special type. For each tetrahedron, we have *two* polyhedra,

$$\{l, m \,|\, n\} \quad \text{and} \quad \{m, l \,|\, n\},$$

which may be called *reciprocal*, since the vertices of each are the centres of the faces of the other. Thus $\{4, 6 \,|\, 4\}$ and $\{6, 4 \,|\, 4\}$ (Figs. i and ii) are reciprocal, while $\{6, 6 \,|\, 3\}$ (Fig. iii) is self-reciprocal.

Now, just as the three angles of a plane triangle are not independent, but add up to $\pi$, so also the six dihedral angles of a tetrahedron are not independent, but satisfy the relation

$$(2.2) \quad \Delta \equiv \begin{vmatrix} 1 & -\cos(ab) & -\cos(ac) & -\cos(ad) \\ -\cos(ba) & 1 & -\cos(bc) & -\cos(bd) \\ -\cos(ca) & -\cos(cb) & 1 & -\cos(cd) \\ -\cos(da) & -\cos(db) & -\cos(dc) & 1 \end{vmatrix} = 0,$$

where $a$, $b$, $c$, $d$ are the four bounding planes. In the case under consideration,

$$\Delta = \left(4\cos^2\frac{\pi}{l}\cos^2\frac{\pi}{m} - \cos^2\frac{\pi}{n}\right)\left(4\sin^2\frac{\pi}{l}\sin^2\frac{\pi}{m} - \cos^2\frac{\pi}{n}\right).$$

After discarding two definitely positive factors, we are left with the condition

$$(2.3) \quad \left(2\cos\frac{\pi}{l}\cos\frac{\pi}{m} - \cos\frac{\pi}{n}\right)\left(2\sin\frac{\pi}{l}\sin\frac{\pi}{m} - \cos\frac{\pi}{n}\right) = 0.$$

By (1.1), the vanishing of the first factor gives the ordinary polyhedron $\{l, m\}$. Hence, to find all the possible skew polyhedra (in three dimensions), we merely have to solve in integers the equation

$$(2.4) \qquad\qquad 2\sin(\pi/l)\sin(\pi/m) = \cos(\pi/n).$$

Apart from the plane-fillings

$$\{3, 6 \,|\, 6\}, \quad \{6, 3 \,|\, 6\}, \quad \{4, 4 \,|\, \infty\},$$

which are already covered by (1.1)*, the only solutions are:

$$\{4, 6 \,|\, 4\}, \quad \{6, 4 \,|\, 4\}, \quad \{6, 6 \,|\, 3\}.$$

These infinite polyhedra may be called "complements" of the cube, octahedron, and tetrahedron, respectively, since each solution $\{l, m \,|\, n\}$ of (2.4) corresponds to a solution $\{l', m' \,|\, n\}$ of (1.1), where

$$1/l + 1/l' = 1/m + 1/m' = \tfrac{1}{2}.$$

We shall see later that "complementary" polyhedra have the same dihedral angles.

For each value of $n$, the relation between $l$ and $m$ may be presented graphically by taking $l$ and $m$ as Cartesian coordinates in a plane. (2.3) is then the equation for a couple of curves, which are shown in Fig. vi when $n = 3$, Fig. vii when $n = 4$. The finite polyhedra $\{l, m\}$ appear as O's on the curve

$$\cos(\pi/l)\cos(\pi/m) = \tfrac{1}{2}\cos(\pi/n),$$

---

* For the plane-fillings we have $1/l + 1/m = \tfrac{1}{2}$, so they can be regarded indifferently as ordinary or skew polyhedra.

while the infinite polyhedra $\{l, m \mid n\}$ appear as dots on the curve

$$\sin(\pi/l) \sin(\pi/m) = \tfrac{1}{2} \cos(\pi/n).$$

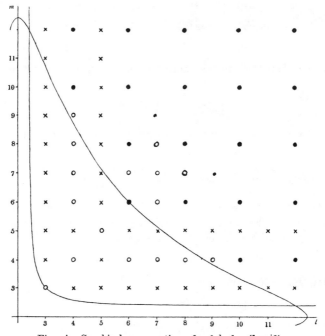

Fig. vi: Graphical enumeration of polyhedra $\{l, m \mid 3\}$.

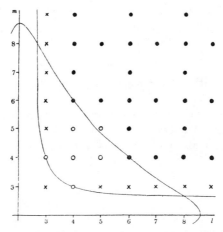

Fig. vii: Graphical enumeration of polyhedra $\{l, m \mid 4\}$.

In the case of the general finite regular polyhedron in four dimensions, the planes $\lambda, \lambda', \mu, \mu'$ have to be replaced by three-spaces through the centre, but the angles between them are still given by (2. 1). They can be regarded as bounding a spherical tetrahedron on the circumscribed hyper-sphere of the polyhedron. As before, the two polyhedra $\{l, m \,|\, n\}$ and $\{m, l \,|\, n\}$ are derivable from the same tetrahedron, and are *reciprocal* in the sense that the vertices, edges, and faces of one correspond to the faces, edges, and vertices of the other. The equation (2.2) has to be replaced by the inequality*

$$\Delta > 0,$$

which reduces to

$$\left(2 \cos \frac{\pi}{l} \cos \frac{\pi}{m} - \cos \frac{\pi}{n}\right)\left(2 \sin \frac{\pi}{l} \sin \frac{\pi}{m} - \cos \frac{\pi}{n}\right) > 0.$$

It is easily seen that the only significant case is when these two factors are positive†. *E.g.*, in Figs. vi and vii, the significant lattice points are those which lie inside the region enclosed by the two curves.

Since we restrict consideration to finite polyhedra in four dimensions, $R$ and $S$ (as we have seen) are rotary reflections; hence their periods, $l$ and $m$, must be *even*. Thus the only admissible solutions are:

$$\{4, 4 \,|\, n\},$$

$$\{4, 6 \,|\, 3\}, \quad \{6, 4 \,|\, 3\},$$

$$\{4, 8 \,|\, 3\}, \quad \{8, 4 \,|\, 3\}.$$

### 3. *Derivation of the polyhedra from uniform polytopes.*

Clearly, the planes or three-spaces $\lambda, \lambda', \mu, \mu'$ are primes of symmetry of the polyhedron, and the reflections in them generate a sub-group of the complete symmetry-group of the polyhedron. ($R$ and $S$ generate a different subgroup.) The vertex C (Fig. v) is the mid-point of the edge $\mu\mu'$ of the tetrahedron $\lambda\lambda'\mu\mu'$, and lies on the bisector of the opposite dihedral angle $(\lambda\lambda')$. The reflections in $\lambda, \lambda'$ transform C into B, C', respectively, while the reflection in $\mu'$ transforms B into D. In fact, Wythoff's construction‡

leads to the uniform polytope $\overset{l}{\underset{\phantom{.}}{\mu}} \begin{array}{c} n \\ \square \\ n \end{array} \overset{l}{\underset{\phantom{.}}{m}}$ whose vertices, edges,

---

* Coxeter, 2, 137.

† When the second factor is negative (and the first positive), the space is Minkowskian instead of Euclidean.

‡ Coxeter, 4, 329.

$\{l\}$'s, and $\{n\}$'s are the vertices, edges, faces, and holes of $\{l, m \mid n\}$. Moreover, the vertex figure of the polytope is an antiprism, whose lateral edges form the vertex figure of the polyhedron.

When $l = m = 4$, the polytope is the double $n$-gonal prism, $[\{n\}, \{n\}]$ or $\{n\} \times \{n\}$, bounded by $2n$ $n$-gonal prisms. The corresponding polyhedron,

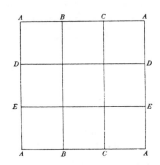

Fig. viii: "Net" of $\{4, 4 \mid 3\}$.

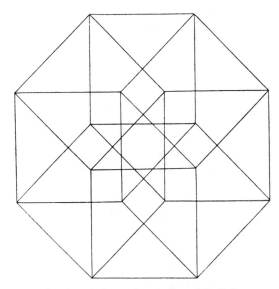

Fig. ix: Orthogonal projection of $\{4, 4 \mid 4\}$.

$\{4, 4 \mid n\}$, can be unfolded into a "net" which consists of $n^2$ squares forming a large square. The opposite sides of this large square are to be identified (or brought into coincidence) in the manner appropriate to the formation of a torus. (See, for instance, Fig. viii, where $n = 3$. The "squared-paper " plane-filling, $\{4, 4\}$, can be regarded as the limiting form as $n$ tends to infinity.) When $n = 4$, the double prism becomes the tessaract or hyper-cube, $\gamma_4$, and the faces of $\{4, 4 \mid 4\}$ are sixteen of the twenty-four squares of this polytope. The vertex figure of $\gamma_4$ is a regular tetrahedron (of edge $\sqrt{2}$), which can be regarded as a digonal antiprism in three ways. Therefore one $\gamma_4$ leads to three $\{4, 4 \mid 4\}$'s. Fig. ix shows the familiar octagonal projection of the tessaract (in which the eight bounding cubes are all plainly visible). In this projection, eight of the twenty-four squares are undistorted, while the remaining sixteen are foreshortened into rhombs. These sixteen squares are the faces of one $\{4, 4 \mid 4\}$.

When $l > 4$, the faces of $\{l, 4 \mid n\}$ are simply all the $\{l\}$'s of the polytope* $t_{1, 2}\{n, \tfrac{1}{2}l, n\}$. In the finite cases, this polytope can be described as the common content of two equal reciprocal regular polytopes $\{n, \tfrac{1}{2}l, n\}$. In particular, $t_{1, 2}\{3, 3, 3\}$ or $t_{1, 2}a_4$ can also be described as a central section of the five-dimensional measure-polytope $\gamma_5$, analogous to the hexagonal section of the cube ($\gamma_3$). The general measure-polytope may be defined as the totality of points $(x_1, x_2, \ldots)$ for which $|x_i| \leqslant 1$; we take its section by the prime $\Sigma x_i = 0$. Thus the vertices of $t_{1, 2}a_4$, and so also of $\{6, 4 \mid 3\}$, are the thirty points

$$(1, 1, 0, -1, -1)$$

(permuted). (See Fig. x.) The polytope is bounded by ten truncated tetrahedra; the polyhedron separates these into two sets of five.

The regular 24-cell, $\{3, 4, 3\}$, may be defined as the totality of points for which

$$|x_i \pm x_j| \leqslant \sqrt{2}\dagger \quad (i, j = 1, 2, 3, 4; \ i < j).$$

Its reciprocal, with vertices $\pm(1, 1, 0, 0)$ (permuted), is then defined by

$$|x_i| \leqslant 1, \quad |x_1 \pm x_2 \pm x_3 \pm x_4| \leqslant 2.$$

Thus the vertices of $t_{1, 2}\{3, 4, 3\}$, and so also of $\{8, 4 \mid 3\}$, are the 288 points

$$\pm(1, \sqrt{2}-1, \sqrt{2}-1, 3-2\sqrt{2}), \quad \pm(2\sqrt{2}-2, 2-\sqrt{2}, 2-\sqrt{2}, 0).$$

The polytope is bounded by forty-eight truncated cubes‡; the polyhedron separates these into two sets of twenty-four.

---

* Coxeter, **4**, 331.

† Or any other positive constant.

‡ Threlfall and Seifert, **1**, 63.

When $l=4$ and $m>4$, the polytope is $t_{0,3}\{n, \tfrac{1}{2}m, n\}$, which can be derived from the regular polytope $\{n, \tfrac{1}{2}m, n\}$ by uniformly shrinking all the bounding $\{n, \tfrac{1}{2}m\}$'s, and inserting $n$-gonal prisms between adjacent pairs of them*. The shrinkage is adjusted so that the lateral faces of the

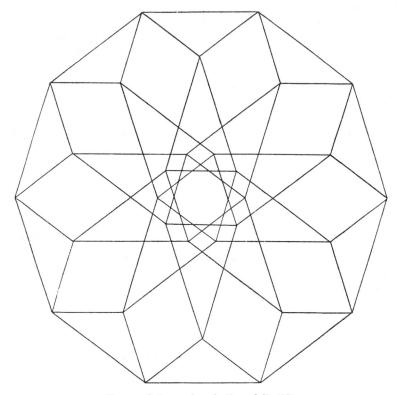

Fig. x: Orthogonal projection of $\{6, 4\,|\,3\}$.

prisms are squares; these squares are then the faces of $\{4, m\,|\,n\}$. It follows that the number of faces of $\{4, m\,|\,n\}$ is $n$ times the number of edges (or plane faces) of $\{n, \tfrac{1}{2}m, n\}$. The vertices of $t_{0,3}\,\alpha_4$, as also of $\{4, 6\,|\,3\}$, are the twenty points

$$(1, 0, 0, 0, -1).$$

---

* Apart from a change of scale, this is equivalent to Mrs. Stott's " expansion " $e_3$. (Stott, **1**, 9.)

The vertices of $t_{0,3}\{3, 4, 3\}$, as also of $\{4, 8\,|\,3\}$, are the 144 points

$$\pm(\sqrt{2},\ 2-\sqrt{2},\ 0,\ 0),\quad \pm(1,\ 1,\ \sqrt{2}-1,\ \sqrt{2}-1).$$

Any two adjacent faces of $\{l, m\,|\,n\}$ belong to a bounding $t_{0,1}\{\frac{1}{2}l, n\}$* of the corresponding polytope. The dihedral angle between them is therefore equal to the dihedral angle of the ordinary polyhedron† $\{\frac{1}{2}l, n\}$, namely

$$2\ \text{arcsin}\ \frac{\cos(\pi/n)}{\sin(2\pi/l)}.$$

In the three-dimensional cases, (2.3) shows that this is equal to

$$2\ \text{arcsin}\ \frac{\cos(\pi/m)}{\sin(\pi/l)}\quad \text{or}\quad 2\ \text{arcsin}\ \frac{\sin(\pi/m)}{\cos(\pi/l)}$$

according as the polyhedron $\{l, m\,|\,n\}$ is ordinary or skew. This explains the fact that the infinite skew polyhedra have the same dihedral angles as the ordinary polyhedra that are their " complements ".

### 4. *The symmetry-groups.*

It is well known that the complete symmetry-group of the ordinary regular polyhedron $\{l, m\}$ is representable on a concentric sphere in such a way that the fundamental region is a spherical triangle of angles $\pi/l$, $\pi/m$, $\pi/2$. The group is generated by the reflections in the sides of this triangle, and has the abstract definition

$$R_1{}^2 = R_2{}^2 = R_3{}^2 = (R_1 R_2)^l = (R_2 R_3)^m = (R_1 R_3)^2 = 1.$$

The rotations

$$R = R_1 R_2,\quad S = R_2 R_3$$

generate a sub-group $\mathfrak{G}$, of order $g$, whose abstract definition is

(4.1) $$R^l = S^m = (RS)^2 = 1.$$

This consists of all *rotations* that are symmetries of the polyhedron. The complete group is derivable from it by adjoining a single reflection, and so is of order $2g$. The area of the fundamental region is $(1/l+1/m-\frac{1}{2})\pi$. Since $2g$ such regions exactly cover the sphere, we have

(4.2) $$2/g = 1/l+1/m-\tfrac{1}{2}.$$

---

* For $l = 4$, an $n$-gonal prism.

† For $l = 4$, of the $n$-gon itself.

Instead of representing the group on a sphere, we can just as well represent it on the surface of the polyhedron itself. The fundamental region is now a plane right-angled triangle $O_1 O_2 O_3$, where

$O_3$ is the centre of a face of the polyhedron,

$O_2$ is the mid-point of a side of this face, and

$O_1$ is one end of this side.

$R_1$, $R_2$, $R_3$ are reflections in the planes joining the centre of the polyhedron to the sides $O_2 O_3$, $O_3 O_1$, $O_1 O_2$ of this triangle. Every face can be divided into $2l$ such triangles, and then each edge of the polyhedron belongs to four of them, and each vertex to $2m$. It is convenient to imagine the triangles to be coloured alternately white and black, so that there are $l$ of either colour in each face, two of either colour at each edge, $m$ of either colour at each vertex, and $g$ of either colour altogether. Hence, if the polyhedron has $f$ faces, $e$ edges, and $v$ vertices,

$$lf = 2e = mv = g,$$

*i.e.*

(4.3) $$f = g/l, \quad e = g/2, \quad v = g/m.$$

Thus (4.2) is equivalent to Euler's theorem

(4.4) $$f + v - e = 2.$$

The faces of a regular *skew* polyhedron can be divided into right-angled triangles in the same manner. Any such triangle is still a fundamental region for the complete symmetry group, but the generating operations which transform the first triangle into its neighbours need no longer be reflections. Actually, $R_1$ and $R_3$ still are reflections (in the planes or three-spaces that we called $\lambda$ and $\mu$), but $R_2$ is the rotation through $\pi$ about the hypotenuse $O_1 O_3$ (or about the plane joining this line to the centre). Hence the sub-group generated by $R$ and $S$ no longer consists solely of rotations. However, since it transforms white triangles into white and black into black, we shall call it the *intrinsic rotation-group* of $\{l, m \,|\, n\}$, and still denote it by $\mathfrak{G}$, and its order by $g$. Both groups now require an extra relation in their abstract definitions, on account of the multiple connectivity of the surface. The formulae (4.3) still hold, but (4.4) has to be replaced by

$$f + v - e = 2 - 2p,$$

where $p$ is the genus; consequently (4.2) becomes

(4.5) $$p = \tfrac{1}{2}g(\tfrac{1}{2} - 1/l - 1/m) + 1.$$

This enables us to find $p$ when $g$ is known, but we have no simple expression for $g$ in terms of $l, m, n$.

Actually, we can find $g$ in any given case by considering the group generated by the reflections in the planes or three-spaces $\lambda, \lambda', \mu, \mu'$. When we denote these reflections by $L, L', M, M'$, (2.1) shows that this group is defined by

(4.6)* $$L^2 = L'^2 = M^2 = M'^2 = (LM)^2 = (L'M')^2$$
$$= (LL')^{\frac{1}{2}l} = (MM')^{\frac{1}{2}m} = (LM')^n = (L'M)^n = 1.$$

Its fundamental region includes two of our right-angled triangles, namely one white and one black, with a common hypotenuse. Hence it is a subgroup of index 2 in the complete group of the polyhedron, so that its order is $g$†. The rotation $R_2$, which interchanges the white and black triangles, is a symmetry of the tetrahedron $\lambda\lambda'\mu\mu'$, and transforms $L, L', M, M'$ into $L', L, M', M$, respectively. Thus we have

$$L = R_1, \quad L' = R_2R_1R_2, \quad M = R_3, \quad M' = R_2R_3R_2.$$

The complete group is derivable from (4.6) by adjoining the involutory operation $R_2$. By direct substitution, its abstract definition is thus

(4.7) $$R_1^2 = R_2^2 = R_3^2 = (R_1R_2)^l = (R_2R_3)^m = (R_1R_3)^2$$
$$= (R_1R_2R_3R_2)^n = 1.$$

Finally, the intrinsic rotation-group $\mathfrak{G}$ can be derived from this (by writing $R_1R_2 = R$, $R_2R_3 = S$) in the form

(4.8) $$R^l = S^m = (RS)^2 = (RS^{-1})^n = 1.$$

The fact that these relations suffice to define $\mathfrak{G}$ may be verified by adjoining the involutory operation $R_2$ (which transforms $R$ and $S$ into their inverses), and so reconstructing (4.7).

In the case of $\{4, 4 \mid n\}$, the group generated by reflections is $[n, 2, n]$ or $[n] \times [n]$, the direct product of two dihedral groups of order $2n$. Hence

---

* Coxeter, 3.
† Therefore $g$ is infinite when $2 \sin(\pi/l) \sin(\pi/m) \leqslant \cos(\pi/n)$ and $l, m$ are even.

$\mathfrak{G}$, defined by*

$$R^4 = S^4 = (RS)^2 = (RS^{-1})^n = 1,$$

is likewise of order $4n^2$. It has a simple representation as a permutation-group of degree $2n$, namely

$$R = (1\ 2\ 3\ 4)(5\ 6\ 7\ 8)(\ldots\ 2n), \quad S = (1\ 2)(3\ 4\ 5\ 6)(\ldots\ 2n);$$

*e.g.*, when $n = 3$†,

$$R = (1\ 2\ 3\ 4)(5\ 6), \qquad S = (1\ 2)(3\ 4\ 5\ 6),$$

and when $n = 4$‡,

$$R = (1\ 2\ 3\ 4)(5\ 6\ 7\ 8), \qquad S = (1\ 2)(3\ 4\ 5\ 6)(7\ 8).$$

In the case of $\{4,\ 6\,|\,3\}$ or $\{6,\ 4\,|\,3\}$, the group generated by reflections is $[3, 3, 3]$, the symmetric group of order 120. Hence $\mathfrak{G}$, defined by

$$R^4 = S^6 = (RS)^2 = (RS^{-1})^3 = 1,$$

is likewise of order 120; and by considering the permutations $(1\ 4\ 2\ 5)$, $(1\ 5)\ (2\ 3\ 4)$, we see that it is again the symmetric group. Geometrically, $[3, 3, 3]$ permutes the five vertices of the regular simplex $\alpha_4$. $\mathfrak{G}$ differs from it in having each *odd* permutation combined with the central inversion.

In the case of $\{4,\ 8\,|\,3\}$ or $\{8,\ 4\,|\,3\}$, the group generated by reflections is $[3, 4, 3]$, of order 1152§. Hence $\mathfrak{G}$, defined by

$$R^4 = S^8 = (RS)^2 = (RS^{-1})^3 = 1,$$

is likewise of order 1152 (although it is a quite different group). The operation

$$(R^2 S^2)^6 = (R_1 . R_2 R_1 R_2 . R_2 R_3 R_2 . R_3)^6 = (L L' M' M)^6$$

is the central inversion; and, by considering the permutations

$$(1\ 8)\ (2\ 7\ 3\ 6)\ (4\ 5), \quad (1\ 6\ 3\ 7\ 4\ 5\ 2\ 8),$$

---

* Burnside, **1**, 419 (III; $a = n$, $b = 0$).
† Miller, **1**, 368 (Degree Six, Order 36, No. 2). The relation $(s_1{}^2 s_2{}^2)^3 = 1$ is superfluous.
‡ Burns, **1**, 208 (Order 64, No. 4). The relation $(s_1 s_2{}^2)^4 = 1$ is superfluous.
§ Goursat, **1**, 87.

we can identify the central quotient group

$$R^4 = S^8 = (RS)^2 = (RS^{-1})^3 = (R^2 S^2)^6 = 1$$

with Dr. Burns's group of order 576, No. 3*.

### 5. *Topological extension of the theory: the general polyhedron* $\{l, m \mid n\}$.

Every regular polyhedron can be interpreted topologically as a *regular map* in the sense of Brahana†, the faces of the polyhedron being the countries of the map. If the faces are $l$-gons, $m$ at each vertex, the map may be built up, face by face, by lettering the vertices, and calling the faces ABC..., ABE..., and so on. *E.g.*, when $l = 4$ and $m = 3$, we have the "topological cube"

ABCD, ABEF, ADGF, BCHE, CDGH, EFGH.

If this abstract construction is carried on in such a way that new letters are introduced whenever possible (*i.e.*, using only the number of sides of the face and the number of faces at a vertex), then the map is automatically regular, and forms a simply-connected surface, finite or infinite according as

$$1/l + 1/m > \text{ or } \leqslant \tfrac{1}{2}.$$

It can be metrically realized as a partition of the sphere, or of the Euclidean or hyperbolic plane, into regular $l$-gons; hence no confusion can arise by using the symbol $\{l, m\}$.

Any other regular map of $l$-gons, $m$ at each vertex, can be derived from $\{l, m\}$ by *identifying* certain edges. The intrinsic rotation-group $\mathfrak{G}$, of order $g = lf = 2e = mv$, may be defined topologically as a permutation group on the edges of the map; namely, as generated by $R$, which cyclically permutes the edges of one face, and $S$, which cyclically permutes the edges at one vertex‡. It is clearly a factor group of the rotation-group of $\{l, m\}$, whose abstract definition is (4.1). The identification of a pair of edges of $\{l, m\}$ can at once be interpreted as an extra relation between $R$ and $S$, and will necessitate the identification of every other pair that are similarly related.

We may, for instance, identify two edges belonging to a chain of edges which leaves, at each vertex, two faces on the left (and $m - 2$ on the right).

---

* Burns, **1**, 210. The central quotient group of [3, 4, 3], on the other hand, is No. 2.
† Brahana, **1**, 269.
‡ Our generators $R$, $S$ differ trivially from Brahana's $S$, $T$; in fact, our $R$, $S$ are his $S$, $S^{-1}T$, while his $S$, $T$ are our $R$, $RS$. Also our $l$, $f$ are his $k$, $n$.

It is easily seen (Fig. xi*) that the operation $RS^{-1}$ takes us one step along such a chain.   Thus the relation

$$(RS^{-1})^n = 1$$

corresponds to the identification of edges which differ by $n$ steps.

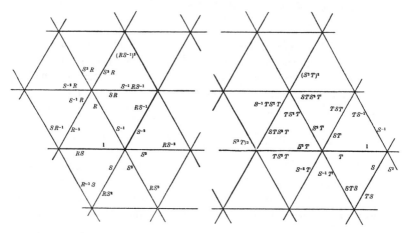

Fig. xi: $R^3 = S^6 = (RS)^2 = 1$.          Fig. xii: $S^6 = T^2 = (ST)^3 = 1$.

We now have an interpretation for the symbol $\{l, m \mid n\}$ regardless of the parity of $l$ and $m$.   § 4 shows that this agrees with our previous interpretation in the case when $l$ and $m$ are both even.

In a family of abstract groups with various periods assigned to certain fixed combinations of generators, it usually happens that the greater periods correspond to greater order†.   Accordingly, we should expect the criterion

(5.1)              $2 \sin(\pi/l) \sin(\pi/m) > \cos(\pi/n)$

(which precisely determines the finiteness of the polyhedron $\{l, m \mid n\}$

---

* In the special case when $l = 3$, (4.1) implies $(RS^{-1})^m = 1$.   But when $l$ and $m$ are both greater than 3, the period of $RS^{-1}$ is infinite.   This follows from (1.1), by considering the Minkowskian polyhedron $\{l, m\}$; for, if $l \geqslant 4$ and $m \geqslant 4$, then $\cos(\pi/n) \geqslant 1$, and $\pi/n$ is a hyperbolic angle.

† This is merely a working rule, and not a demonstrable theorem, as the following example shows : although $\{6, 6 \mid 3\}$ is infinite, we shall see later that $\{6, 7 \mid 3\}$ and $\{7, 7 \mid 3\}$ are finite.

when $l$ and $m$ are even) to retain some significance when the parity of $l$ and $m$ is unrestricted. Actually, we find that it fails in two respects.

First, for certain values of $l$, $m$, $n$, satisfying the criterion (as well as for certain values violating it), the group *collapses*, in the sense that the relations (4.8) imply $R = S = 1$. *E.g.*, there is no polyhedron $\{4, 5 \mid 3\}$. Secondly, for certain values of $l$, $m$, $n$, violating the criterion, the group remains finite. In fact, the order of the group, considered as a function of $l, m, n$, has a tendency to take, when $l$ or $m$ is odd, a smaller value than that which we should expect by interpolation from the values when $l$ and $m$ are even. (Since the value when $l$ and $m$ are even is determined by the volume of a spherical tetrahedron, this interpolation could be given a precise meaning in terms of Schläfli functions.) However, we can say this: *whenever the group is infinite, the criterion is violated, i.e.*, whenever the criterion is satisfied, the group either is finite or collapses. In Figs. vi and vii, we mark the known finite polyhedra as O's, the known infinite polyhedra as dots, and the known cases of collapse as crosses. We observe that the O's are all inside or "just outside" the region of validity of the criteria. (Unmarked lattice points correspond to groups which have not yet been investigated.)

When $l$ or $m = 3$, we must have $n = m$ or $l$ (respectively); any other value of $n$ causes collapse (although, if the assigned value of $n$ is a multiple of its proper value, the collapse will be merely partial).

*When $m > 3$, and $l$ and $n$ are both even* (or when $l > 3$, and $m$ and $n$ are both even), *the criterion* (5.1) *holds perfectly*; *e.g.*, $\{4, 7 \mid 4\}$ and $\{5, 6 \mid 4\}$ are infinite. For (4.8) can be written in the form

$$(5.2) \qquad S^m = T^2 = (ST)^l = (S^2 T)^n = 1,$$

which has been thoroughly investigated elsewhere* in the case when $l$ and $n$ are even.

By putting $S^2$ in place of $S$ in (5.2), we see that $\{l, 5 \mid n\}$ *and* $\{n, 5 \mid l\}$ *have the same group*. In particular, $\{5, 5 \mid 3\}$ has the same group as $\{3, 5 \mid 5\}$ or $\{3, 5\}$, namely the icosahedral group. $\{5, 5 \mid 3\}$ can, in fact, be realized metrically as the *great dodecahedron*† $\{5, \frac{5}{2}\}$ (Fig. xiii). All other polyhedra $\{n, 5 \mid 3\}$, $\{5, n \mid 3\}$ are impossible, since they would have the same group as $\{3, 5 \mid n\}$.

---

* Coxeter, 5, 284.

† *Cf.* Brahana and Coble, 1, 14 (Fig. VII), The other three Kepler-Poinsot polyhedra give nothing fresh. In fact, $\{\frac{5}{2}, 5\}$, $\{\frac{5}{2}, 3\}$, $\{3, \frac{5}{2}\}$, are topologically identical with $\{5, \frac{5}{2}\}$, $\{5, 3\}$, $\{3, 5\}$, respectively.

Apart from

$$\{3, m \,|\, n\}, \quad \{m, 3 \,|\, n\} \quad (n \neq m),$$

$$\{n, 5 \,|\, 3\}, \quad \{5, n \,|\, 3\} \quad (n \neq 5),$$

there are no known cases of collapse.

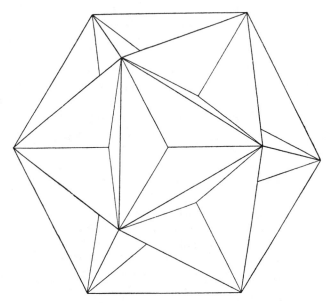

Fig. xiii: The great dodecahedron, $\{5, 5 \,|\, 3\}$.

The remaining polyhedra that satisfy (5.1) are

$$\{4, 5 \,|\, 4\}, \quad \{4, 7 \,|\, 3\}, \quad \{4, 5 \,|\, 5\},$$

and their reciprocals. The last of these has, of course, the same group as $\{5, 5 \,|\, 4\}$, which does not satisfy (5.1).

For $\{4, 5 \,|\, 4\}$ and $\{5, 4 \,|\, 4\}$, ⑤ is*

$$S^5 = T^2 = (ST)^4 = (S^2 T)^4 = 1,$$

of order 160. $\{4, 5 \,|\, 4\}$ can be realized metrically in five dimensions, by taking half the squares of the measure-polytope $\gamma_5$. Fig. xiv shows an isometric projection of the vertices and edges of $\gamma_5$, in which the 80 squares appear as rhombs of two kinds: 40 of angle $\frac{1}{5}\pi$, and 40 of angle $\frac{2}{5}\pi$. Either

---

* Coxeter, 5, 284.

set of 40 are projections of the faces of a $\{4, 5 \mid 4\}$. (This partition of the 80 squares is not really unique, but depends on the plane of projection.)

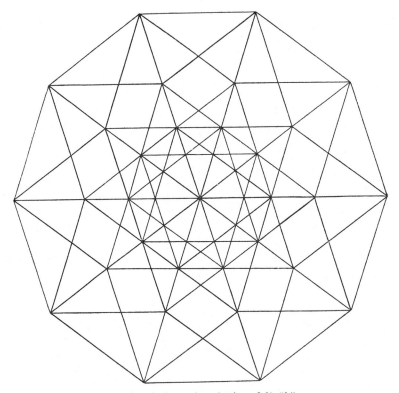

Fig. xiv: Orthogonal projection of $\{4, 5 \mid 4\}$.

For $\{4, 5 \mid 5\}$, $\{5, 4 \mid 5\}$, and $\{5, 5 \mid 4\}$, $\mathfrak{G}$ is

$$S^5 = T^2 = (ST)^4 = (S^2T)^5 = 1$$

or*

$$R^5 = S^5 = (RS)^2 = (RS^{-1})^4 = 1,$$

the simple group of order 360 (*i.e.*, the alternating group of degree six). Fig. xv shows a conformal representation of $\{4, 6\}$ *qua* partition of the hyperbolic plane, with letters to indicate the identifications that produce $\{4, 6 \mid 3\}$ (*cf.* Fig. viii).

---

*Todd and Coxeter, **1**, 31 (5); Brahana, **1**, 274.

For $\{4, 7|3\}$ and $\{7, 4|3\}$, $\mathfrak{G}$ is*

$$S^7 = T^2 = (ST)^4 = (S^2 T)^3 = 1,$$

the simple group of order 168. This is a special case of

$$S^7 = T^2 = (ST)^l = (S^2 T)^3 = 1.$$

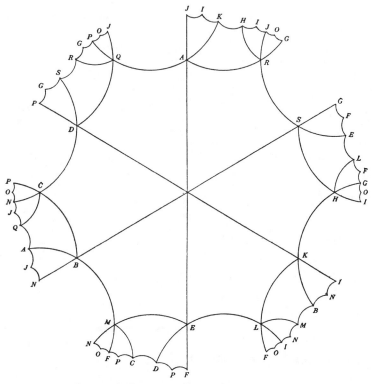

Fig. xv: The topological polyhedron $\{4, 6|3\}$.

On putting $R^3$ for $S$, this becomes

$$R^7 = T^2 = (RT)^3 = (R^3 T)^l = 1,$$

or, since $R^2 TR = R(TRT)^{-1} = RT^{-1} R^{-1} T$,

$$R^7 = T^2 = (RT)^3 = (RT^{-1} R^{-1} T)^l = 1.$$

---

*Burnside, **1**, 422. $S_7 = S^2$, $S_2 = T$.

In the form

$$S^3 = T^2 = (ST)^7 = (S^{-1} T^{-1} ST)^l = 1,$$

this family of groups has been extensively studied by Brahana and Sinkov[*]. By putting $l = 6$ and then $l = 7$, we see that, for the three polyhedra

$$\{6, 7 \,|\, 3\}, \quad \{7, 6 \,|\, 3\}, \quad \{7, 7 \,|\, 3\},$$

$\mathfrak{G}$ is the simple group of order 1092.

There remain, in Fig. vi, one pair of lattice points that are "only just outside the region", namely the points corresponding to the polyhedra

$$\{4, 9 \,|\, 3\}, \quad \{9, 4 \,|\, 3\}.$$

After enormous labour, I succeeded in enumerating the 272 co-sets of a cyclic subgroup of order 9, thus proving that the group

$$S^9 = T^2 = (ST)^4 = (S^2 T)^3 = 1$$

is of order 2448. Then Sinkov completed its identification with the simple group of that order, by citing the linear fractional substitutions

$$S = \left(7 \, \frac{x+1}{x+4}\right), \quad T = (-x) \pmod{17}.$$

With the help of (4.3), we can now write out Table I. In describing the various groups, we use the symbols $S_n$ and $A_n$ to denote the symmetric and alternating groups of degree $n$.

## 6. *A further extension*: *the polyhedron* $\{q_1, m \,|\, q_2, q_3, \ldots\}$.

In §5, we derived the topological polyhedron $\{l, m \,|\, n\}$ from $\{l, m\}$ by identifying two edges that belong to a certain "chain". We may generalize our results by using a chain of edges which leaves, at each vertex, (say) $j$ on the right (and $m-j$ on the left). When $j = 1$ this is a face; when $j = 2$ it is a "hole". We define a generalized skew polyhedron

$$(6.1) \qquad\qquad \{q_1, m \,|\, q_2, q_3, \ldots\}$$

---

[*] Brahana, 2; Sinkov, 1.

by identifying points $q_j$ steps along the $j$-th chain. In the most general case we would allow $j$ to take all values from 1 to $[\frac{1}{2}m]$. In certain cases, such a skew polyhedron can be realized metrically in Euclidean space of $m$ (or fewer) dimensions. Its vertex figure is then a skew $m$-gon whose sides, first diagonals, second diagonals, etc., are of lengths

$$2 \cos (\pi/q_1), \quad 2 \cos (\pi/q_2), \quad 2 \cos (\pi/q_3), \quad \dots .$$

Since these generalized skew polyhedra do not necessarily occur in reciprocal pairs, it is convenient to replace $R$ by $TS^{-1}$, as in (5.2). A glance at Fig. xii makes it clear that the group takes the form

$$S^m = T^2 = (S^j T)^{q_j} = 1 \quad (j = 1, 2, \dots),$$

or, in terms of $R$ and $S$,

$$R^{q_1} = S^m = (RS)^2 = (RS^{-1})^{q_2} = (RS^{-2})^{q_3} = \dots = 1.$$

When all the $q$'s are even, the only finite polyhedra of this kind are*

$$\{4, \, m \,|\, 4^{[\frac{1}{2}m]-1}\}$$

and                $$\{4, \, m \,|\, 4^{\frac{1}{2}m-2}, 2p\} \quad (m \text{ even}).$$

These can both be realized metrically in Euclidean $m$-space, the former by squares from the measure-polytope $\gamma_m$, and the latter by squares from the generalized prism (or "rectangular product") of $\frac{1}{2}m$ $2p$-gons:

$$[\{2p\}, \, \{2p\}, \, \dots] \quad \text{or} \quad \{2p\}^{\frac{1}{2}m}.$$

The polyhedron $\{4, \, m \,|\, 4^{[\frac{1}{2}m]-1}\}$, with $g = 2^m m$, has $2^{m-2} m$ faces (squares), $2^{m-1} m$ edges, and $2^m$ vertices (viz., all the edges and vertices of $\gamma_m$). Accordingly, its genus is $2^{m-3}(m-4)+1$. (The cases $m = 4$ and $m = 5$ were considered earlier. See Figs. ix, xiv.) The properties of $\{4, \, m \,|\, 4^{\frac{1}{2}m-2}, 2p\}$ ($m$ even) can be deduced similarly from the fact that $g = (2p)^{\frac{1}{2}m} m$.

The vertex figure of $\{4, \, m \,|\, 4^{[\frac{1}{2}m]-1}\}$ is a skew $m$-gon whose sides and all kinds of diagonals are of length $\sqrt{2}$. In fact, the vertex figure of $\gamma_m$ is a regular simplex $\alpha_{m-1} \sqrt{2}$, whose Petrie polygon† is the vertex figure of the

---

* Coxeter, 5, 283, (4.2), (4.5). We write $4^n$ for a row of $n$ 4's.
† Coxeter, 1, 203.

polyhedron. The vertex figure of $\{4, m \mid 4^{\frac{1}{2}m-2}, 2p\}$ ($m$ even) differs from this in having all pairs of *opposite* vertices distant $2 \cos (\pi/2p)$. When we make $p$ infinite, this becomes the Petrie polygon of the cross-polytope $\beta_{\frac{1}{2}m} \sqrt{2}$. Now, this cross-polytope is the vertex figure of $\delta_{\frac{1}{2}m+1}$, the net of measure-polytopes, *i.e.*, the infinite polytope whose vertices are all the lattice points in $\frac{1}{2}m$ dimensions. Hence, writing $r$ for $\frac{1}{2}m$, we have the infinite skew polyhedron

$$\{4, \ 2r \mid 4^{r-2}\},$$

whose faces belong to $\delta_{r+1}$. This can be regarded as a generalization of $\{4, 4\}$ and $\{4, 6 \mid 4\}$.

There is also a generalization of $\{6, 3\}$ and $\{6, 4 \mid 4\}$: the infinite polyhedron

$$\{6, \ m \mid 4^{[\frac{1}{2}m]-1}\}.$$

Its faces are all the hexagons of the uniform $(m-1)$-dimensional space-filling* whose graph consists of an $m$-gon with every vertex ringed. The remaining plane faces of this space-filling are squares, which form, when $m = 5$, another infinite polyhedron:

$$\{4, \ 5 \mid 6\}.$$

When the parity of the $q$'s is unrestricted, very little is known about the possible polyhedra (6.1). One line of investigation is suggested by the fact that

$$\{6, \ 6 \mid 3, \ 4\}$$

can be realized metrically by taking all the equatorial hexagons of all the bounding cuboctahedra of the "truncated tesseract" $t_1 \gamma_4$. The corresponding group is simply

$$S^6 = T^2 = (S^2 T)^3 = (S^3 T)^4 = 1,$$

since these relations imply $(ST)^6 = 1$. This is the "hyper-pyritohedral" group† $[(3, 3)', 4]$, of order 192, since it is derivable from $[3, 3, 4]$‡ by putting

$$S = R_1 R_2 R_4, \quad T = R_1 R_3.$$

---

* Coxeter, **4**, 334.
† Coxeter, **5**, 295.
‡ *I.e.*,

$$R_1{}^2 = R_2{}^2 = R_3{}^2 = R_4{}^2 = (R_1 R_2)^3 = (R_2 R_3)^3 = (R_3 R_4)^4 = (R_1 R_3)^2 = (R_1 R_4)^2 = (R_2 R_4)^2 = 1.$$

The hexagons cut one another diagonally, producing "false vertices". This polyhedron, therefore, is a kind of generalized Kepler-Poinsot polyhedron. Four-dimensional space admits nineteen reciprocal pairs of such skew star-polyhedra; but they lie beyond the scope of the present work.

In conclusion, we mention the case when none of the $q$'s are specified save $q_1$ and $q_3$, *i.e.*, the case of the polyhedron*

$$\{l, m\,|, q\}   (m \geqslant 6)$$

with $q$-gonal "second holes". The group is

(6.2)          $S^m = T^2 = (ST)^l = (S^3\,T)^q = 1,$

or          $R^l = S^m = (RS)^2 = (RS^{-2})^q = 1,$

or          $R^l = T^2 = (RT)^m = (R^2\,TRT)^q = 1.$

When $l = 3$, this last definition becomes†

$$R^3 = T^2 = (RT)^m = (R^{-1}\,TRT)^q = 1.$$

In particular, one of Brahana's results‡ shows that $\{3, 8\,|, 6\}$ is infinite.

When $m = 7$, we write $S^2$ for $S$ in (6.2), obtaining

$$S^7 = T^2 = (ST)^q = (S^2\,T)^l = 1.$$

Thus $\{l, 7\,|, q\}$ *has the same group as* $\{q, 7\,|\,l\}$. $\{3, 7\,|, 6\}$ and $\{3, 7\,|, 7\}$ provide the extraordinary phenomenon of two polyhedra which have the same number of vertices, edges, and faces (and therefore the same genus), *and the same group*, but which still are distinct.

When $m$ is even, $R$ and $S^2$ generate a self-conjugate sub-group of index 2:

$$R^l = S'^{\frac{1}{2}m} = (RS')^l = (RS'^{-1})^q = 1,$$

so that, when we put $s_1 = R^{-1}$, $s_2 = RS'$,

$$s_1{}^l = s_2{}^l = (s_1 s_2)^{\frac{1}{2}m} = (s_1{}^2 s_2)^q = 1.$$

---

* $m = 3$ implies $q = 2$; $m = 4$ implies $q = l$; and $\{l, 5\,|, q\}$ is the same as $\{l, 5\,|q\}$.
† Brahana, 2.
‡ Brahana, 3, 901.

**59**

It seems likely that the only non-trivial cases when this group is finite are those considered by G. A. Miller[*], and one immediate consequence of these, namely:

$$l = 3, \quad \tfrac{1}{2}m = q = 4,$$

$$l = 3, \quad \tfrac{1}{2}m = 4, \quad q = 5,$$

$$l = 3, \quad \tfrac{1}{2}m = 5, \quad q = 4,$$

$$l = 4, \quad \tfrac{1}{2}m = q = 3.$$

In the first case, Miller gives the permutations

$$s_1 = (abc)\,(def), \quad s_2 = (aeh)\,(cdg).$$

Since $S$ transforms $s_1$ into $s_2$, and $S^2 = S' = s_1 s_2$, we have

$$R = (acb)\,(dfe), \quad S = (agbcdehf).$$

Hence the group of $\{3, 8\,|\, 4\}$ is the "extended 168-group"[†], *i.e.*, the group of all linear fractional transformations modulo 7. By putting $s_1' = S'$, $s_2' = S'^{-1}R$, we obtain the 168-group itself in the alternative form

$$s_1'^4 = s_2'^4 = (s_1' s_2')^3 = (s_1'^2 s_2')^3 = 1,$$

with $\qquad\qquad s_1' = (abdh)\,(cefg), \quad s_2' = (ahfd)\,(bcge).$

In this case, $s_1'$ is transformed into $s_2'$ by $(afcebd)$; so the group of $\{4, 6\,|\, 3\}$ is again the extended 168-group.

Du Val has pointed out that the polyhedron $\{3, 7\,|\, 4\}$ can be metrically realized (albeit singularly) in seven dimensions. Its faces are the fifty-six triangles of the regular simplex $a_7$; its edges and vertices are those of the simplex, each counted three times. The six triangles that meet at an edge of $a_7$ have to be paired in a special manner.

---

[*] Miller, 2.

[†] Van der Waerden calls this group $PGL\,(2, 7)$, $PSL\,(2, 7)$ being the 168-group itself, which Dickson calls $LF(2, 7)$.

TABLE I.  Finite polyhedra $\{l, m|n\}$.

| Polyhedron | $f$ | $e$ | $v$ | $p$ | $\mathfrak{G}$ | $g$ |
|---|---|---|---|---|---|---|
| $\{3, 3\,\lvert\,3\} = \{3, 3\}$ | 4 | 6 | 4 | 0 | $A_4 \sim LF(2, 3)$ | 12 |
| $\{3, 4\,\lvert\,4\} = \{3, 4\}$ | 8 | 12 | 6 | 0 | $\left.\begin{array}{c}\\\\\end{array}\right\}$ $S_4$ | 24 |
| $\{4, 3\,\lvert\,4\} = \{4, 3\}$ | 6 | 12 | 8 | 0 | | |
| $\{3, 5\,\lvert\,5\} = \{3, 5\}$ | 20 | 30 | 12 | 0 | | |
| $\{5, 3\,\lvert\,5\} = \{5, 3\}$ | 12 | 30 | 20 | 0 | $A_5 \sim LF(2, 5`)$ | 60 |
| $\{5, 5\,\lvert\,3\} = \{5, \tfrac{5}{2}\}$ | 12 | 30 | 12 | 4 | | |
| $\{4, 4\,\lvert\,n\}$ | $n^2$ | $2n^2$ | $n^2$ | 1 | | $4n^2$ |
| $\{4, 5\,\lvert\,4\}$ | 40 | 80 | 32 | 5 | | 160 |
| $\{5, 4\,\lvert\,4\}$ | 32 | 80 | 40 | 5 | | |
| $\{4, 6\,\lvert\,3\}$ | 30 | 60 | 20 | 6 | $S_5$ | 120 |
| $\{6, 4\,\lvert\,3\}$ | 20 | 60 | 30 | 6 | | |
| $\{4, 7\,\lvert\,3\}$ | 42 | 84 | 24 | 10 | $LF(2, 7)$ | 168 |
| $\{7, 4\,\lvert\,3\}$ | 24 | 84 | 42 | 10 | | |
| $\{4, 5\,\lvert\,5\}$ | 90 | 180 | 72 | 10 | $A_6$ | 360 |
| $\{5, 4\,\lvert\,5\}$ | 72 | 180 | 90 | 10 | | |
| $\{5, 5\,\lvert\,4\}$ | 72 | 180 | 72 | 19 | | |
| $\{4, 8\,\lvert\,3\}$ | 288 | 576 | 144 | 73 | | 1152 |
| $\{8, 4\,\lvert\,3\}$ | 144 | 576 | 288 | 73 | | |
| $\{6, 7\,\lvert\,3\}$ | 182 | 546 | 156 | 105 | | |
| $\{7, 6\,\lvert\,3\}$ | 156 | 546 | 182 | 105 | $LF(2, 13)$ | 1092 |
| $\{7, 7\,\lvert\,3\}$ | 156 | 546 | 156 | 118 | | |
| $\{4, 9\,\lvert\,3\}$ | 612 | 1224 | 272 | 171 | $LF(2, 17)$ | 2448 |
| $\{9, 4\,\lvert\,3\}$ | 272 | 1224 | 612 | 171 | | |
| $\{7, 8\,\lvert\,3\}$ | 1536 | 5376 | 1344 | 1249 | | 10752 |
| $\{8, 7\,\lvert\,3\}$ | 1344 | 5376 | 1536 | 1249 | | |

TABLE II.  Finite polyhedra $\{l, m\,\lvert, q\}$.

| Polyhedron | $f$ | $e$ | $v$ | $p$ | $\mathfrak{G}$ | $g$ |
|---|---|---|---|---|---|---|
| $\{3, 6\,\lvert, q\}$ | $2q^2$ | $3q^2$ | $q^2$ | 1 | | $6q^2$ |
| $\{3, 2q\,\lvert, 3\}$ | $2q^2$ | $3q^2$ | $3q$ | $\tfrac{1}{2}(q-1)(q-2)$ | | $6q^2$ |
| $\{3, 7\,\lvert, 4\}$ | 56 | 84 | 24 | 3 | $LF(2, 7)$ | 168 |
| $\{3, 8\,\lvert, 4\}$ | 112 | 168 | 42 | 8 | $\left.\begin{array}{c}\\\end{array}\right\}$ $PGL(2, 7)$ | 336 |
| $\{4, 6\,\lvert, 3\}$ | 84 | 168 | 56 | 15 | | |
| $\{3, 7\,\lvert, 6\}$ | 364 | 546 | 156 | 14 | $\left.\begin{array}{c}\\\end{array}\right\}$ $LF(2, 13)$ | 1092 |
| $\{3, 7\,\lvert, 7\}$ | 364 | 546 | 156 | 14 | | |
| $\{3, 8\,\lvert, 5\}$ | 720 | 1080 | 270 | 46 | | 2160 |
| $\{3, 10\,\lvert, 4\}$ | 720 | 1080 | 216 | 73 | | |
| $\{4, 6\,\lvert, 2\}$ | 12 | 24 | 8 | 3 | $S_4 \times S_2$ | 48 |
| $\{5, 6\,\lvert, 2\}$ | 24 | 60 | 20 | 9 | $A_5 \times S_2$ | 120 |
| $\{3, 11\,\lvert, 4\}$ | 2024 | 3036 | 552 | 231 | $LF(2, 23)$ | 6072 |
| $\{3, 7\,\lvert, 8\}$ | 3584 | 5376 | 1536 | 129 | | 10752 |
| $\{3, 9\,\lvert, 5\}$ | 12180 | 18270 | 4060 | 1016 | $LF(2, 29) \times A_5$ | 36540 |

*References.*

Andreini, 1.　" Sulle reti di poliedri regolari e semiregolari ", *Mem. della Soc. ital. delle Sci.* (3), 14 (1905), 75–110.

Brahana, 1.　" Regular maps and their groups ", *American J. of Math.*, 49 (1927), 268–284.

————, 2.　" Certain perfect groups generated by two operators of orders two and three ", *American J. of Math.*, 50 (1928), 345–356.

————, 3.　" On the groups generated by two operators of orders two and three whose product is of order eight ", *American J. of Math.*, 53 (1931), 891–901.

Brahana and Coble, 1.　" Maps of twelve countries with five sides with a group of order 120 containing an ikosahedral subgroup ", *American J. of Math.*, 48 (1926), 1–20.

Burns, 1.　" Abstract definitions of groups of degree eight ", *American J. of Math.*, 37 (1915), 195–214.

Burnside, 1.　" Theory of groups of finite order " (Cambridge, 1911).

Cauchy, 1.　" Recherches sur les polyèdres ", *Journal de l'Ecole Polyt.*, 9 (1813), 68–86.

Coxeter, 1.　" The densities of the regular polytopes ", *Proc. Camb. Phil. Soc.*, 27 (1931), 201–211.

————, 2.　" The polytopes with regular-prismatic vertex figures ", *Proc. London Math. Soc.* (2), 34 (1932), 126–189.

————, 3.　" The complete enumeration of finite groups of the form

$$R_i{}^2 = (R_i R_j)^{k_{ij}} = 1 \text{ ",}$$

*Journal London Math. Soc.*, 10 (1935), 21–25.

————, 4.　" Wythoff's construction for uniform polytopes ", *Proc. London Math. Soc.* (2), 38 (1935), 327–339.*

————, 5.　" The abstract groups $R^m = S^m = (R^j S^j)^{p_j} = 1$, $S^m = T^2 = (S^j T)^{2p_j} = 1$, and $S^m = T^2 = (S^{-j} T S^j T)^{p_j} = 1$ ", *Proc. London Math. Soc.* (2), 41 (1936), 278–301.

Goursat, 1.　" Sur les substitutions orthogonales et les divisions régulières de l'espace ", *Ann. Sci. de l'Ecole Norm. Sup.* (3), 6 (1889), 9–102.

Miller, 1.　" Abstract definitions of all the substitution groups whose degrees do not exceed seven ", *American J. of Math.*, 33 (1911), 363–372.

————, 2.　" Groups generated by two operators of order three whose product is of order four ", *Bull. Amer. Math. Soc.*, 26 (1920), 361–369.

Sinkov, 1.　" A set of defining relations for the simple group of order 1092 ", *Bull. Amer. Math. Soc.*, 41 (1935), 237–240.

Stott, 1.　" Geometrical reduction of semiregular from regular polytopes and space fillings ", *Ver. der K. Akad. van Wet.* (Amsterdam) (1), 11.1 (1910).

Threlfall and Seifert, 1.　" Topologische Untersuchung der Diskontinuitätsbereiche endlicher Bewegungsgruppen des dreidimensionalen sphärischen Raumes ", *Math. Annalen*, 104 (1931), 1–70.

Todd and Coxeter, 1.　" A practical method for enumerating co-sets of a finite abstract group ", *Proc. Edinburgh Math. Soc.* (2), 5 (1936), 26–34.

*Reference "Coxeter 4" is Chapter 3 of this book.

# 6

# SELF-DUAL CONFIGURATIONS AND REGULAR GRAPHS

Reprinted by permission of the
American Mathematical Society from Bulletin 56.

1. **Introduction.** A configuration ($m_c$, $n_d$) is a set of $m$ points and $n$ lines in a plane, with $d$ of the points on each line and $c$ of the lines through each point; thus $cm = dn$. Those permutations which preserve incidences form a group, "the group of the configuration." If $m = n$, and consequently $c = d$, the group may include not only *symmetries* which permute the points among themselves but also *reciprocities* which interchange points and lines in accordance with the principle of duality. The configuration is then "self-dual," and its symbol ($n_d$, $n_d$) is conveniently abbreviated to $n_d$. We shall use the same symbol for the analogous concept of a configuration in three dimensions, consisting of $n$ points lying by $d$'s in $n$ planes, $d$ through each point.

With any configuration we can associate a diagram called the *Menger graph* [13, p. 28],* in which the points are represented by dots or "nodes," two of which are joined by an arc or "branch" whenever the corresponding two points are on a line of the configuration. Unfortunately, however, it often happens that two different configurations have the same Menger graph. The present address is concerned with another kind of diagram, which represents the configuration uniquely. In this *Levi graph* [32, p. 5], we represent the points and lines (or planes) of the configuration by dots of two colors, say "red nodes" and "blue nodes," with the rule that *two nodes differently colored are joined whenever the corresponding elements of the configuration are incident.* (Two nodes of the same color are never joined.) Thus the Levi graph for ($m_c$, $n_d$) has $m$ red nodes and $n$ blue nodes, with each red node joined to $c$ blue nodes and each blue node joined to $d$ red nodes, so that there are $cm = dn$ branches altogether.

As simple instances in two dimensions we have the triangle $3_2$, whose Levi graph is a hexagon with red and blue vertices occurring

---

*For bibliography, see pages 147–49.

alternately; and the complete quadrangle $(4_3, 6_2)$, whose Levi graph may be regarded as a tetrahedron with a red node at each vertex and a blue node at the midpoint of each edge.

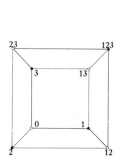

FIG. 1. The cube $\gamma_3$.

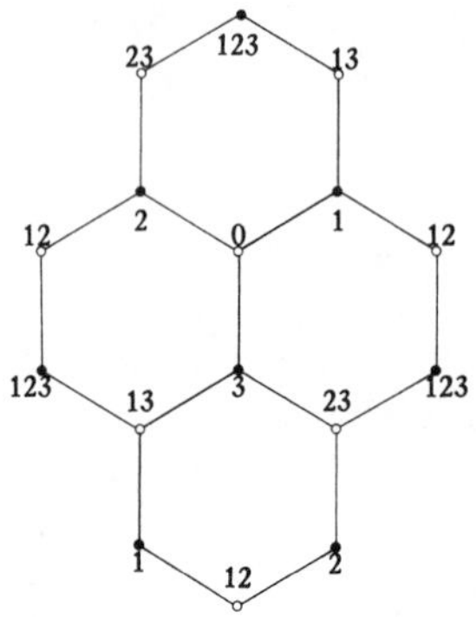

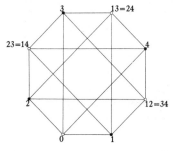

FIG. 2. The regular map $\{6, 3\}_{2,0}$.

As instances in three dimensions we have the tetrahedron $4_3$, whose Levi graph (Fig. 1 or 3) consists of the vertices and edges of a cube (with the vertices colored alternately red and blue); and Möbius's configuration $8_4$, whose Levi graph (Fig. 5) consists of the vertices and edges of a four-dimensional hypercube $\gamma_4$ [12, p. 123], as we shall see in §10. This $8_4$ may be regarded as a pair of tetrahedra so placed that each vertex of either lies in a face-plane of the other.

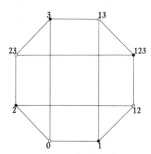

FIG. 3. A distorted cube.

FIG. 4. $\gamma_4/2$.

The symbols at the vertices in Figs. 1 and 5 represent Cartesian coordinates in the following manner: 0 is the origin, and every other symbol indicates which of the three or four coordinates have the value 1, while the rest are zero. Thus two vertices are joined by an edge whenever their symbols are derived from each other by inserting or removing a single digit. (Here 0 does not count as a digit, but rather as the absence of digits.) A vertex is red or blue according as its symbol contains an even or odd number of digits (thus 0 is red). The group of symmetries and reciprocities of Möbius's configuration is of order $4! \, 2^4 = 384$; for it is isomorphic with the symmetry group of the hypercube, and this consists of the permutations of the digits 1, 2, 3, 4, combined with reflections $R_1$, $R_2$, $R_3$, $R_4$ which act on the symbols as follows: $R_i$ adds a digit $i$ whenever that digit is not yet present, and removes $i$ whenever it is present; for example, $R_3$ transforms the square 0 1 13 3 into the same square with its vertices named in the reverse order.

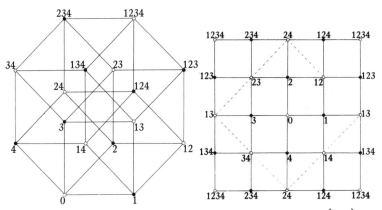

FIG. 5. The hypercube $\gamma_4$.    FIG. 6. The regular maps $\{4, 4\}_{4,0}$ and $\{4, 4\}_{2,2}$.

It is possible (in three ways) to select 16 of the 24 square faces of the hypercube so as to form the regular skew polyhedron $\{4, 4 | 4\}$ [11, p. 43]. Topologically, this is a regular map on a torus, as in Fig. 6 (with peripheral elements identified as marked). Fig. 4 is named $\gamma_4/2$ because it can be derived from the hypercube $\gamma_4$ by identifying opposite elements, as in the passage from spherical to elliptic space. When applied to Fig. 6, this identification yields a map of eight (instead of sixteen) squares on a torus, given by identifying opposite sides of the square drawn in broken lines. Comparing Fig. 4 with Fig. 3, we see that this graph of 8 nodes and 16 branches may be regarded as a cube with opposite vertices joined, though its group (of order $2 \cdot 4!^2 = 1152$) is much larger than that of the cube itself. It is the Levi graph for a rather trivial configuration consisting of four points on a line and four planes through the same line.

Fig. 2 shows that the Levi graph for the tetrahedron (Fig. 1) forms another regular map on a torus: it decomposes the surface into four hexagons [13, p. 25]. If we look for these hexagons (such as 0 1 12 123 23 3) in the original cube, we find that they are Petrie polygons: regular skew polygons whose various pairs of adjacent sides belong to different faces. Conversely, the map has its own Petrie polygons (such as 0 1 13 3), which are the faces of the cube.

This investigation is justified by the general principle that interesting configurations are represented by interesting graphs. In practice, either subject is liable to throw light on the other. A known property of Kummer's configuration $16_6$ led to the discovery of a new regular skew polyhedron in six dimensions, as we shall see in §13. On the other hand, a certain graph, described long ago by an electrical engineer, was found to represent a highly symmetrical configuration $12_3$ in the real projective plane (§8).

**2. Regular graphs.** To be precise, a *graph* is a collection of $N_0$ objects called *nodes*, $N_1$ pairs of which satisfy a certain symmetric relation. Such a pair of nodes is called a *branch* (or rather, the two nodes are said to be *joined* by a branch). Since we shall never allow two branches to join the same two nodes, it is convenient to regard the $N_0$ nodes as points and the $N_1$ branches as line-segments joining pairs of the points. If the number of branches at a node is the same for all nodes, say $d$, the graph is said to be of *degree d*, and we see that

$$2N_1 = dN_0.$$

An ordered sequence of $s$ distinct branches, consecutively adjacent (forming a continuous path from one node to another in a definite direction) is called an *s-arc*; for example, the 1-arcs are the $2N_1$ directed branches. A graph is said to be connected if every two nodes are joined by some $s$-arc. We shall be concerned solely with connected graphs. A closed sequence of $m$ distinct branches, consecutively adjacent (forming a continuous path from a node to itself in $m$ steps, $m > 2$) is called an *m-circuit*. The smallest value of $m$ for which an $m$-circuit occurs is called the *girth* of the graph; for example, the graph of vertices and edges of the cube has degree 3 and girth 4. It is often possible to find an $N_0$-circuit that includes all the nodes of the graph (and $N_0$ of the $N_1$ branches); this is called a *Hamiltonian* circuit [12, p. 8], and enables us to draw the graph as an $N_0$-gon with $N_1 - N_0$ diagonals inserted. (See Figs. 3, 4, 9, 12, 15, 18, 22, 23.)

A graph of degree $d$ evidently has $d - 1$ times as many $s$-arcs as $(s-1)$-arcs, provided $1 < s < m$. Thus the number of $s$-arcs is

$$(d - 1)^{s-1} 2N_1 = (d - 1)^{s-1} dN_0 \qquad (1 < s < m).$$

Those permutations of the nodes which leave them joined as before form a group, "the group of the graph." A graph is said to be *s-regular* if its *s*-arcs are all alike while its (*s*+1)-arcs are not all alike, that is, if the group is transitive on the *s*-arcs but not on the (*s*+1)-arcs. Thus the cube is 2-regular (not 3-regular, since a 3-arc may belong either to a face or to a Petrie polygon). For an *s*-regular graph of degree 3, the two ways in which a given *s*-arc may be extended to an (*s*+1)-arc are essentially different; so the subgroup leaving the *s*-arc invariant is of order 1, and the order of the whole group is equal to the number of *s*-arcs, namely

$$2^s N_1.$$

But when $d > 3$, the final branch of an (*s*+1)-arc may sometimes be chosen in several equivalent ways; so we can only say that the order is *divisible* by $(d-1)^{s-1} 2N_1$.

The above definitions differ only very slightly from those proposed by Tutte [47], who proved that, for every *s*-regular graph of girth *m*,

$$s \leqq m/2 + 1.$$

He called the graph an *m-cage* in the case of maximum regularity

$$s = [m/2 + 1].$$

Restricting consideration to the case when $d = 3$, he found a cage for each value of *m* up to 8, except 7, and proved that there are no other cages (of degree 3).

The 3-cage is the complete graph with four nodes, that is, the graph of vertices and edges of a tetrahedron. The 4-cage is the Thomsen graph, which has six nodes 1, 2, 3, 4, 5, 6, and nine branches *ij*, where *i* is even and *j* odd [22, p. 403; 5, p. 35]. The 5-cage is the Petersen graph, whose ten nodes may be denoted by the unordered pairs of digits 1, 2, 3, 4, 5, with the rule that two nodes are joined whenever their symbols have no common digit [28, p. 194]. The 6-cage and 8-cage are more complicated graphs which we shall describe in §§4 and 9.

The Petersen graph may conveniently be drawn as a large pentagon surrounding a small pentagram, with corresponding vertices joined [19, p. 222]. Accordingly, let us denote it by

$$\{5\} + \left\{ \frac{5}{2} \right\}$$

and try to generalize this procedure. The symbol $\{n\}+\{n/d\}$ will denote an $n$-gon $A_0A_1 \cdots A_{n-1}$ and a star $n$-gon $B_0B_dB_{2d}B_{3d} \cdots$, with $A_iB_i$ joined. The star polygon $\{n/d\}$ exists whenever $n/d$ is a fraction in its lowest terms with $n > 2d$ [12, p. 93]. Thus the number of different kinds of $n$-gon is $\phi(n)/2$, and the values of $n$ for which there is just one star $n$-gon are given by

$$\phi(n) = 4,$$

namely $n = 5$, 8, 10, 12. On examining Figs. 11, 17, and 21, we find that $\{8\}+\{8/3\}$ and $\{12\}+\{12/5\}$ are 2-regular, while $\{10\}+\{10/3\}$ is 3-regular. The last is the 3-regular graph described by Tutte [47, p. 460]. But all three of these graphs were described earlier by Foster [18, p. 315].

The regularity of such a graph is easily computed by writing the mark 0 at one node, 1 at each adjacent node, 2 at any further nodes adjacent to these, and so on. Then all the $s$-arcs emanating from the initial node are marked $012 \cdots s$.

The complete graph with $n+1$ nodes is evidently 2-regular. This graph, whose group is symmetric, may be described as the graph of vertices and edges of an $n$-dimensional simplex. More generally, the graph of vertices and edges of an $n$-dimensional regular polytope $(n > 2)$ is 2-regular whenever the vertex figure is a simplex [12, p. 128] but only 1-regular otherwise. However, the regularity is occasionally increased by identifying opposite vertices so as to obtain a regular "honeycomb" covering elliptic $(n-1)$-space. The case of the four-dimensional hypercube $\gamma_4$ has already been mentioned. The other instance is the dodecahedron, which reduces to the Petersen graph [30, p. 69; 12, p. 51, Fig. 3.6E with $ij = ji$].

We saw in §1 how every configuration can be represented in a unique manner by a "Levi graph." Clearly, the configuration and graph are isomorphic: the group of the configuration (including reciprocities as well as symmetries) is the same as the group of the graph. In particular, *if a configuration $n_3$ has an $s$-regular Levi graph $(s > 1)$, the order of its group is*

$$2^sN_1 = 2^s3n.$$

3. **Regular maps.** It was proved by Petersen [38, p. 420] that every graph can be embedded into a surface so as to cover the surface with a map of $N_2$ non-overlapping simply-connected regions (polygons) whose boundaries consist of the $N_1$ branches and $N_0$ nodes of the graph. Since a polyhedron is such a map, it is natural to call the regions *faces*, the branches *edges*, and the nodes *vertices*. The em-

bedding is by no means unique, but there must be at least one for which the characteristic

$$- N_0 + N_1 - N_2$$

in minimum. When the minimum characteristic is $-2$, the graph is said to be *planar*, since it can be embedded into the ordinary plane or sphere. The simplest graphs of characteristic $-1$ are the Thomsen graph and the complete 5-point (or four-dimensional simplex). The corresponding maps are derived from the hexagonal prism and pentagonal antiprism by identifying antipodal elements. A famous theorem of Kuratowski, elegantly proved by Whitney [50], states that a necessary and sufficient condition for a graph to be planar is that it have no part homeomorphic to either of these two special graphs. (In the case of a graph of degree 3, only the Thomsen graph needs to be mentioned.)

Those permutations of the elements of a map which preserve the incidences form a group $\mathfrak{g}$, "the group of the map." It is interesting to compare this with $\mathfrak{G}$, the group of the graph formed by the vertices and edges alone. Clearly, $\mathfrak{g}$ is a subgroup of $\mathfrak{G}$, and any operation of $\mathfrak{G}$ not belonging to $\mathfrak{g}$ yields another map of the same kind in which the same vertices and edges are differently distributed into faces. Hence the graph can be embedded into the same kind of surface in a number of ways equal to the index of $\mathfrak{g}$ in $\mathfrak{G}$. In particular, if $\mathfrak{g}$ is the whole of $\mathfrak{G}$, the embedding is unique.

A map is said to be *regular* if its group includes the cyclic permutation of the edges belonging to any one face and also the cyclic permutation of the edges that meet at any one vertex of this face [7]. Thus the group must be transitive on the vertices, and on the edges, and on the faces. Such a map is "of type $\{p, q\}$" if $p$ edges belong to a face, and $q$ to a vertex. The *dual* map, whose edges cross those of the original map, is then of type $\{q, p\}$.

The above cyclic permutations or "rotations," R and S, satisfy the relations

$$R^p = S^q = (RS)^2 = 1,$$

and generate a group of order $4N_1/\epsilon$, where $\epsilon = 2$ or 1 according as the surface is orientable or unorientable ("two-sided" or "one-sided") [8, p. 408]. When $\epsilon = 1$ this is the whole group of the map, but when $\epsilon = 2$ it may be either the whole group or a subgroup of index 2. The latter case arises when the map is symmetrical by reflection as well as by rotation, that is, when there is an operation that interchanges two adjacent faces without altering their common vertices.

The *Petrie polygon* of a regular map is an *h*-circuit of edges, such that every two consecutive edges, but no three, belong to a face [12, pp. 24, 61, 90, 108]. A particular specimen of this polygon is determined by any two adjacent sides of any face; therefore the various specimens are all alike, and we are justified in speaking of *the* Petrie polygon of the map. The number *h* is easily seen to be $\epsilon$ times the period of the "translation" $R^2S^2$ [13, Fig. xi]. Since every edge belongs to two such circuits, the Petrie polygons themselves form a regular map of type $\{h, q\}$ having the same $N_1$ edges, the same $N_0$ vertices, and the same group, though covering a different surface. The relation between the two maps is evidently symmetric: the faces of either are the Petrie polygons of the other. (See Figs. 1 and 2.)

Every map of type $\{p, q\}$ can be derived by making suitable identifications in a simple-connected "universal covering map" [46, p. 8] which we call simply "the map $\{p, q\}$." This is, in general, infinite, and we naturally think of it as covering the Euclidean or hyperbolic plane [43, p. 162]. In fact, since the Euler-Poincaré characteristic

$$K = -N_0 + N_1 - N_2 = \left(1 - \frac{2}{p} - \frac{2}{q}\right)N_1$$

$$= [(p-2)(q-2) - 4]\frac{N_1}{pq}$$

cannot be negative unless $(p-2)(q-2) < 4$, the only finite universal regular maps $(K = -2)$ are

$$\{p, 2\}, \quad \{2, p\}, \quad \{3, 3\}, \quad \{4, 3\}, \quad \{3, 4\}, \quad \{5, 3\}, \quad \{3, 5\}.$$

The first of these is the partition of the sphere into two hemispheres by a *p*-gon along the equator; and the corresponding graph is merely the *p*-gon, $\{p\}$. The second, being the partition of the sphere into *p* lunes by *p* meridians, is irrelevant to our discussion, as we are not interested in graphs having two nodes joined by more than one branch. (However, Tutte includes $\{2, 3\}$ in his treatment [47], calling it the 2-cage.)

The remaining spherical $\{p, q\}$'s are the five Platonic solids, for which

$$N_1 = 2pq/[4 - (p-2)(q-2)], \qquad h = (4N_1 + 1)^{1/2} - 1$$

[12, pp. 5, 13, 19]. Of these, all save the tetrahedron $\{3, 3\}$ have pairs of antipodal elements [12, p. 91]. Identifying such pairs, we obtain the four regular maps in the elliptic plane $(K = -1)$:

$$\{4, 3\}/2, \qquad \{3, 4\}/2, \qquad \{5, 3\}/2, \qquad \{3, 5\}/2.$$

When regarded as graphs, the first is the complete 4-point, like $\{3, 3\}$; the second is merely a triangle with repeated sides; the third is the Petersen graph, as we saw in §2; and the fourth is the complete 6-point (or five-dimensional simplex).

Regular maps on the torus $(K = 0)$ are more interesting [13]. There is a map $\{4, 4\}_{b,c}$ for any non-negative integers $b$ and $c$ (with $b \geqq c$ to avoid repetition). This is derived from the ordinary plane tessellation $\{4, 4\}$, whose vertices have integral Cartesian coordinates, by identifying all points of the lattice generated by the vectors $(b, c)$ and $(-c, b)$. Similarly, $\{3, 6\}_{b,c}$ is derived from the infinite triangular tessellation $\{3, 6\}$, whose vertices have integral coordinates referred to oblique axes inclined at $\pi/3$, by identifying all points of the lattice generated by the vectors $(b, c)$ and $(-c, b+c)$. Thus $\{4, 4\}_{b,c}$ has $b^2+c^2$ vertices, while $\{3, 6\}_{b,c}$ has $b^2+bc+c^2$. Finally $\{6, 3\}_{b,c}$ is the dual of $\{3, 6\}_{b,c}$, and has $2(b^2+bc+c^2)$ vertices. These maps are symmetrical by reflection whenever $b = c$ or $bc = 0$ (as in Figs. 2, 6, 14, 20), but only by rotation in other cases (for example, Fig. 8).

It often happens that the map of minimum characteristic for a given graph is not regular, although a regular map is obtained by allowing a higher characteristic. For instance, the Thomsen graph and the complete 5-point, which form irregular maps on the elliptic plane, form regular maps on the torus, namely

$$\{6, 3\}_{1,1} \quad \text{and} \quad \{4, 4\}_{2,1}$$

[13, Figs. iv and ii].

Regarded as a graph, $\{4, 4\}_{b,c}$ is 1-regular if $b+c > 4$, since there are two obviously different types of 2-arc, one of which belongs to a 4-circuit while the other does not. For the same reason, $\{4, 4\}_{3,0}$ is 1-regular. But the straight 2-arc does belong to a 4-circuit if $b+c = 4$; so it is not surprising to find that $\{4, 4\}_{4,0}$ and $\{4, 4\}_{3,1}$ are 2-regular, while $\{4, 4\}_{2,2}$ is 3-regular (Fig. 6). Also $\{4, 4\}_{2,1}$, being the complete 5-point, is 2-regular. The girth of $\{4, 4\}_{b,c}$ is 4 if $b+c > 3$, but only 3 if $b+c = 3$.

All the graphs $\{3, 6\}_{b,c}$ are 1-regular, of girth 3 and degree 6. But *there are infinitely many 2-regular graphs of degree 3*. In fact, $\{6, 3\}_{b,c}$ is 2-regular whenever $b+c > 3$; but $\{6, 3\}_{2,1}$ is 4-regular (Fig. 8), while $\{6, 3\}_{1,1}$ and $\{6, 3\}_{3,0}$ are 3-regular (Fig. 14). The girth of $\{6, 3\}_{b,c}$ is 6 if $b+c > 2$, but only 4 if $b+c = 2$ (as in Fig. 2 or the Thomsen graph).

When $b$ and $c$ are coprime, we obtain a Hamiltonian circuit for $\{4, 4\}_{b,c}$ by taking a "straight path" along the edges [13, p. 27]. When this circuit is drawn as a regular polygon $\{b^2+c^2\}$, the remaining branches of the graph form a star polygon

$$\left\{\frac{b^2 + c^2}{bb' + cc'}\right\}$$

having the same $b^2+c^2$ vertices. Here $b'$ and $c'$ are given by the congruence

$$bc' - cb' \equiv \pm 1 \ (\mathrm{mod} \ b^2 + c^2)$$

with whichever sign makes $bb'+cc' < (b^2+c^2)/2$. In particular, $\{4, 4\}_{b,1}$ can be drawn as a $\{b^2+1\}$ with its inscribed $\{(b^2+1)/b\}$ (given by $b'=1$, $c'=0$).

Foster has drawn a pentagram $\{5\}$ with its inscribed pentagram $\{5/2\}$, and a decagon $\{10\}$ with its inscribed decagram $\{10/3\}$, as instances of "symmetrical geometrical circuits" [18, p. 316, Fig. 10]. These 2-regular graphs are now seen to be $\{4, 4\}_{2,1}$ and $\{4, 4\}_{3,1}$.

Similarly, the graph $\{3, 6\}_{b,c}$ can be drawn, when $b$ and $c$ are coprime, as a $\{b^2+bc+c^2\}$ with two inscribed star polygons of different densities.

The Petrie polygon of $\{4, 4\}_{b,c}$ is an $h$-circuit, where $h$ is twice the period of $R^2S^2$ in the group

$$R^4 = S^4 = (RS)^2 = (RS^{-1})^b(R^{-1}S)^c = 1$$

[13, p. 25]. Since $(RS^{-1})^b(R^{-1}S)^b = (RS^{-1}R^{-1}S)^b = (R^2S^2)^b$, we have $h = 2b$ if $b = c$; but whenever $b \neq c$,

$$h = 2(b^2 + c^2)/(b, c).$$

(The denominator is the greatest common divisor.) Similarly, for $\{3, 6\}_{b,c}$ or $\{6, 3\}_{b,c}$, $h$ is twice the period of $R^2S^2$ in

$$R^3 = S^6 = (RS)^2 = (RS^{-2})^b(R^{-1}S^2)^c = 1,$$

namely

$$h = 2(b^2 + bc + c^2)/(b, c)$$

for all values of $b$ and $c$ [8, p. 418]. In particular, when $(b, c)=1$, $\{4, 4\}_{b,c}$ and $\{3, 6\}_{b,c}$ have the peculiarity that the number of sides of the Petrie polygon is twice the number of vertices of the whole map, which means that the Petrie polygon is *singular* (as Petrie himself noticed in 1931); for example, the Petrie polygon of $\{4, 4\}_{2,1}$ is a singular decagon such as 0213243041 [13, Fig. i].

The Petrie polygon of $\{6, 3\}_{b,c}$ has the same number of sides as that of the dual $\{3, 6\}_{b,c}$, but is never singular. In fact, since $\{6, 3\}_{b,c}$ has $2(b^2+bc+c^2)$ vertices, its Petrie polygon is Hamiltonian when $(b, c) = 1$. In this case the graph can be drawn as a regular $2(b^2+bc+c^2)$-gon with certain pairs of vertices joined; for example, the Thomsen graph $\{6, 3\}_{1,1}$ can be drawn as a hexagon with opposite vertices joined, and $\{6, 3\}_{2,1}$ can be drawn as in Fig. 9.

On the other hand, $\{4, 4\}_{,b,c}$ and $\{3, 6\}_{b,c}$ have Hamiltonian Petrie polygons when $(b, c) = 2$, except in the single case of $\{4, 4\}_{2,2}$ (Fig. 6). The "red" vertices of Fig. 20 (that is, those marked with capital letters) form a $\{3, 6\}_{2,2}$ whose Petrie polygons are dodecagons such as

$$A_1B_2C_3A_4B_1C_2A_3B_4C_1A_2B_3C_4.$$

Thus the graph of vertices and edges of $\{3, 6\}_{2,2}$ can be drawn as a dodecagon $\{12\}$, a dodecagram $\{12/5\}$, and a pair of hexagons, all having the same twelve vertices. (This is the Menger graph for Fig. 19.)

The three maps $\{4, 4\}_{2,2}$, $\{6, 3\}_{3,0}$, and $\{6, 3\}_{1,1}$ have $h = p$; so it is not surprising to find that in each case the derived map formed by the Petrie polygons is isomorphic to the original map. For instance, the Thomsen graph forms a map $\{6, 3\}_{1,1}$ having faces

$$123456, \qquad 143652, \qquad 163254$$

and Petrie polygons

$$123654, \qquad 143256, \qquad 163452,$$

or vice versa.

Two regular maps of positive characteristic will be considered in §§5 and 7.

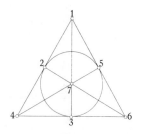

FIG. 7. Fano's $7_3$.

**4. Fano's $7_3$.** The regular map $\{6, 3\}_{2,1}$ first arose in the work of Heawood on generalizations of the four-color problem [21, p. 333]. Since it consists of seven hexagons, each contiguous to all the others, it demonstrates the fact that as many as seven colors may be needed for coloring a map on a torus. But Heawood proved also that seven colors will suffice for every map on a torus; so the "seven-color problem" was completely solved.

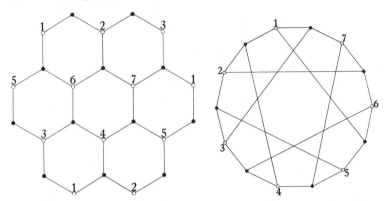

FIG. 8. The regular map $\{6, 3\}_{2,1}$.            FIG. 9. Another view of the 6-cage.

Let alternate vertices of $\{6, 3\}_{2,1}$ be numbered 1, 2, 3, 4, 5, 6, 7, as in Fig. 8. Each of the remaining seven vertices is then determined by the three to which it is joined; so we have a system of seven triples

$$124, \qquad 235, \qquad 346, \qquad 457, \qquad 561, \qquad 672, \qquad 713,$$

derived from one another by cyclic permutation of the digits. If the two types of vertices are the red and blue nodes of a Levi graph, representing the points and lines of a configuration $7_3$, the seven triples show which sets of points are collinear. Every three collinear points are the diagonal points of the complete quadrangle formed by the remaining four points. This state of affairs is indicated in Fig. 7, where the circle is to be regarded as a seventh line. Its impossibility in the usual systems of geometry is often taken as an axiom [**17**, p. 115; **49**, p. 45].

But such a configuration occurs in many finite geometries; for example, in $PG(2, 2)$ where it is the whole plane [**17**, p. 114; **49**, p. 202]. This means that each point has three coordinates (not all zero) belonging to the field of residue-classes modulo 2, namely 0 and 1 with the rule

$$1 + 1 = 0,$$

and three points are collinear whenever their coordinates in each position have sum zero. Such a coordinate symbol may be regarded as a mark in the Galois field $GF[2^3]$, and the simple rule is that three points are collinear whenever their marks have sum zero. Let the modulus of the Galois field be the irreducible cubic

$$\lambda^3 + \lambda + 1$$

or 1011. Then the successive powers of the primitive root $\lambda$ or 10 are

$$\lambda = 010, \ \lambda^2 = 100, \ \lambda^3 = 011, \ \lambda^4 = 110, \ \lambda^5 = 111, \ \lambda^6 = 101, \ \lambda^7 = 001.$$

These are the proper coordinate symbols for the points previously named 1, 2, 3, 4, 5, 6, 7; and the rule for collinearity is a consequence of the identity $\lambda^r(1+\lambda+\lambda^3) = 0$.

The Levi graph for $PG(2, 2)$ may be drawn as in Fig. 9 by placing the numbers 1, 2, $\cdots$, 7 at alternate vertices of a regular 14-gon, joining the vertex between 1 and 2 to 4, that between 2 and 3 to 5, and so on. The peripheral 14-gon appears in $\{6, 3\}_{2,1}$ as a Petrie polygon, which is Hamiltonian since 2 and 1 are coprime.

The blue node 124 belongs to six 2-arcs (as defined in §2): two through each of the red nodes 1, 2, 4. The only nodes not used up in this manner are the four red nodes 3, 5, 6, 7. This remark suggests a third construction for the graph. Regard these four red nodes as forming a tetrahedron. Take a blue node on each of the six edges; for example, a blue node 137 on the edge 37. Join the two blue nodes on each pair of opposite edges, and take a red node on each join; for example, a red node 1 on the join of 137 and 156. Finally, join these three red nodes 1, 2, 4 to a seventh blue node 124.

The relation between the finite projective geometry and its Levi graph shows at once that the graph is a 6-cage (that is, 4-regular, of girth 6). For, each 6-circuit of the graph represents a triangle of the geometry, and each 4-arc represents a sequence

    point-line-point-line-point   or   line-point-line-point-line,

consecutively incident. Since every such sequence is part of a triangle, all the 4-arcs are alike, and the graph is 4-regular. The group of the graph, being the group of collineations and correlations of $PG(2, 2)$, is of order 336 with a simple subgroup of index 2. The order may also be computed by setting $s = 4$ and $n = 7$ in the expression $2^s 3n$ at the end of §2.

Most of these ideas can be generalized at once to $PG(2, p^n)$ for any prime-power $p^n$. Singer [42, p. 378] observed that the

$$q = p^{2n} + p^n + 1$$

points in this finite plane may be so ordered that the $i$th point is represented by the mark $\lambda^i$, where $\lambda$ is a primitive root of $GF[p^{3n}]$ while $\lambda^q$ is a primitive root of the subfield $GF[p^n]$ to which the coordinates of the points belong. The condition for three points $\lambda^a$, $\lambda^b$, $\lambda^c$ to be collinear is

$$\alpha\lambda^a + \beta\lambda^b + \gamma\lambda^c = 0,$$

where $\alpha$, $\beta$, $\gamma$ belong to the $GF[p^n]$; therefore multiplication by $\lambda$ represents a collineation.

For the same reason as before, the Levi graph for this configuration $q_{p^n+1}$ is 4-regular, of grith 6; so we may call it a 6-cage of degree $p^n+1$. Singer's cyclic rule provides a Hamiltonian circuit, enabling us to draw the graph as a $\{2q\}$ with certain diagonals inserted.

Since there are $N_0 = 2q$ nodes, the number of 4-arcs is

$$(d - 1)^3 dN_0 = 2p^{3n}(p^n + 1)q,$$

whereas the order of the group of collineations and correlations of $PG(2, p^n)$ is known to be

$$2np^{3n}(p^{2n} - 1)(p^{3n} - 1) = 2n(p^n - 1)^2 p^{3n}(p^n + 1)q$$

[48, p. 368]. Since a 4-arc represents the three sides and two vertices (or vice versa) of a triangle, it follows that the collineations leaving a triangle entirely invariant form a subgroup of order $n(p^n - 1)^2$. The factor

$$(p^n - 1)^2 = q - 3p^n$$

is simply the number of points not lying on any side of a given triangle, that is, the number of ways in which the triangle can be completed into a quadrangle. This leaves just $n$ for the order of the subgroup leaving a quadrangle entirely invariant, namely, the group of automorphisms of the Galois field, which is the cyclic group of order $n$ generated by the transformation

$$x \longrightarrow x^p.$$

Generalizing $7_3$ in another direction, we construct a graph of $n$ red and $n$ blue nodes, such that the red nodes are numbered $0, 1, \cdots, n-1$ while the blue nodes are joined to the triples

$$0 \ 1 \ r, \qquad 1 \ 2 \ r+1, \qquad \cdots \pmod{n}$$

with $2 < r < n/2$. This inequality ensures that no two triples shall contain more than one common member. The case $n = 8$ ($r = 3$) will be considered in §5.

When $n = 9$ or 10, it is immaterial whether we take $r = 3$ or 4; for the transformation $x \rightarrow 5x + 4$ changes the triple 013 (mod 9) into 401, while $x \rightarrow 3x + 1$ changes 013 (mod 10) into 140. These systems form configurations $9_3$ and $10_3$ in the real projective plane: in the notation of Kantor [26; 27] they are the $9_3$ of type (C) and the $10_3$ of type (A). The Levi graphs can be drawn in a manner analogous to Figs. 9 and 12. But these cases, where the graphs are not even 1-regular, seem less interesting than the configurations of Pappus and Desargues (Kantor's $9_3$ of type (A) and $10_3$ of type (B)) which we shall discuss in §§6 and 7.

On the other hand, an infinite family of 2-regular graphs of this cyclic nature can be obtained by considering the maps $\{6, 3\}_{b,c}$, where $b$ and $c$ are coprime. When the vertices of any map $\{6, 3\}_{b,c}$ are colored alternately red and blue, those of either color evidently form a map $\{3, 6\}_{b,c}$. (When the $\{6, 3\}_{b,c}$ arises as the Levi graph for a configuration, the $\{3, 6\}_{b,c}$ is the corresponding Menger graph.) The two colors occur alternately along any Petrie polygon of the $\{6, 3\}_{b,c}$, and the red vertices of the Petrie polygon form a "straight path" in the red $\{3, 6\}_{b,c}$. When $b$ and $c$ are coprime, the Petrie polygon of $\{6, 3\}_{b,c}$ is Hamiltonian; so also is the straight path in $\{3, 6\}_{b,c}$, as we remarked in §3. Assigning consecutive numbers to the vertices of this straight path, we obtain a triple system of the above kind with

$$n = b^2 + bc + c^2$$

[13, p. 27]. In terms of $b'$ and $c'$, given by the congruence

$$bc' - cb' \equiv 1 \pmod{n},$$

we find $r = bb' + bc' + cc'$ or $n + 1 - (bb' + bc' + cc')$, whichever is smaller. Thus, in the case of $\{6, 3\}_{b,1}$ we have simply

$$n = b^2 + b + 1, \quad b' = b - 1, \quad c' = 1, \quad r = n + 1 - (b^2 + 1) = b + 1;$$

so the triples are

$$0 \ 1 \ b+1, \quad 1 \ 2 \ b+2, \quad \cdots \pmod{b^2 + b + 1}.$$

### 5. The Möbius-Kantor $8_3$. The triple system

$$0 \ 1 \ 3, \quad 1 \ 2 \ 4, \quad 2 \ 3 \ 5, \quad \cdots \pmod{2p}$$

provides a combinatorial solution for Möbius's problem of finding a pair of simple $p$-gons, each inscribed in the other [36, p. 446]. In fact, the vertices of the $p$-gon $0 \ 2 \ 4 \ \cdots$ would lie on the respective sides 13, 35, $\cdots$ of the $p$-gon $1 \ 3 \ 5 \ \cdots$, and the vertices of the latter would lie on the sides 24, 46, $\cdots$ of the former. For a realization of this combinatorial scheme in the projective plane, Kantor considered an arbitrary $p$-gon 0 2 4 $\cdots$ and a tentative position for the point 1 on the side 24. This determines further points on the remaining sides, connected by a chain of perspectivities

$$1 \overset{0}{\overline{\wedge}} 3 \overset{2}{\overline{\wedge}} 5 \cdots$$

which leads eventually to a point 1' on the original side 24. The problem is solved if this 1' coincides with 1; so the given $p$-gon 0 2 4 $\cdots$ provides two solutions whenever the projectivity $1 \overline{\wedge} 1'$ is hyperbolic [26, pp. 916–917; 27, pp. 1291–1291]. Thus a solution is always possible in the complex plane, though not necessarily in the real plane.

In the case of quadrangles ($p=4$), writing 8 instead of 0, we have the triple system

1 2 4,    2 3 5,    3 4 6,    4 5 7,    5 6 8,    6 7 1,    7 8 2,    8 1 3

and the perspectivities

$$1 \overset{8}{\overline{\wedge}} 3 \overset{2}{\overline{\wedge}} 5 \overset{4}{\overline{\wedge}} 7 \overset{6}{\overline{\wedge}} 1'.$$

Möbius proved that in this simplest case the projectivity $1 \overline{\wedge} 1'$ is necessarily elliptic; hence the configuration $8_3$ cannot be constructed in the real plane. However, in the complex plane, in terms of the cube roots of unity

$$\omega = (-1 + 3^{1/2}i)/2 \quad \text{and} \quad \omega^2 = (-1 - 3^{1/2}i)/2,$$

the eight points may be taken to be

1 (1, 0, 0),    2 (0, 0, 1),    3 ($\omega$, −1, 1),    4 (−1, 0, 1),

5 (−1, $\omega^2$, 1),    6 (1, $\omega$, 0),    7 (0, 1, 0),    8 (0, −1, 1),

so that the eight lines are their polars with respect to the conic

$$x^2 + y^2 = z^2,$$

namely

782 [1, 0, 0],    671 [0, 0, 1],    568 [−$\omega$, 1, 1],    457 [1, 0, 1],

346 [1, −$\omega^2$, 1],    235 [1, $\omega$, 0],    124 [0, 1, 0],    813 [0, 1, 1].

The configuration $8_3$ cannot constitute the whole of a finite geometry as $7_3$ does; for the eight points occur in pairs of *opposites* (such as 1 and 5) which do not belong to lines, and the eight lines occur in pairs of opposites (such as 124 and 568) which have no common point. Nevertheless, eight points and eight lines having the desired incidences can be found in some finite geometries such as $PG(2, 3)$ or, more conveniently, $EG(2, 3)$. This is the affine geometry in which the two coordinates of a point belong to the field of residue-classes modulo 3, namely 0, 1, 2 with the rules

$$1 + 2 = 0, \quad 2 + 2 = 1, \quad 2 \times 2 = 1,$$

and three points are collinear whenever their coordinates in both positions have sum zero. Such a coordinate symbol may be regarded

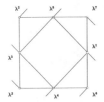

FIG. 10. The Möbius-Kantor $8_3$.

as a mark in the Galois field $GF[3^2]$, and the simple rule is that three points are collinear whenever their marks have sum zero.[2] Let the modulus of the Galois field be the irreducible quadratic

$$\lambda^2 + \lambda + 2$$

or 112. Then the successive powers of the primitive root $\lambda$ or 10 are

$$\lambda = 10, \lambda^2 = 21, \lambda^3 = 22, \lambda^4 = 02, \lambda^5 = 20, \lambda^6 = 12, \lambda^7 = 11, \lambda^8 = 01.$$

These are the proper coordinate symbols for the points previously named 1, 2, 3, 4, 5, 6, 7, 8; and the rule for collinearity is a consequence of the identity $\lambda^r(1+\lambda+\lambda^3) = 0$, which holds since

$$\lambda^3 + \lambda + 1 = (\lambda + 2)(\lambda^2 + \lambda + 2).$$

The only point of the geometry not present in the configuration is the origin 00. (See Fig. 10, where the points are represented in the Euclidean plane as if the coordinate residue 2 were the ordinary number $-1$. This representation naturally obscures the collinearity of such points as $\lambda^4, \lambda^5, \lambda^7$.)

The corresponding Levi graph, of 16 nodes and 24 branches, has a Hamiltonian circuit whose alternate vertices are 1 2 3 4 5 6 7 8, enabling us to draw it as in Fig. 12 and to see at a glance that it is 2-regular.

---

[2] The generalization to $EG(2, p^n)$ has been worked out by Bose [6, pp. 3–5].

Each node of either color is joined by 2-arcs to six of the remaining seven nodes of that color; for example, the red node 1 is joined to 2, 4, 6, 7, 8, 3, but not to 5. Thus the nodes (and consequently also the branches) occur in pairs of opposites, such as 1 and 5. The operation of interchanging opposites is indicated in the notation by adding or subtracting 4, or multiplying by $\lambda^4 = 2$, or making a half-turn about the origin of the finite affine geometry. In the complex projective plane, with the above choice of coordinates, it is the harmonic homology with center $(\omega, \omega^2, 1)$ and axis $[\omega, \omega^2, -1]$.

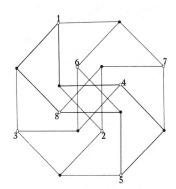

FIG. 11. The 2-regular graph $\{8\} + \{8/3\}$.

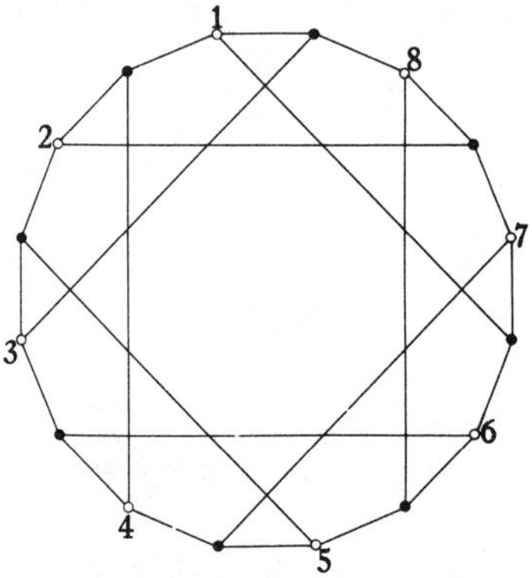

FIG. 12. Another view of $\{8\} + \{8/3\}$.

The same graph has already been mentioned in §2 qua $\{8\}$ $+ \{8/3\}$. In Fig. 11, the central $\{8/3\}$ has been twisted to exhibit the fact that the branches of such a graph can be found among the edges of the four-dimensional hypercube (Fig. 5). This $\{8/3\}$ and the peripheral $\{8\}$ are projections of two Petrie polygons of the hypercube [12, p. 223], and it is easy to pick out four further Petrie polygons, one of which has been drawn in heavy lines. Naming their alternate vertices, we thus find six 8-circuits:

$$\begin{array}{ccc} 1\ 3\ 5\ 7, & 1\ 2\ 5\ 6, & 1\ 4\ 5\ 8, \\ 2\ 4\ 6\ 8, & 3\ 4\ 7\ 8, & 2\ 3\ 6\ 7. \end{array}$$

They correspond to the three ways in which the $8_3$ can be regarded as a pair of simple quadrangles, each inscribed in the other. They form a regular map of six octagons, each sharing two opposite sides with another. The characteristic is

$$- N_0 + N_1 - N_2 = -16 + 24 - 6 = 2.$$

Referring to Threlfall's catalogue of such maps [46, p. 44] we find this listed as "Nr. 2." Since it is symmetrical by reflection, its group is of order $2 \cdot 6 \cdot 8 = 96$. But the group of the graph (which is also the group of symmetries and reciprocities of the configuration) has order $2^5 3n = 96$. So in this case the configuration and the map are isomorphic: no symmetry is lost by embedding the graph into a surface. In other words, the embedding is unique, unlike the embedding of the 6-cage into the torus (which can be done in eight different ways).

Returning to the complex projective plane, we see that the conic

$$x^2 + y^2 = z^2$$

determines a polarity which appears in the map as a half-turn about the mid-point of either of the two opposite edges 4 457, 8 813. (The conic touches line 457 at point 4, and line 813 at point 8.) Since there are altogether twelve pairs of opposite edges, we can immediately infer the existence of twelve such conics, each touching two opposite lines of the configuration at two opposite points, while the remaining six points and six lines form two self-polar triangles.

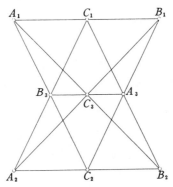

FIG. 13. Pappus's $9_3$.

**6. Pappus's $9_3$.** In the analytic foundations of general projective geometry of two dimensions we define a *point* as a triple of numbers $(x, y, z)$, not all zero, with the understanding that $(x\lambda, y\lambda, z\lambda)$ is the same point for any $\lambda \neq 0$ [24, p. 176]. Similarly we define a *line* $[X, Y, Z]$ which is the same as $[\lambda X, \lambda Y, \lambda Z]$; and we say that the point $(x, y, z)$ and line $[X, Y, Z]$ are *incident* if

$$Xx + Yy + Zz = 0.$$

These "numbers" may be elements of any division ring. It follows from the above definition that the general point collinear with $(x, y, z)$ and $(x', y', z')$ may be expressed as $(x\lambda+x', y\lambda+y', z\lambda+z')$. One of the most striking passages in the development is Hilbert's observation that the commutativity of multiplication is equivalent to Pappus's Theorem, which may be stated as follows:

*If two triangles are doubly perspective they are triply perspective.*

We say that two triangles $A_1A_2A_3$ and $B_1B_3B_2$ are *perspective* if the lines $A_1B_1$, $A_2B_3$, $A_3B_2$ all pass through one point, say $C_1$. We say that they are *doubly* perspective if the same kind of relation holds after the vertices of one triangle have been cyclically permuted, namely if either

$$A_1B_3,\ A_2B_2,\ A_3B_1\ \text{all pass through one point } C_2$$

or

$$A_1B_2,\ A_2B_1,\ A_3B_3\ \text{all pass through one point } C_3$$

[49, p. 100]. Finally, they are *triply* perspective if all three of these relations hold, as in Fig. 13.

Clearly, we can choose a coordinate system so that the points

$$A_1, \qquad A_2, \qquad A_3, \qquad C_1, \qquad B_1, \qquad B_3, \qquad B_2$$

are

$$(1, 0, 0),\ \ (0, 1, 0),\ \ (0, 0, 1),\ \ (1, 1, 1),\ \ (x, 1, 1),\ \ (1, y, 1),\ \ (1, 1, z).$$

Then the lines $A_1B_3$, $A_2B_2$, $A_3B_1$, $A_1B_2$, $A_2B_1$, $A_3B_3$ are

$$[0,\ -1,\ y],\ [z,\ 0,\ -1],\ [-1,\ x,\ 0],\ [0,\ z,\ -1],\ [-1,\ 0,\ x],\ [y,\ -1,\ 0].$$

Thus

$A_3B_1$ passes through the point $C_2 = A_1B_3 \cdot A_2B_2 = (1, yz, z)$   if   $xyz = 1$,

and

$A_3B_3$ passes through the point $C_3 = A_1B_2 \cdot A_2B_1 = (xz, 1, z)$   if   $yxz = 1$.

This shows that Pappus's Theorem holds if and only if

$$xy = yx.$$

The two triply perspective triangles and their centers of perspective form the *Pappus configuration* $9_3$, whose incidences are summarized in the statement that the points $A_i$, $B_j$, $C_k$ are collinear whenever

$$i + j + k \equiv 0 \pmod 3$$

[31, p. 106; 29, p. 17].

In its more orthodox form, Pappus's Theorem asserts the collinearity of the cross joins

$$A_3 = B_1C_2 \cdot B_2C_1, \qquad B_3 = C_1A_2 \cdot C_2A_1, \qquad C_3 = A_1B_2 \cdot A_2B_1$$

of three points $A_1$, $B_1$, $C_1$ on one line with three points $A_2$, $B_2$, $C_2$ on another. In Möbius's configuration $8_3$, which we discussed in §5, we find three points 2, 8, 7 on one line, and three points 3, 4, 6 on another, while two of the intersections of cross joins are 1 and 5. According to Pappus's Theorem, the three lines 15, 26, 37 must be concurrent. Since the two triads may be rearranged as 8, 7, 2 and 6, 3, 4, the three lines 15, 26, 48 are likewise concurrent. Hence:

*In any geometry satisfying Pappus's Theorem, the four pairs of opposite points of $8_3$ are joined by four concurrent lines.*

The result is a special Pappus configuration in which the two "triply perspective triangles" have collapsed to form sets of *collinear* points: 235 and 671. The centers of perspective form a third degenerate triangle: 480, So altogether we have a configuration

$$(9_4, \ 12_3)$$

of nine points lying by threes on twelve lines, four through each point. Collinear points are indicated by the rows, columns, diagonals, and "broken diagonals" of the square matrix

$$\begin{matrix} 2 & 8 & 7 \\ 5 & 0 & 1 \\ 3 & 4 & 6 \end{matrix}$$

[35, p. 335].

Returning to the general case, we find that the Pappus configuration $9_3$ may be regarded in six ways as a cycle of three "Hessenberg" triangles, each inscribed in the next. Three of the ways are

$$\begin{array}{c|c|c} A_3B_1C_1 & A_1B_3C_1 & A_1B_1C_3 \\ A_1B_2C_2 & A_2B_1C_2 & A_2B_2C_1 \\ A_2B_3C_3 & A_3B_2C_3 & A_3B_3C_2 \end{array}$$

and the other three can be derived from these by the consistent interchange of two of the suffix numbers, say 1 and 2. The corresponding Levi graph, of 18 nodes and 27 branches, turns out to be 3-regular. Each of the eighteen Hessenberg triangles appears as a 6-circuit, and the nine of them displayed above form a map of nine hexagons on a torus, as in Fig. 14, which we recognize as $\{6, 3\}_{3,0}$. Thus the graph can be embedded into the torus in two distinct ways, in agreement with the fact that the group of the graph has order $2^3 \cdot 3 \cdot 9 = 216$, while that of the map has order $2 \cdot 6 \cdot 9 = 108$.

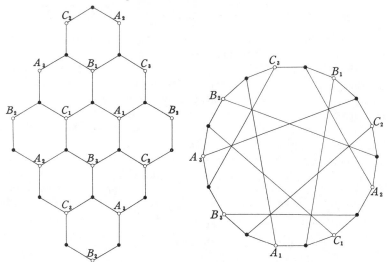

FIG. 14. The regular map $\{6, 3\}_{3,0}$.

FIG. 15. Another view of the Pappus graph.

Kommerell [29, p. 19] observed that each Hessenberg triangle determines an involutory reciprocity for which that triangle is "self-polar" while the other two triangles in the same cycle are interchanged. The reciprocity is not, in general, a polarity; but it will be if these other two triangles are in perspective [14, p. 66]. In one of the two maps $\{6, 3\}_{3,0}$, the Hessenberg triangle appears as a hexagon and the reciprocity appears as a half-turn about the center of that hexagon; in the other map the same triangle appears as a Petrie polygon (zig-zag) and the reciprocity appears as a reflection.

After locating the Hamiltonian circuit

$$A_3 \; B_3 \; A_1 \; C_1 \; A_2 \; C_2 \; B_1 \; C_3 \; B_2,$$

we can draw the graph as in Fig. 15.

7. **Desargues' $10_3$.** As long ago as 1846, Cayley remarked that the ten lines and ten planes determined by five points of general position in projective 3-space meet an arbitrary plane in a *Desargues configuration* (Fig. 16) consisting of ten points

| 12 | 23 | 34 | 45 | 15 | 13 | 24 | 35 | 14 | 25 |
|----|----|----|----|----|----|----|----|----|----|

and ten lines

| 345 | 145 | 125 | 123 | 234 | 245 | 135 | 124 | 235 | 134 |

with the rule that the point $ij$ is incident with the line $ijk$ [9, p. 318].

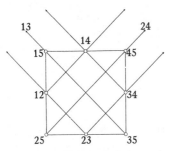

FIG. 16. Desargues' $10_3$.

This configuration can be regarded in ten ways as a pair of triangles in perspective; for example, the triangles

14 24 34   and   15 25 35

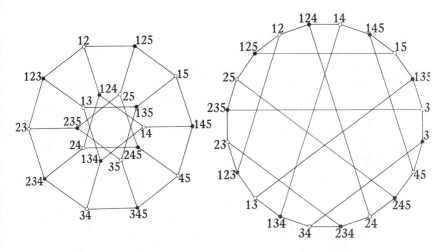

FIG. 17. The 3-regular graph
$\{10\} + \{10/3\}$.

FIG. 18. Another view of the
Desargues graph.

are in perspective from the point 45 and from the line 123, and are interchanged by a homology which appears in this notation as the transposition (4 5). Since the symmetric group of order 120 is generated by four transpositions, the collineation group of the configuration is generated by four such homolgies. K. G. C. von Staudt [44, p. 134] proved that the ten points and ten lines are interchanged by a certain polarity. Adjoining this to the collineation group of order 120, we obtain a group of collineations and correlations of order 240, which is abstractly the direct product of the symmetric group with the group of order 2.

We see in Fig. 17 that the Levi graph, of twenty nodes and thirty lines, is

$$\{10\} + \left\{\frac{10}{3}\right\},$$

in the notation of §2. Since this is 3-regular, its group has order

$$2^3 \; 30 = 240.$$

The $10_3$ can be regarded in six ways as consisting of a pair of simple pentagons, consecutive vertices of either lying on alternate sides of the other; for example

| 12 | 23 | 34 | 45 | 51 |
|----|----|----|----|----|
| 14 | 42 | 25 | 53 | 31 |

[49, p. 51]. This instance may be associated with the cycles (12345), (14253), and we can use in a similar fashion five other pairs of cycles. Corresponding to these twelve pentagons, the graph contains twelve 10-circuits; for example, the cycle (12345) provides the 10-circuit

| 12 | 123 | 23 | 234 | 34 | 345 | 45 | 451 | 51 | 512. |
|----|-----|----|-----|----|-----|----|-----|----|------|

From the cycle (12345) we derive five others by shifting each digit in turn two places on (or back). This systematic rule gives us one cycle from each of the six pairs, namely

$$(12345), \quad (23145), \quad (13425), \quad (12453), \quad (14235), \quad (12534).$$

The six corresponding 10-circuits form a regular map of six decagons, each sharing two opposite sides with another; for example, the 10-circuits arising from (12345) and (23145) share the sides

$$123 \; 23 \quad \text{and} \quad 145 \; 45.$$

This map has characteristic $-20+30-6=4$; but the surface is unorientable. The remaining six 10-circuits form another such map, and the faces of either map are the Petrie polygons of the other.

In either map, the cyclic permutations of three digits, which generate the icosahedral (alternating) group, appear as trigonal rotations about vertices, while von Staudt's polarity appears as a half-turn about the center of any face (or rather, about the centers of all faces simultaneously).

It is interesting to compare this with another regular map of six decagons on the same surface, namely the map formed by the six Petrie polygons of the regular dodecahedron $\{5, 3\}$. Again each decagon shares two opposite sides with another, and again the group is the direct product of the icosahedral group with the group of order 2. We have here a remarkably perfect instance of maps that are not isomorphic but "allomorphic" [43, p. 101]. The difference is seen by considering the girth of the graph of vertices and edges. In the last case, where the vertices and edges are simply those of the dodecahedron itself, the girth is 5; but the girth of $\{10\} + \{10/3\}$ is 6.

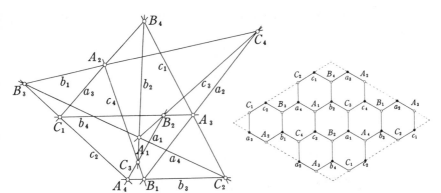

FIG. 19. A regular $12_3$.

FIG. 20. The regular map $\{6, 3\}_{2,2}$.

**8. A regular $12_3$.** After proving the impossibility of a pair of real quadrangles, each inscribed in the other, Möbius remarked: "Auf Vielecke von mehreren Seiten habe ich die Untersuchung nicht ausgedehnt" [36, p. 446]. As we observed in §5, the problem of finding a pair of mutually inscribed $p$-gons was solved for the complex projective plane by Kantor in 1881, using the invariant points of a certain projectivity $1 \bar{\wedge} 1'$ on the side 24 of a given $p$-gon $024 \cdots$. Möbius had shown, in 1828, that this projectivity has no real invariant points when $p = 4$. Since all quadrangles are projectively equivalent, we cannot remedy this state of affairs by modifying the initial quadrangle. But two $p$-gons with $p > 4$ are not, in general, projectively equivalent; so it is natural to expect a suitable choice of the initial $p$-gon to render the above projectivity hyperbolic. When $p = 5$ this expectation is realized by the "non-Desarguesian" $10_3$: the (A) of Kantor [27, pp. 1292, 1314] which is the VII of Martinetti [33a, p. 17]. It can hardly be doubted that greater values of $p$ will serve just as well. However, the corresponding Levi graphs are only 0-regular. The higher regularity of the graph for $8_3$ is a consequence of the fact that this special configuration can be regarded *in three different ways* as a pair of mutually inscribed quadrangles. To achieve this refinement when $p > 4$ we must abandon the cyclic scheme.

Fig. 19 shows a $12_3$ which can be regarded in three ways as a pair of hexagons, each inscribed in the other:

$$A_1 \ B_2 \ C_1 \ A_2 \ B_1 \ C_2 \quad \text{and} \quad A_3 \ B_4 \ C_3 \ A_4 \ B_3 \ C_4,$$
$$A_1 \ B_3 \ C_1 \ A_3 \ B_1 \ C_3 \quad \text{and} \quad A_2 \ B_4 \ C_2 \ A_4 \ B_2 \ C_4,$$
$$A_1 \ B_4 \ C_1 \ A_4 \ B_1 \ C_4 \quad \text{and} \quad A_2 \ B_3 \ C_2 \ A_3 \ B_2 \ C_3.$$

The configuration is self-dual; for, if the twelve lines are denoted by

$$a_1 = A_4B_2C_3, \qquad b_1 = A_2B_3C_4, \qquad c_1 = A_3B_4C_2,$$
$$a_2 = A_3B_1C_4, \qquad b_2 = A_1B_4C_3, \qquad c_2 = A_4B_3C_1,$$
$$a_3 = A_2B_4C_1, \qquad b_3 = A_4B_1C_2, \qquad c_3 = A_1B_2C_4,$$
$$a_4 = A_1B_3C_2, \qquad b_4 = A_3B_2C_1, \qquad c_4 = A_2B_1C_3,$$

then the twelve points are

$$A_1 = a_4 \cdot b_2 \cdot c_3, \qquad B_1 = a_2 \cdot b_3 \cdot c_4, \qquad C_1 = a_3 \cdot b_4 \cdot c_2,$$
$$A_2 = a_3 \cdot b_1 \cdot c_4, \qquad B_2 = a_1 \cdot b_4 \cdot c_3, \qquad C_2 = a_4 \cdot b_3 \cdot c_1,$$
$$A_3 = a_2 \cdot b_4 \cdot c_1, \qquad B_3 = a_4 \cdot b_1 \cdot c_2, \qquad C_3 = a_1 \cdot b_2 \cdot c_4,$$
$$A_4 = a_1 \cdot b_3 \cdot c_2, \qquad B_4 = a_3 \cdot b_2 \cdot c_1, \qquad C_4 = a_2 \cdot b_1 \cdot c_3.$$

The triads of collinear points comprise twelve of the twenty-four possible arrangements $A_iB_jC_k$ with $i, j, k$ all different. The remaining twelve triads form triangles; for instance $A_2B_4C_3$ is a triangle with sides $b_2$, $c_4$, $a_3$. Hence the Levi graph (consisting of twelve red nodes $A_1, \cdots$, twelve blue nodes $a_1, \cdots$, and thirty-six branches $A_1b_2, \cdots$) contains twelve 6-circuits such as

$$A_2a_3B_4b_2C_3c_4,$$

and can be embedded into the torus to make a map of twelve hexagons which we recognize as $\{6, 3\}_{2,2}$ (Fig. 20).

The symmetries and reciprocities of the configuration form a group of order 144. We can obtain this number by considering either the graph or the map. Since the graph is 2-regular, its group has order

$$2^2 36 = 144;$$

and since the map is symmetrical by reflection, its group has order four times the number of edges.

The Petrie polygons of $\{6, 3\}_{2,2}$ are dodecagons such as

$$A_1b_2C_3c_4B_1a_2A_3b_4C_1c_2B_3a_4 \quad \text{and} \quad a_1B_2c_3C_4b_1A_2a_3B_4c_1C_2b_3A_4,$$

which can be put together to form $\{12\}+(12/5)$, as in Fig. 21.

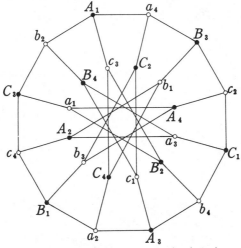

FIG. 21. The 2-regular graph $\{12\}+\{12/5\}$.

When Tutte was told about this graph, he discovered a Hamiltonian circuit, enabling us to draw it as a regular 24-gon with appropriate diagonals inserted (Fig. 22).

The configuration $12_3$ may be regarded as a set of three triangles

$$A_1A_2A_3, \quad B_2B_3B_1, \quad C_3C_1C_2$$

in perspective by pairs from centers $A_4$, $B_4$, $C_4$, while the vertices lie on three lines

$$A_1B_3C_2, \quad A_3B_2C_1, \quad A_2B_1C_3.$$

A special case was described by Hesse, Salmon, and Zacharias [51, p. 149], namely the case when the vertices also lie on the three lines

$$A_1B_1C_1, \quad A_2B_2C_2, \quad A_3B_3C_3,$$

while the centers of perspective lie on a line $A_4B_4C_4$. We now have a configuration $(12_4, 16_3)$ of twelve points lying by threes on sixteen lines, four through each point. Calling the four new lines

$$d_1, \qquad d_2, \qquad d_3, \qquad d_4,$$

we see that the twelve points are expressible in the form

$$a_i \cdot b_j \cdot c_k \cdot d_l,$$

where the suffixes *ijkl* run over the twelve odd permutations of 1234.

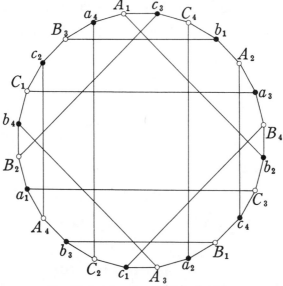

FIG. 22. Another view of $\{12\} + \{12/5\}$.

For comparison with Zacharias's notation, we observe that he numbered the lines from 1 to 16 in the order

$$d_1 \quad b_2 \quad c_3 \quad a_4 \quad c_4 \quad a_3 \quad d_2 \quad b_1 \quad a_2 \quad c_1 \quad b_4 \quad d_3 \quad b_3 \quad d_4 \quad a_1 \quad c_2.$$

Hesse constructed this $(12_4, 16_3)$ by taking the points of contact of the twelve tangents to an elliptic cubic curve that can be drawn from three collinear points of the curve. To make it real we must use a bipartite cubic with the three collinear points all on the odd branch. This is most easily seen by regarding the cubic as the locus of the point

$$(\wp u, \wp' u, 1)$$

where the elliptic function has the real period $2\omega$ and the imaginary period $2\omega'$, so that points on the odd branch are given by real values of the parameter $u$ while points on the oval are given by real values plus the imaginary half-period $\omega'$. Three collinear points on the odd branch may be taken to have real parameters $-2\alpha$, $-2\beta$, $-2\gamma$, where

$$\alpha + \beta + \gamma \equiv 0 \ (\text{mod } 2\omega),$$

and then the parameters of the twelve points of contact are as follows:

| | | | | | | | |
|---|---|---|---|---|---|---|---|
| $A_1$ | $\alpha$ | $A_2$ | $\alpha + \omega$ | $A_3$ | $\alpha + \omega'$ | $A_4$ | $\alpha + \omega + \omega'$ |
| $B_1$ | $\beta$ | $B_2$ | $\beta + \omega'$ | $B_3$ | $\beta + \omega + \omega'$ | $B_4$ | $\beta + \omega$ |
| $C_1$ | $\gamma$ | $C_2$ | $\gamma + \omega + \omega'$ | $C_3$ | $\gamma + \omega$ | $C_4$ | $\gamma + \omega'$ |

The sets of collinear points are now evident. (In one way of drawing the figure, we find the points $A_1B_1C_1A_2B_4C_3$ in that order on the odd branch, and $A_4B_3C_2A_3B_2C_4$ on the oval.) Feld [**17a**, p. 553] considered the case when $\alpha$, $\beta$, $\gamma$ are 0 and $\pm 2\omega/3$, so that $A_1$, $B_1$, $C_1$ are the three real inflexions while the rest are the nine real sextactic points.

Zacharias discovered that a Euclidean specialization of this $(12_4, 16_3)$ is provided by Morley's celebrated figure of a triangle $A_1A_2A_3$ with its angle trisectors and their intersections. Morley's theorem states that the trisectors

$$A_1B_1C_1, \qquad A_2B_1C_3, \qquad A_2B_2C_2, \qquad A_3B_2C_1, \qquad A_3B_3C_3, \qquad A_1B_3C_2$$

form an equilateral triangle $B_2B_3B_1$ by their first intersections and another triangle $C_3C_1C_2$ (usually *not* equilateral) by their second intersections. Zacharias observed that these two triangles are in perspective with $A_1A_2A_3$ and with each other, from collinear centers $C_4$, $B_4$, $A_4$.

9. **The Cremona-Richmond** $15_3$. Six points of general position in real projective 4-space—say 1, 2, 3, 4, 5, 6—determine fifteen further points such as $P_{12}$: the intersection of the line 12 with the hyperplane 3456. These lie by threes on fifteen lines such as $P_{12}P_{34}P_{56}$: the common line of the three hyperplanes

$$3456, \qquad 1256, \qquad 1234$$

[**39**, p. 131]. This configuration $15_3$ in four dimensions was chosen by Baker as a frontispiece for both volumes II and IV of his *Principles of geometry*. A $15_3$ in two or three dimensions can be derived by projection. Dualizing the three-dimensional version, we obtain a set of

15 lines and 15 planes which can be identified with the configuration formed by the tritangent planes of a nodal cubic surface [15].

The $15_3$ in two dimensions is self-dual. For, instead of associating the points with the fifteen pairs of six symbols 1, 2, 3, 4, 5, 6 and the lines with the fifteen *synthemes* such as (12, 34, 56) we can just as well associate the lines with the fifteen pairs of six other symbols

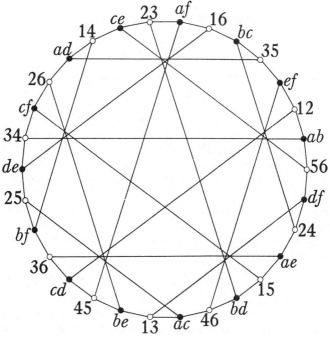

FIG. 23. Tutte's 8-cage (the most regular of all graphs).

$a, b, c, d, e, f$ and the points with fifteen synthemes such as $(ab, cd, ef)$. Richmond's way of doing this is indicated on the Levi graph in Fig. 23. Thus the group of the configuration is the group of automorphisms of the symmetric group of degree six [8, p. 210] and its order is

$$2 \cdot 6! = 1440.$$

Martinetti cited this $15_3$ as the simplest instance of a configuration containing *no triangles* [33, p. 174; cf. 41, p. 44]. The corresponding property of the Levi graph is that its girth is not 6 but 8. Since the order of the group is

$$1440 = 2^* \; 3 \cdot 15,$$

the regularity is $s = 5$. Hence this graph is Tutte's 8-cage, the most regular of all graphs [**47**, p. 460]. Tutte's own description of it can be reconciled with ours as follows. Regard the eight red nodes

| 15 | 36 | 25 | 46 |
|----|----|----|----|
| 26 | 45 | 16 | 35 |

as forming a cube (or two interpenetrating tetrahedra: 15 25 35 45 and 16 26 36 46). Take a blue node on each of the twelve edges; for instance, a blue node *cf* on the edge 15 26, and a blue node *de* on the opposite edge 25 16. Join the blue nodes on each pair of opposite edges, and take a red node on each join; for instance, a red node 34 on the join of *cf* and *de*, and a red node 12 on the join of *cd* and *ef*. Join the two red nodes thus associated with each set of four parallel edges, and take a blue node on each join; for instance, a blue node *ab* on the join of 12 and 34. Finally, join the three blue nodes thus associated with the three dimensions, namely *ab*, *ce*, *df*, to a fifteenth red node 56.

There is apparently no regular embedding for this graph. We must be content with a map of one decagon and ten octagons. Naming alternate vertices, we find a decagon

$$13 \ 24 \ 35 \ 14 \ 25$$

surrounded by octagons 16 24 36 45, 12 35 16 34, and others derived from these two by cyclic permutation of the digits 1, 2, 3, 4, 5 (leaving 6 unchanged). The surface is easily seen to be unorientable, of characteristic

$$- 30 + 45 - (1 + 10) = 4.$$

The decagon represents, in the configuration, a pentagon that is self-polar for the reciprocity $(1 \ d)(2 \ b)(3 \ a)(4 \ f)(5 \ c)(6 \ e)$. Thus Richmond's notation could be improved by writing

$$a, b, c, d, e, f \quad \text{in place of his} \quad d, b, a, f, c, e.$$

On the other hand, the configuration contains another pentagon that is self-polar for the "natural" reciprocity

$$(1 \ a)(2 \ b)(3 \ c)(4 \ d)(5 \ e)(6 \ f)$$

without any change of notation.

The 8-cage is not only the Levi graph for the $15_3$, but also represents, in a different way, the $60_{15}$ known as *Klein's configuration* [**45**, p. 447; **25**, p. 42]. Klein's sixty points and sixty planes are the vertices and faces of fifteen tetrahedra whose pairs of opposite edges are the directrices of the fifteen linear congruences determined by pairs of six mutually apolar linear complexes (or six mutually commutative null polarities) in complex projective 3-space. The fifteen red nodes of the 8-cage represent the pairs of directrices (or the fifteen congruences) while the fifteen blue nodes represent the tetrahedra.

**10. Möbius's $8_4$.** Before resuming the discussion of Möbius's mutually inscribed tetrahedra (§1), let us mention a few other self-dual configurations of points and planes in real projective 3-space. The simplest kind is the $n_n$ consisting of $n$ points on a line and $n$ planes through the same line. Since each point is incident with every plane, the Levi graph consists of $n$ red nodes all joined to each of $n$ blue nodes. Since this is 3-regular, of girth 4, we call it a 4-cage of degree $n$. When $n = 2$ it is the square $\{4\}$. When $n = 3$ it is the Thomsen graph $\{6, 3\}_{1,1}$. When $n = 4$ it is $\gamma_4/2$ (Fig. 4) or $\{4, 4\}_{2,2}$ (Fig. 6).

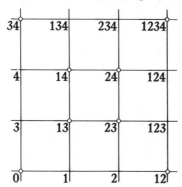

FIG. 24. Gallucci's version of Möbius's $8_4$.

Let 1, 2, 3 denote three planes forming a trihedron with vertex 0, and let 23, 31, 12 be arbitrary points on the three edges, forming a fourth plane 123. We now have a tetrahedron $4_3$, whose Levi graph is the cube $\gamma_3$ (Fig. 1). Continuing in the same manner, let 4 denote a fourth plane through 0, meeting planes 1, 2, 3 in three lines on which we take further arbitrary points 14, 24, 34. In this way we obtain three new planes 234, 341, 412, which intersect in an eighth point 1234. By a theorem of Möbius [36, p. 443], this point 1234 lies on the plane 123, so that we have altogether a configuration $8_4$, whose Levi graph is the four-dimensional hypercube $\gamma_4$ (Fig. 5).

Möbius deduced his theorem from the observation that the quadrangular relation $Q(A'B'C', F'G'H')$ for six collinear points implies $Q(F'G'H', A'B'C')$. This is a corollary of Pappus's Theorem, which

Baker showed to be equivalent to the theorem that if three skew lines all intersect three other skew lines, any transversal to the first set of three intersects any transversal to the second set [1, pp. 47–51]. The following argument relates Möbius's theorem to the transversal theorem directly [cf. **20**, p. 76; **4**, p. 144].

The line joining points 13 and 14 is the line of intersection of planes 1 and 134; likewise the join of 23 and 24 is the intersection of 2 and 234 (see Fig. 24). Hence the transversal from 0 to these two lines is the intersection of planes 1 and 2, and the transversal from 34 is the intersection of 134 and 234. Similarly, the transversal from 0 to the two lines 13 23 and 14 24 is $3 \cdot 4$, and the transversal from 12 is $123 \cdot 124$. The "transversal theorem" asserts that the two lines $134 \cdot 234$ and $123 \cdot 124$ intersect, while Möbius's theorem asserts that the four planes 234, 134, 124, 123 are concurrent; these two statements are clearly equivalent.

The following analytic proof of Möbius's theorem indicates its dependence on the commutative law. Choose the tetrahedron of reference so that the four points and four planes

| 23, | 31, | 12, | 0, |
|-----|-----|-----|-----|
| 1, | 2, | 3, | 123, |

are

| $(1, 0, 0, 0)$, | $(0, 1, 0, 0)$, | $(0, 0, 1, 0)$, | $(0, 0, 0, 1)$, |
|-----|-----|-----|-----|
| $[1, 0, 0, 0]$, | $[0, 1, 0, 0]$, | $[0, 0, 1, 0]$, | $[0, 0, 0, 1]$. |

Take the unit plane $[1, 1, 1, 1]$ to pass through the line $123 \cdot 4$, so that the plane 4 is $[1, 1, 1, 0]$. Then the points 14, 24, 34 are of the form

$$(0, 1, -1, a), \qquad (-1, 0, 1, b), \qquad (1, -1, 0, c),$$

and the planes 234, 314, 124 are

$$[0, c, -b, 1], \qquad [-c, 0, a, 1], \qquad [b, -a, 0, 1].$$

These and 123 all pass through the point $(a, b, c, 0)$, which is the desired point of concurrence 1234. (The incidence of 124 and 1234 requires $ba = ab$.)

Benneton [**3**, p. 30] makes a different choice of unit plane so as to exhibit the four planes and four points

| 234, | 314, | 124, | 4, |
|------|------|------|-----|
| 14, | 24, | 34, | 1234 |

in the form

$$[0,\ c,\ -b,\ a], \qquad [-c,\ 0,\ a,\ b], \qquad [b,\ -a,\ 0,\ c], \qquad [a,\ b,\ c,\ 0],$$
$$(0,\ c,\ -b,\ a), \qquad (-c,\ 0,\ a,\ b), \qquad (b,\ -a,\ 0,\ c), \qquad (a,\ b,\ c,\ 0).$$

Representing both the point $(x,\ y,\ z,\ t)$ and the plane $[x,\ y,\ z,\ t]$ by the quaternion $t+xi+yj+zk$, he neatly expresses all the points and planes of the configuration in the form

$$1,\quad i,\quad j,\quad k,\quad V,\quad iV,\quad jV,\quad kV,$$

where $V$ denotes the pure quaternion $ai+bj+ck$.

In §1 we described the reflections $R_1, R_2, R_3, R_4$ of the hypercube, which add a digit to every symbol lacking it, and remove the same digit from every symbol containing it. We now see that the corresponding reciprocities of Möbius's $8_4$ may be described as follows: $R_1, R_2, R_3$ have the effect of multiplying every quaternion on the left by $i, j, k$, respectively, while $R_4$ has the effect of multiplying on the right by $V$. It follows that these four mutually commutative reciprocities can be extended to the whole space as *null polarities* [25, p. 38].

When the $8_4$ is represented by its Levi graph, $\gamma_4$, the two mutually inscribed tetrahedra appear as two opposite cells $\gamma_3$, transformed into each other by the reflection $R_4$. Since $\gamma_4$ has four pairs of opposite cells, we see clearly that the $8_4$ can be regarded *in four different ways* as a pair of mutually inscribed tetrahedra, transformed into each other by a null polarity.

**11. Cox's $(2^{d-1})_d$.** After constructing Möbius's $8_4$ as above, let 5 be a fifth plane through 0, meeting planes 1, 2, 3, 4 in four lines on which we take further arbitrary points 15, 25, 35, 45. Leaving out the five planes, one by one, we obtain five points

$$2345, \qquad 3451, \qquad 4512, \qquad 5123, \qquad 1234.$$

By a theorem of Homersham Cox [10, p. 67], these five points lie in one plane 12345, so that we have altogether a configuration $16_5$, whose Levi graph is the five-dimensional hypercube $\gamma_5$ [12, p. 244]. By symmetry, we merely have to prove that the first four of the five points are coplanar; and this follows from the dual of Möbius's theorem as applied to the four points 15, 25, 35, 45 in the plane 5. (We simply add the digit 5 to all the symbols occurring in Möbius's theorem; that is, we apply the reciprocity $R_5$. Cf. Richmond [39a].)

Continuing in this manner we obtain Cox's general theorem, to the effect that $d$ concurrent planes 1, 2, $\cdots$, $d$, with arbitrary points on their lines of intersection, determine a configuration of $2^{d-1}$ points and $2^{d-1}$ planes, with $d$ points on each plane and $d$ planes through each point, so that the Levi graph is the $d$-dimensional hypercube $\gamma_d$. Thus the group of symmetries and reciprocities is of order

$$2^d \ d!.$$

Instead of *arbitrary* points (12, 13, and so on) on the lines of intersection of pairs of the $d$ given planes, we may take the points determined on these lines by an arbitrary sphere through the point of concurrence 0. The $2^{d-1}$ planes intersect the sphere in $2^{d-1}$ circles [**37**, p. 271]. By stereographic projection we obtain Clifford's configuration of $2^{d-1}$ points and $2^{d-1}$ circles, with $d$ points on each circle and $d$ circles through each point. These again are represented by the $2^d$ vertices of $\gamma_d$.

**12. Kummer's $16_6$.** Let the five initial planes of Cox's $16_5$ be represented by the quaternions

$$i, \quad j, \quad k, \quad V = ai + bj + ck, \qquad V' = a'i + b'j + c'k.$$

If the reciprocity $R_1$ were a null polarity, it would transform the plane $V'$ into $iV'$ or $(0, c', -b', a')$, which is a definite point on the line of intersection of planes 1 and 5, whereas Cox would take his point 15 to be arbitrary on that line, say $(0, c', -b', \alpha)$. Hence, for the general $16_5$ the reciprocities $R_1, \cdots, R_5$ are *not* null polarities; and the same remark holds for any greater value of $d$.

On the other hand, the special case when $R_1, \cdots, R_5$ *are* null polarities is interesting in a different way. $R_1, R_2, R_3$ are the operations of multiplying on the left by $i, j, k$, while $R_4$ and $R_5$ are the operations of multiplying on the right by two anticommutative pure quaternions (or orthogonal vectors) $V$ and $V'$. Writing

$$V = qiq^{-1}, \qquad V' = qjq^{-1},$$

we can use a coordinate transformation to change the symbols of the planes

$$1, \qquad 2, \qquad 3, \qquad 4, \qquad 5$$

from

$$i, \qquad j, \qquad k, \qquad qiq^{-1}, \qquad qjq^{-1}$$

to

$$iq, \qquad jq, \qquad kq, \qquad qi, \qquad qj.$$

Then the initial point 0 is not 1 but $q$, and the reciprocities $R_1, R_2, R_3$

have the same effect as before, while $R_4$ and $R_5$ are the operations of multiplying on the right by $i$ and $j$. Applying the product $R_1R_2R_3R_4R_5$, we find that the plane 12345 is $kjiqij = qk$, which passes through the initial point $q$. (This is in marked contrast with the general $16_5$, where the point 0 and plane 12345 are not incident.)

Applying the null polarities $R_1, \cdots, R_5$ in turn, we find that the incidence of 0 and 12345 implies the incidence of 1 and 2345, of 12 and 345, and so on. Thus the special configuration is really not a $16_5$ but a $16_6$; and it is natural to use the symbol 6 for 12345, 16 for 2345, 126 as an alternative for 345, and so on, in agreement with the classical notation for the sixteen *nodes* and sixteen *tropes* of Kummer's quartic surface [25, p. 16]. We now have six mutually commutative null polarities $R_1, \cdots, R_6$: the first three are left-multiplications by $i, j, k$, and the last three are right-multiplications by $i, j, k$ [see **1a**, p. 138; **3**, p. 38]. The essential difference between this $16_6$ and Cox's $32_6$ is that now $R_1R_2R_3R_4R_5R_6$ is the identity.

We have seen that the Levi graph for Cox's $16_5$ is the five-dimensional hypercube $\gamma_5$. The effect of the new incidences is to insert new branches joining the pairs of opposite vertices (as in the derivation of Fig. 4 from Fig. 3). More symmetrically, we can take the six-dimensional hypercube $\gamma_6$ (which represents $32_6$) and identify pairs of opposite elements. Thus

*The Levi graph for Kummer's $16_6$ is $\gamma_6/2$.*

In other words, after coloring the sixty-four vertices of $\gamma_6$ alternately red and blue, we can say that the sixteen pairs of opposite red vertices represent the sixteen nodes of Kummer's surface, while the sixteen pairs of opposite blue vertices represent the sixteen tropes. The thirty-two red vertices by themselves belong to the semi-regular polytope $h\gamma_6$ [12, p. 158] ,whose connection with Kummer's $16_6$ has already been pointed out by Du Val [16, p. 65].

The most convenient coordinates for the vertices of $\gamma_6$ (of edge 2) are

$$(\pm 1, \pm 1, \pm 1, \pm 1, \pm 1, \pm 1),$$

with an even number of minus signs for a red vertex and an odd number of minus signs for a blue vertex. The 240 squares such as

$$(\pm 1, \pm 1, 1, 1, 1, 1)$$

represent the 120 *Kummer lines* [25, p. 77]; the 160 cubes such as

$$(\pm 1, \pm 1, \pm 1, 1, 1, 1)$$

represent the 80 *Rosenhain tetrads* [25, p. 78]; the sixty $\gamma_4$'s such as

$$(\pm 1,\ \pm 1,\ \pm 1,\ \pm 1,\ 1,\ 1)$$

represent thirty[3] Möbius configurations $8_4$; and the twelve cells $\gamma_6$ represent the six null polarities.

Each of the 160 cubes is obtained by fixing the signs of three of the six coordinates and leaving the remaining three ambiguous. Since the fixing of signs can be done in 8 ways (for the same choice of three coordinates), we thus obtain 8 cubes which together exhaust the 64 vertices of $\gamma_6$, and which occur in four pairs of opposites. Any two of these four pairs of opposite $\gamma_3$'s belong to a pair of opposite $\gamma_4$'s. Finally, the three coordinates whose signs were fixed can be chosen in

$$\binom{6}{3} = 20$$

ways. We thus verify the well known fact that there are twenty ways of regarding the $16_6$ as a set of four Rosenhain tetrahedra any two of which form a Möbius configuration $8_4$ [see **34** or **2**].

Besides these peripheral elements of $\gamma_6$, there are also 120 large inscribed cubes (of edge $2^{3/2}$ instead of 2) such as

$$(1,\ 1,\ 1,\ 1,\ 1,\ 1)(1,\ 1,\ 1,\ 1,\ -1,\ -1)$$

$$(-1,\ -1,\ 1,\ 1,\ 1,\ 1)(-1,\ -1,\ 1,\ 1,\ -1,\ -1)$$

$$(1,\ 1,\ -1,\ -1,\ 1,\ 1)(1,\ 1,\ -1,\ -1,\ -1,\ -1)$$

$$(-1,\ -1,\ -1,\ -1,\ 1,\ 1)(-1,\ -1,\ -1,\ -1,\ -1,\ -1),$$

which represent the 60 Göpel tetrads of points and the 60 Göpel tetrads of planes [**25**, p. 79].

**13. The skew icosahedron.** In the above list of significant subsets of Kummer's $16_6$, we have not mentioned the *Weber hexads* [**25**, p. 80] such as 0, 12, 23, 34, 45, 51. The corresponding vertices of $\gamma_6$ are

$$0,$$

| 12, | | 23, | | 34, | | 45, | | 51, |
|-----|---|-----|---|-----|---|-----|---|-----|
| | 1236, | | 2346, | | 3456, | | 4516, | | 5126, |

$$123456;$$

or, in terms of coordinates,

---

[3] Blaschke [**4**, p. 151] found only fifteen Möbius configurations; but that was surely a mistake. After the removal of any one $8_4$, the remaining points and planes of the $16_6$ form another $8_4$.

$$(1, 1, 1, 1, 1, 1),$$

$$(-1, -1, 1, 1, 1, 1), (1, -1, -1, 1, 1, 1), \cdots, (-1, 1, 1, 1, -1, 1),$$

$$(-1, -1, -1, 1, 1, -1), (1, -1, -1, -1, 1, -1), \cdots,$$

$$(-1, -1, 1, 1, -1, -1),$$

$$(-1, -1, -1, -1, -1, -1).$$

These twelve points in Euclidean 6-space form twenty equilateral triangles of side $2^{3/2}$, such as

0 12 23,    12 23 1236,    23 1236 2346,    and    1236 2346 123456,

which are the faces of a *regular skew icosahedron* (Fig. 25). This re-

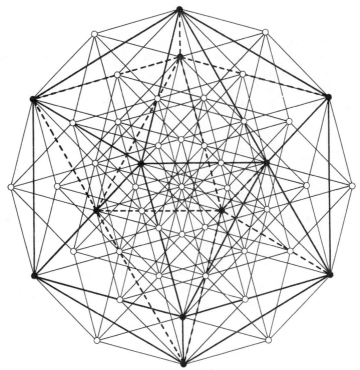

FIG. 25. $\gamma_6$ and the skew icosahedron.

markable surface is just as symmetrical as the Platonic icosahedron [**12**, p. 52], but its angles are different. In fact, the "vertex figure"

12 23 34 45 51,

instead of being a regular plane pentagon of angle 108°, is a regular skew pentagon of angle 90° (since 12 23 34 41 is a square). Similarly, the dihedral angle is reduced from about 138° 12' to 109° 28', since the two adjacent faces

$$0 \ 12 \ 23 \quad \text{and} \quad 0 \ 23 \ 34$$

belong to the octahedron 0 12 23 34 41 1234.

From the fifteen points 12, 13, $\cdots$, 56 we can pick out a rectangular skew pentagon such as 12 23 34 45 51 in seventy-two ways; hence there are seventy-two skew icosahedra for each vertex of $\gamma_6$, and

$$72 \cdot 64/12 = 384$$

of them altogether, representing the 192 Weber hexads of nodes and the 192 Weber hexads of tropes.

**14. Self-reciprocal polytopes in the Euclidean 4-space.** Although the Levi graph has been defined only for configurations of points and lines, or of points and planes, the idea can evidently be extended, for example, to configurations of points and hyperplanes in more than three dimensions. We shall be content to describe two instances: the configurations $5_4$ and $24_6$ formed by the vertices and cells of the four-dimensional polytopes $\{3, 3, 3\}$ and $\{3, 4, 3\}$ [12, pp. 120, 148].

The former is the four-dimensional simplex $\alpha_4$. In the Levi graph its five vertices are represented by five red nodes $A$, $B$, $C, D, E$, and the respectively opposite cells by five blue nodes $a$, $b$, $c$, $d$, $e$. Red and blue nodes are joined whenever their letters disagree. Thus the group is of order 240: the direct product of the symmetric group of degree 5 and the group of order 2.

The five letters can be arranged as a cycle in twelve ways, corresponding to the twelve Petrie polygons of the simplex [12, p. 225]. Every four consecutive vertices of the Petrie polygon belong to a cell; and by placing the symbol for the cell between the middle two of the four vertices, we obtain a symmetrical Hamiltonian circuit of the graph. In this manner the Petrie polygon $ABCDE$ yields the circuit

$$A \ d \ B \ e \ C \ a \ D \ b \ E \ c,$$

which we may rewrite as

$$0 \ 1 \ 2 \ 3 \ 4 \ 5 \ 6 \ 7 \ 8 \ 9,$$

so that the even and odd digits represent the red and blue nodes, or the vertices and cells of the simplex, and the symbols for a vertex and its opposite cell differ by 5. Thus the graph can be drawn as a regular decagon 0123456789 with its inscribed decagram 0369258147 [see **18**, p. 315, Fig. 10, IV-11-A]. This 2-regular graph of degree 4 and girth 4 can be embedded into the torus to form the map $\{4, 4\}_{3,1}$ (see §3), which has ten square faces

$$0143, \ 1254, \ 2365, \ \cdots, \ 9032.$$

The embedding can be done in 6 ways, since the group of the map is of order 40. This order arises as

$$10 \cdot 2 \cdot 2,$$

since the ten symbols can be cyclically permuted or reversed, or multiplied by 3 (mod 10) so as to interchange the decagon and decagram.

A more symmetrical way of drawing the graph is in Euclidean 4-space, where we may take two equal regular simplexes in reciprocal positions and join each vertex of either to the four nearest vertices of the other. (This is analogous to the two tetrahedra of Kepler's *stella octangula* which can be joined by the edges of a cube.) Since the regular four-dimensional simplex has the Schläfli symbol $\{3, 3, 3\}$, an appropriate notation for this aspect of the graph is

$$2\{3, 3, 3\}.$$

Similarly, $2\{3, 4, 3\}$ denotes the 1-regular graph of degree 6 and girth 4 whose 48 nodes are the vertices of two reciprocal 24-cells $\{3, 4, 3\}$, each vertex of either being joined to the six vertices of an octahedral cell of the other. This is the Levi graph for the polytope $\{3, 4, 3\}$ qua configuration $24_6$. Robinson [**40**, p. 44] called it $\delta_4 + \delta_4'$, and remarked that its 144 branches are the sides of 18 concentric octagons in different planes. By identifying opposite elements we derive

$$2\{3, 4, 3\}/2:$$

the Levi graph for Stephanos's configuration $12_6$ consisting of three "desmic" tetrahedra in real projective 3-space [**45; 25**]. In fact, the Cartesian coordinates for the 24 vertices of $\{3, 4, 3\}$, being the constituents of Hurwitz's 24 quaternion units

$$\pm i, \ \pm j, \ \pm k \quad \text{and} \quad (\pm 1 \pm i \pm j \pm k)/2,$$

are the same as the homogeneous coordinates for the 12 points of the $12_6$ [see **23**, p. 170; **3**, p. 35].

By Pappus's Theorem as stated in §6, if a triangle $A_1A_2A_3$ is in perspective with $B_1B_3B_2$ and with $B_3B_2B_1$, it is also in perspective with $B_2B_1B_3$. Stephanos's desmic tetrahedra come from the analogous theorem that if a tetrahedron $A_1A_2A_3A_4$ is in perspective with $B_2B_1B_4B_3$ and $B_3B_4B_1B_2$ and $B_4B_3B_2B_1$, it is also in perspective with $B_1B_2B_3B_4$. However, the resemblance is superficial, as the permutations of the $B$'s are no longer cyclic. In one respect this three-dimensional theorem is simpler: instead of depending on Pappus's Theorem, it can be deduced from the "propositions of incidence" alone. In fact, the given perspectivities imply that the line $A_iB_j$ intersects $A_jB_i$, whence $A_iB_i$ intersects $A_jB_j$. Thus the four lines $A_iB_i$ all intersect one another, and since they are not all coplanar they must be concurrent.

By a simple change of notation, from $B_1B_2B_3$ to $B_2B_3B_1$, we obtain a rule for sets of collinear points that is identical with the rule for Hesse's $(12_4, 16_3)$ which we considered in §8. Hence, as Zacharias observed [51, p. 152], this plane configuration can be derived from Stephanos's $12_6$ by projection from an arbitrary point in the 3-space.

SUMMARY

| Configuration $n_d$ | Space | Group order | Graph or map | Girth $m$ | Regularity $s$ | Number of $s$-arcs $2(d-1)^{s-1}dn$ |
|---|---|---|---|---|---|---|
| Fano's $7_3$ | $PG(2,2)$ | 336 | $\{6,3\}_{2,1}$ | 6 | 4 | 336 |
| Möbius-Kantor $8_3$ | Complex plane | 96 | $\{8\}+\{8/3\}$ | 6 | 2 | 96 |
| Pappus's $9_3$ | Real plane | 216 | $\{6,3\}_{3,0}$ | 6 | 3 | 216 |
| Desargues' $10_3$ | Real plane | 240 | $\{10\}+\{10/3\}$ | 6 | 3 | 240 |
| The new $12_3$ | Real plane | 144 | $\{12\}+\{12/5\}$ or $\{6,3\}_{2,2}$ | 6 | 2 | 144 |
| Cremona-Richmond $15_3$ | Real plane | 1440 | Tutte's 8-cage | 8 | 5 | 1440 |
| $3_3$ | Real 3-space | 72 | $\{6,3\}_{1,1}$ | 4 | 3 | 72 |
| $4_4$ | Real 3-space | 1152 | $\{4,4\}_{2,2}$ | 4 | 3 | 288 |
| Tetrahedron $4_3$ | Real 3-space | 48 | $\{6,3\}_{2,0}$ | 4 | 2 | 48 |
| Möbius's $8_4$ | Real 3-space | 384 | $\{4,4\}_{4,0}$ | 4 | 2 | 192 |
| Cox's $(2^{d-1})_d$ | Real 3-space | $2^d d!$ | $\gamma_d$ | 4 | 2 | $2^d d(d-1)$ |
| Kummer's $16_6$ | Real 3-space | 23040 | $\gamma_6/2$ | 4 | 2 | 960 |
| Stephanos's $12_6$ | Real 3-space | 1152 | $2\{3,4,3\}/2$ | 4 | 1 | 144 |
| Simplex $5_4$ | Real 4-space | 240 | $\{4,4\}_{2,1}$ | 4 | 2 | 120 |

BIBLIOGRAPHY

1. H. F. Baker, *Principles of geometry*, vol. 1, 2d ed., Cambridge, 1929.

1a. ———, ibid. vol. 3, Cambridge, 1934.

2. ———, *Note to the preceding paper by C. V. H. Rao*, Proc. Cambridge Philos. Soc. vol. 42 (1946) pp. 226–229.

**3.** G. Benneton, *Configurations harmoniques et quaternions*, Ann. École Norm. (3) vol. 64 (1947) pp. 1–58.

**4.** W. Blaschke, *Projektive Geometrie*, Wolfenbüttel-Hannover, 1947.

**5.** W. Blaschke and G. Bol, *Geometrie der Gewebe*, Berlin, 1938.

**6.** R. C. Bose, *An affine analogue of Singer's theorem*, J. Indian Math. Soc. N.S vol. 6 (1942) pp. 1–15.

**7.** H. R. Brahana, *Regular maps and their groups*, Amer. J. Math. vol. 49 (1927) pp. 268–284.

**8.** W. Burnside, *Theory of groups of finite order*, 2d ed., Cambridge, 1911.

**9.** A. Cayley, *Sur quelques théorèmes de la géométrie de position* (1846), Collected Mathematical Papers, vol. 1, 1889, pp. 317–328.

**10.** H. Cox, *Application of Grassmann's Ausdehnungslehre to properties of circles*, Quart. J. Math. vol. 25 (1891) pp. 1–71.

**11.** H. S. M. Coxeter, *Regular skew polyhedra in three and four dimensions*, Proc. London Math. Soc. (2) vol. 43 (1937) pp. 33–62.*

**12.** ———, *Regular polytopes*, New York, 1963.

**13.** ———, *Configurations and maps*, Reports of a Mathematical Colloquium (2) vol. 8 (1948) pp. 18–38.

**14.** ———, *The real projective plane,* Cambridge, 1961.

**15.** L. Cremona, *Teoremi stereometrici, dei quali si deducono le proprietà dell'esagrammo di Pascal*, Memorie della Reale Accademia dei Lincei vol. 1 (1877) pp. 854–874.

**16.** P. Du Val, *On the directrices of a set of points in a plane*, Proc. London Math. Soc. (2) vol. 35 (1933) pp. 23–74.

**17.** G. Fano, *Sui postulati fondamentali della geometria proiettiva in uno spazio lineare a un numero qualunque di dimensioni*, Giornale di Matematiche vol. 30 (1892) pp. 106–132.

**17a.** J. M. Feld, *Configurations inscriptible in a plane cubic curve*, Amer. Math. Monthly vol. 43 (1936) pp. 549–555.

**18.** R. M. Foster, *Geometrical circuits of electrical networks*, Transactions of the American Institute of Electrical Engineers vol. 51 (1932) pp. 309–317.

**19.** R. Frucht, *Die Gruppe des Petersen'schen Graphen und der Kantensysteme der regulären Polyeder*, Comment. Math. Helv. vol. 9 (1937) pp. 217–223.

**20.** G. Gallucci, *Studio della figura delle otto rette e sue applicazioni alla geometria del tetraedro ed alla teoria della configurazioni*, Rendiconto dell'Accademia delle Scienze fisiche e matematiche (Sezione della Società reale di Napoli) (3) vol. 12 (1906) pp. 49–79.

**21.** P. J. Heawood, *Map-colour theorem*, Quart. J. Math. vol. 24 (1890) pp. 332–338.

**22.** L. Henneberg, *Die graphische Statik der starren Körper*, Encyklopädie der Mathematischen Wissenschaften vol. 4.1 (1908) pp. 345–434.

**23.** E. Hess, *Weitere Beiträge zur Theorie der räumlichen Configurationen*, Verhandlungen den K. Leopoldinisch-Carolinischen Deutschen Akademie Naturforscher vol. 75 (1899) pp. 1–482.

**24.** W. V. D. Hodge and D. Pedoe, *Methods of algebraic geometry*, vol. 1, Cambridge, 1947.

**25.** R. W. H. T. Hudson, *Kummer's quartic surface*, Cambridge, 1905.

**26.** S. Kantor, *Über die Configurationen (3, 3) mit den Indices 8, 9 und ihren Zusammenhang mit den Curven dritter Ordnung*, Sitzungsberichte der Mathematisch-Naturwissenschaftliche Classe der K. Akademie der Wissenschaften, Wien vol. 84.1 (1882) pp. 915–932.

27. ——, *Die Configurationen* (3, 3)$_{10}$, ibid. pp. 1291–1314.

28. D. König, *Theorie der endlichen und unendlichen Graphen*, Leipzig, 1936.

29. K. Kommerell, *Die Pascalsche Konfiguration* 9$_3$, Deutsche Mathematik vol. 6 (1941) pp. 16–32.

30. A. Kowalewski, W. R. *Hamilton's Dodekaederaufgabe als Buntordnungsproblem*, Sitzungsberichte der Mathematisch-Naturwissenschaftliche Klasse der K. Akademie der Wissenschaften, Wien vol. 126.2a (1917) pp. 67–90.

31. F. W. Levi, *Geometrische Konfigurationen*, Leipzig, 1929.

32. ——, *Finite geometrical systems*, Calcutta, 1942.

33. V. Martinetti, *Sopra alcune configurazioni piane*, Annali di Matematica (2) vol. 14 (1887) pp. 161–192.

33a. ——, *Sulle configurazioni piane* $\mu_3$, Annali di Matematica (2) vol. 15 (1888) pp. 1–26.

34. ——, *Alcune considerazioni sulla configurazione di Kummer*, Rend. Circ. Mat. Palermo vol. 16 (1902) pp. 196–203.

35. G. A. Miller, H. F. Blichfeldt, and L. E. Dickson, *Theory and applications of finite groups*, New York, 1916.

36. A. F. Möbius, *Kann von zwei dreiseitigen Pyramiden eine jede in Bezug auf die andere um- und eingeschrieben zugleich heissen?* (1828), Gesammelte Werke, vol. 1, 1886, pp. 439–446.

37. F. Morley and F. V. Morley, *Inversive geometry*, Boston, 1933.

38. J. Petersen, *Les 36 officiers*, Annuaire des Mathématiciens (1902) pp. 413–427.

39. H. W. Richmond, *On the figure of six points in space of four dimensions*, Quart. J. Math. vol. 31 (1900) pp. 125–160.

39a. ——, *On the configuration of Homersham Cox*, J. London Math. Soc. vol. 16 (1941) pp. 105–112.

40. G. de B. Robinson, *On the orthogonal groups in four dimensions*, Proc. Cambridge Philos. Soc. vol. 27 (1931) pp. 37–48.

41. A. Schoenflies, *Ueber die regelmässigen Configurationen* $n_3$, Math. Ann. vol. 31 (1888) pp. 43–69.

42. J. Singer, *A theorem in finite projective geometry and some applications to number theory*, Trans. Amer. Math. Soc. vol. 43 (1938) pp. 377–385.

43. D. M. Y. Sommerville, *An introduction to the geometry of n dimensions*, London, 1929.

44. K. G. C. von Staudt, *Geometrie der Lage*, Nürnberg, 1847.

45. C. Stephanos, *Sur les systèmes desmiques de trois tétraèdres*, Bull. Sci. Math. (2) vol. 3 (1879) pp. 424–456.

46. W. Threlfall, *Gruppenbilder*, Abhandlungen der Mathematisch-Physischen Klasse der Sächsischen Akademie der Wissenschaften vol. 41.6 (1932) pp. 1–59.

47. W. T. Tutte, *A family of cubical graphs*, Proc. Cambridge Philos. Soc. vol. 43 (1947) pp. 459–474.

48. O. Veblen, *Collineations in a finite projective geometry*, Trans. Amer. Math. Soc. vol. 8 (1907) pp. 366–368.

49. O. Veblen and J. W. Young, *Projective geometry*, vol. 1, Boston, 1910.

50. H. Whitney, *Planar graphs*, Fund. Math. vol. 21 (1933) pp. 73–84.

51. M. Zacharias, *Untersuchungen über ebene Konfigurationen* (12$_4$, 16$_3$), Deutsche Mathematik vol. 6 (1941) pp. 147–170.

*Reference 11 is Chapter 5 of this book.

# 7

# TWELVE POINTS IN *PG*(5,3) WITH 95040 SELF-TRANSFORMATIONS

Reprinted with permission,
from the *Proceedings of the Royal Society*,
A, Vol. 247 (1958).

The title of this paper could have been 'Geometry in five dimensions over $GF(3)$' (cf. Edge 1954), or 'The geometry of the second Mathieu group', or 'Duads and synthemes', or 'Hexastigms', or simply 'Some thoughts on the number 6'. The words actually chosen acknowledge the inspiration of the late H. F. Baker, whose last book (Baker 1946) develops the idea of duads and synthemes in a different direction. The special property of the number 6 that makes the present development possible is the existence of an outer automorphism for the symmetric group of this degree. The consequent group of order 1440 is described abstractly in §1, topologically in §2, and geometrically in §§3 to 7. The kernel of the geometrical discussion is in §5, where the chords of a non-ruled quadric in the finite projective space $PG(3, 3)$ are identified with the edges of a graph having an unusually high degree of regularity (Tutte 1958). It is seen in §4 that the ten points which constitute this quadric can be derived very simply from a 'hexastigm' consisting of six points in $PG(4, 3)$ (cf. Coxeter 1958). The connexion with Edge's work is described in §6. Then §7 shows that the derivation of the quadric from a hexastigm can be carried out in two distinct ways, suggesting the use of a second hexastigm in a different 4-space. It is found in §8 that the consequent configuration of twelve points in $PG(5, 3)$ can be divided into two hexastigms in 66 ways. The whole set of 132 hexastigms forms a geometrical realization of the Steiner system $\mathfrak{S}(5, 6, 12)$, whose group is known to be the quintuply transitive Mathieu group $M_{12}$, of order 95040. Finally, §9 shows how the same 5-dimensional configuration can be regarded (in 396 ways) as a pair of mutually inscribed simplexes, like Möbius's mutually inscribed tetrahedra in ordinary space of 3 dimensions.

## 1. SYLVESTER'S DUADS AND SYNTHEMES

Sylvester (1844, p. 92)*noticed that six digits 1, 2, 3, 4, 5, 6 form fifteen pairs of *duads* such as 12 (the same as 21), which can be taken together in fifteen sets of three such as (12, 34, 56), called *synthemes*. (It is understood that (12, 34, 56) is the same as (12, 56, 34) or (34, 12, 56), etc.) Duads and synthemes are in (3, 3) correspondence: each syntheme consists of three duads, and each duad belongs to three synthemes. The synthemes occur in six families of five, each exhausting all the fifteen duads. 'And it will be observed that every two families have one, and only one, syntheme in common between them' (Sylvester 1861, p. 268). Table 1, (cf. Baker 1930, p. 221) exhibits the six families in its rows, and again in its columns, Naming them $a, b, c, d, e, f$, we can distinguish each syntheme by a pair of letters (one for the row and one for the column).

Thus the synthemes of numbers correspond to duads of letters, and the synthemes of letters to duads of numbers. For instance, the various occurrences of '12' in table 1 serve to identify this duad of numbers with the syntheme $(ae, bc, df)$. In fact, an equally valid table, resembling table 1, could be made by interchanging numbers and letters in the natural order

$$(1\,a) \quad (2\,b) \quad (3\,c) \quad (4\,d) \quad (5\,e) \quad (6\,f). \tag{1.1}$$

Richmond (1900, p. 133) remarked: 'It does not appear that this reciprocity can be followed out to any result of value.' This defeatist attitude was probably due to the amount of arbitrariness in naming Sylvester's six families of synthemes. For

*For references, see page 165.

instance, our choice of $a$, $b$, $c$, $d$, $e$, $f$ differs from Richmond's, yet both notations exhibit the same kind of reciprocity.

The symmetric group $\mathfrak{S}_6$ permutes the numbers and letters in corresponding but different ways; e.g. (in our notation) the permutations

$$(1\ 2\ 3\ 4\ 5), \quad (1\ 2\ 3\ 4\ 5\ 6), \quad (1\ 2)$$

of the numbers correspond to the permutations

$$(a\ b\ c\ d\ e), \quad (a\ e\ f)\ (b\ d), \quad (a\ e)\ (b\ c)\ (d\ f)$$

of the letters, and the permutations

$$(1\ 5)\ (2\ 3)\ (4\ 6), \quad (1\ 4)\ (2\ 6)\ (3\ 5), \quad (1\ 3)\ (2\ 4)\ (5\ 6), \quad \ldots$$

of the numbers (see table 1) correspond to the transpositions

$$(a\ b), \quad (a\ c), \quad (a\ d), \quad \ldots$$

of the letters. Richmond apparently failed to connect this idea with the fact that the symmetric group of degree six (unlike every other symmetric group) possesses outer automorphisms. Of the 720 outer automorphisms, 36 are involutory (one for each subgroup of order five in $\mathfrak{S}_6$, as we shall see in § 2). Richmond's reciprocity and ours are simply two of these thirty-six involutory automorphisms.

TABLE 1. DUADS AND SYNTHEMES

|   | $a$ | $b$ | $c$ | $d$ | $e$ | $f$ |
|---|-----|-----|-----|-----|-----|-----|
| $a$ | — | 15, 23, 46 | 14, 26, 35 | 13, 24, 56 | 12, 36, 45 | 16, 25, 34 |
| $b$ | 15, 23, 46 | — | 12, 34, 56 | 14, 25, 36 | 16, 24, 35 | 13, 26, 45 |
| $c$ | 14, 26, 35 | 12, 34, 56 | — | 16, 23, 45 | 13, 25, 46 | 15, 24, 36 |
| $d$ | 13, 24, 56 | 14, 25, 36 | 16, 23, 45 | — | 15, 26, 34 | 12, 35, 46 |
| $e$ | 12, 36, 45 | 16, 24, 35 | 13, 25, 46 | 15, 26, 34 | — | 14, 23, 56 |
| $f$ | 16, 25, 34 | 13, 26, 45 | 15, 24, 36 | 12, 35, 46 | 14, 23, 56 | — |

## 2. TUTTE'S 8-CAGE

Any (3, 3) correspondence (between two sets of the same number of objects) may conveniently be represented as an even graph of degree 3 whose vertices represent the objects while the edges indicate which pairs of objects correspond. Table 1 establishes a (3, 3) correspondence between the fifteen duads of numbers and the fifteen duads of letters. In this case the graph, having 30 vertices and 45 edges, is the 5-regular 8-cage of Tutte (1958), shown in figures 1 and 2. It is said to be 5-*regular* because all the paths formed by 5 successive edges are equivalent under the group of automorphisms of the graph. (This, as we have seen, is the group of automorphisms of $\mathfrak{S}_6$, of order 1440.) For instance, in spite of any appearance to the contrary, the two paths

$$13\ ad\ 24\ be\ 35\ ac \quad \text{and} \quad 13\ ce\ 46\ ab\ 15\ cf$$

are exactly alike. On the other hand, the longer paths

$$13\ ad\ 24\ be\ 35\ ac\ 26 \quad \text{and} \quad 13\ ad\ 24\ be\ 35\ ac\ 14$$

are essentially different, as the former can be completed (by adding $bf$) to form an 8-circuit or octagon, whereas the latter cannot.

The graph is said to be of *girth* 8 because it contains octagons but no shorter circuits. It is called the 8-*cage* because it has the highest possible regularity for its girth.

The aspect shown in figure 1 arises from the existence of a *Hamiltonian* circuit which includes all the 30 vertices. The crossings of the 15 'diagonal' edges must, of

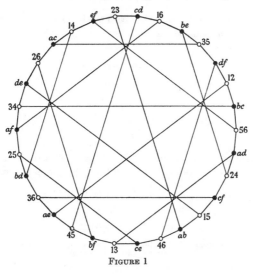

FIGURE 1

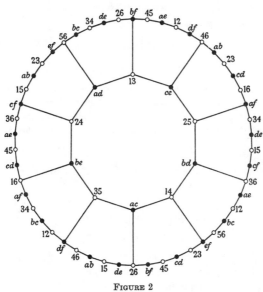

FIGURE 2

course, be ignored. Such crossings can be avoided by embedding the graph in a suitable surface to form a *map*. Since a map made up entirely of $N$ octagons would have $4N$ edges, the simplest possible embedding of Tutte's graph (which has 45 edges) is a map of ten octagons and one decagon. The surface is non-orientable, of characteristic

$$30 - 45 + 11 = -4$$

(Coxeter 1950, p. 443). In figure 2 it has been cut and spread out flat, the way one unfolds a torus into a rectangle. Such an embedding of the graph can be done in 144 ways, two for each decagon; e.g. the octagon

$$13\ ad\ 56\ bc\ 34\ de\ 26\ bf$$

(in figure 2) could be replaced by

$$13\ ad\ 56\ ef\ 23\ cd\ 45\ bf$$

without altering the central decagon

$$13\ ad\ 24\ be\ 35\ ac\ 14\ bd\ 25\ ce.$$

The group of the map (a subgroup of index 144 in the group of the graph) is simply the 'rotation' group of the decagon, namely the cyclic group $\mathfrak{C}_{10}$ generated by the permutation

$$S = (1\ d\ 2\ e\ 3\ a\ 4\ b\ 5\ c)\ (6f),$$

whose square,     $$S^2 = (1\ 2\ 3\ 4\ 5)\ (a\ b\ c\ d\ e),\qquad(2\cdot1)$$

generates one of the subgroups $\mathfrak{C}_5$ of $\mathfrak{S}_6$. The fifth power,

$$S^5 = (1\ a)\ (2\ b)\ (3\ c)\ (4\ d)\ (5\ e)\ (6f),$$

is the reciprocity $(1\cdot1)$, which appears in figure 2 as the half-turn about the centre. The same reciprocity arises when we replace $S$ by

$$S' = S^{-3} = (1\ b\ 3\ d\ 5\ a\ 2\ c\ 4\ e)\ (6f),$$

so that     $$S'^2 = S^4 = (1\ 3\ 5\ 2\ 4)\ (a\ c\ e\ b\ d);$$

but now the decagon is     $$12\ bc\ 34\ de\ 15\ ab\ 23\ cd\ 45\ ae.$$

We thus see that the subgroup $\{S^2\}$ of order five in $\mathfrak{S}_6$ is associated with two decagons (and four maps) and with a single reciprocity $S^5 = S'^5$, that is, an involutory outer automorphism of $\mathfrak{S}_6$. In this way, the 36 subgroups $\mathfrak{C}_5$ in $\mathfrak{S}_6$ yield 36 reciprocities; e.g. the six reciprocities that interchange 6 and $f$ are

$$(1\ a)\ (2\ b)\ (3\ c)\ (4\ d)\ (5\ e)\ (6f),$$
$$(1\ e)\ (2\ d)\ (3\ c)\ (4\ b)\ (5\ a)\ (6f),$$
$$(1\ d)\ (2\ c)\ (3\ b)\ (4\ a)\ (5\ e)\ (6f),$$
$$(1\ c)\ (2\ b)\ (3\ a)\ (4\ e)\ (5\ d)\ (6f),$$
$$(1\ b)\ (2\ a)\ (3\ e)\ (4\ d)\ (5\ c)\ (6f),$$
$$(1\ a)\ (2\ e)\ (3\ d)\ (4\ c)\ (5\ b)\ (6f).$$

The effect of the first reciprocity on figure 2 has already been described. The remaining five appear in figure 1 as reflexions in the five obvious lines of symmetry. The last comes from the subgroup generated by $(2\ 3\ 4\ 5\ 6)\ (f\ e\ d\ c\ b).$

The following interpretation of duads and synthemes was suggested by B. H. Neumann.

The symmetric group $\mathfrak{S}_6$ itself contains thirty odd permutations of period two: fifteen transpositions such as (1 2), and fifteen products of three disjoint transpositions such as (1 2) (3 4) (5 6). Representing these thirty permutations by the vertices of the 8-cage, we see that each vertex is equal to the product of its three neighbours.

### 3. RICHMOND'S HEXASTIGM

In a projective 4-space, a *hexastigm* (Richmond 1900, p. 127) is the figure formed by six points *1, ..., 6* (no five in a 3-space), joined in sets of two, three, and four by

$$15\ \textit{edges} \qquad 12, ..., 56,$$
$$20\ \textit{faces} \qquad 123, ..., 456,$$
$$15\ \textit{solids} \qquad 1234, ..., 3456.$$

Each edge meets the opposite solid in one of 15 *diagonal points* such as

$$P_{12} = 12.3456.$$

Three edges that have no common vertex can be chosen in fifteen ways. Being three skew lines, they have a unique transversal; e.g. the transversal to *12, 34, 56* is the line
$$P_{12}P_{34}P_{56} = 1234.1256.3456.$$

Each diagonal point belongs to three transversals; e.g. $P_{12}$ lies not only on $P_{12}P_{34}P_{56}$ but also on $P_{12}P_{35}P_{46}$ and $P_{12}P_{36}P_{45}$. Thus the fifteen diagonal points and fifteen transversals form a configuration $15_3$: three points on each line and three lines through each point. These are the points

$$A, B, C, \quad A', B', C', \quad P, Q, R, \quad P', Q', R', \quad L, M, N$$

of Baker (1940, pp. 113–16); our hexastigm *123456* is his *FGHF'G'H'*.

Regarding the entries of table 1 as transversals, we see that each row or column consists of five of which together use up all the fifteen points. Baker showed that five such lines are *associated* in the sense that every plane meeting four of them meets the fifth as well. Denoting the six pentads of associated lines by the letters *a, ..., f*, we may use table 1 to represent the transversals and diagonal points by duads and synthemes of letters (whereas we previously represented them by synthemes and duads of numbers).

The remaining sixty joins of diagonal points will be found to consist of four lines through each of fifteen *harmonic points* $Q_{ij}$; for instance, the four lines $P_{1k}P_{2k}$ ($k = 3, 4, 5, 6$) all pass through the point $Q_{12}$, which is the harmonic conjugate of $P_{12}$ with respect to *1* and *2*.

The ten pairs of opposite faces of the hexastigm meet in ten points 'of minor importance' (Richmond 1900, p. 128) such as

$$P_{123} = P_{456} = 123.456.$$

The forty-five lines joining pairs of these ten points consist of three lines through each of the fifteen harmonic points; e.g. the lines

$$P_{145}P_{245}, \quad P_{135}P_{235}, \quad P_{134}P_{234}$$

all pass through $Q_{12}$. The same forty-five lines consist of three meeting each of the fifteen transversals; e.g. the lines

$$P_{134}P_{156}, \quad P_{123}P_{356}, \quad P_{125}P_{345}$$

all meet $P_{12}P_{34}P_{56}$.

These results can be verified at once by means of Möbius's 'barycentric calculus' (Whitehead 1898, pp. 136–43; Baker 1929, p. 71). Assuming the syzygy

$$1+2+3+4+5+6 = 0, \tag{3.1}$$

we have
$$P_{ij} = i+j, \quad P_{jkl} = j+k+l$$

(so that $P_{12}+P_{34}+P_{56} = 0$ and $P_{456} = -P_{123}$),

$$P_{ik}-P_{jk} = i-j = Q_{ij},$$
$$P_{ikl}-P_{jkl} = i-j = Q_{ij}, \tag{3.2}$$

$$P_{134}-P_{156} = P_{34}-P_{56}, \quad P_{123}-P_{356} = P_{12}-P_{56}, \quad P_{125}-P_{345} = P_{12}-P_{34}.$$

## 4. THE HEXASTIGM IN $PG(4, 3)$

When the co-ordinate field is restricted to $GF(3)$, so that our hexastigm *123456* is in a finite space $PG(4,3)$, the ten 'minor' points $P_{jkl}$ all lie in a hyperplane (or 'prime'); i.e. they belong to a $PG(3,3)$. To see this, we observe that, in $GF(3)$, the relations (3·1) and (3·2) imply

$$Q_{16}+Q_{26}+Q_{36}+Q_{46}+Q_{56} = 0.$$

These five harmonic points thus lie in the 3-space spanned by any four of them. Since
$$Q_{ij} = Q_{i6}-Q_{j6},$$

all the fifteen harmonic points lie in this same hyperplane. Since

$$P_{jkl} = Q_{j6}+Q_{k6}+Q_{l6},$$

the ten 'minor' points do likewise.

Among the forty-five lines joining pairs of these ten points, the three that meet a transversal all meet it in the same point, namely, the point where this transversal meets the hyperplane. For instance, since

$$P_{34}-P_{56} = P_{34}+(P_{12}+P_{34}) = P_{12}-P_{34}$$

and
$$P_{12}-P_{56} = P_{12}+(P_{12}+P_{34}) = -(P_{12}-P_{34}),$$

the three expressions at the end of § 3 all refer to the same point, which we naturally denote by the syntheme (12, 34, 56).

Thus the forty-five joins not only pass by threes through the fifteen harmonic or 'negative' points such as $Q_{12}$, but also pass by threes through fifteen 'positive' points such as (12, 34, 56). Each of them contains one point from each set of fifteen; e.g. the join $P_{134}P_{234}$ contains

$$P_{134}-P_{234} = Q_{12} \quad \text{and} \quad P_{134}+P_{234} = (12, 34, 56).$$

(The barycentric calculus is particularly well suited to geometry over $GF(3)$, in which every line contains just four points; for the remaining two points on a line $AB$ are simply $A \pm B$.) The ten 'minor', fifteen 'negative' and fifteen 'positive' points exactly account for all the $1 + 3 + 3^2 + 3^3 = 40$ points in the 3-space.

Disregarding the ten 'minor' points, we have a *graph* consisting of forty-five lines joining pairs of thirty points, each point lying on three of the lines. The 'duad and syntheme' notation serves to identify this graph with the 8-cage described in § 2.

### 5. The non-ruled quadric in $PG(3, 3)$

In the finite projective 4-space $PG(4, 3)$, we may set up a system of homogeneous co-ordinates (consisting of the marks 0 and $\pm 1$ of the Galois field) by choosing *12345* as simplex of reference while *6* is the 'unit point' $(1, 1, 1, 1, 1)$ or

$$(-1, -1, -1, -1, -1)$$

(for better agreement with $(3 \cdot 1)$).

The planes *123* and *456*, namely

$$x_4 = x_5 = 0 \quad \text{and} \quad x_1 = x_2 = x_3,$$

meet in the point $(1, 1, 1, 0, 0)$, which is $P_{123}$. The ten points of this kind constitute the non-ruled quadric or 'ellipsoid'

$$x_1^2 + x_2^2 + x_3^2 + x_4^2 + x_5^2 = 0$$

in the 3-space $x_1 + x_2 + x_3 + x_4 + x_5 = 0$ (cf. Edge 1954, p. 274). This quadric is clearly non-ruled, since it meets its tangent plane $x_1 + x_2 + x_3 = 0$ only at the 'point of contact' $(1, 1, 1, 0, 0)$. It 'separates' the remaining thirty points of $PG(3, 3)$ into the fifteen *negative* points such as

$$Q_{12} = (1, -1, 0, 0, 0), \quad Q_{56} = (1, 1, 1, 1, -1),$$

for which $\Sigma x^2 = -1$, and the fifteen *positive* points ($\Sigma x^2 = 1$) such as

$$(12, 34, 56) = (1, 1, -1, -1, 0),$$

for which table 1 suggests an alternative symbol involving a duad of letters, so that

$$(12, 34, 56) = Q_{bc} \tag{5·1}$$

and analogously

$$Q_{23} = (ab, cd, ef). \tag{5·2}$$

In terms of the quadric, the conclusion of § 4 may be expressed as follows: *The forty-five chords and the thirty points not on the quadric are the edges and vertices of an 8-cage* (Tutte 1958).

Moreover, each of the ten points on the quadric is joined to the other nine by nine chords which are nine of the thirteen lines through the given point. The remaining four lines through this point are the tangents there, forming a flat pencil in the tangent plane; e.g. the three remaining points on each of the four tangents at $P_{123}$ are

$$Q_{12} Q_{13} Q_{23}, \quad Q_{45} Q_{46} Q_{56},$$
$$Q_{ac} Q_{af} Q_{cf}, \quad Q_{bd} Q_{be} Q_{de}.$$

Following Edge (1954, p. 272), we call the first two *negative* tangents and the last two *positive* tangents. This description exhausts the thirteen points in the plane, in agreement with the well-known geometrical fact that the point of contact of a tangent plane is joined to every other point in the plane by a tangent line. Incidentally, it is natural to assign two further symbols to the point of contact in accordance with the equations

$$P_{123} = P_{456} = P_{acf} = P_{bde} \tag{5.3}$$

(cf. Richmond 1900, p. 132, table III).

## 6. EDGE'S PENTAGONS

If we wish to represent the six vertices of our hexastigm in $PG(4, 3)$ by geometric objects in the subspace $PG(3, 3)$, it is natural to represent the vertex *1* by the set of five points

$$Q_{12}, \quad Q_{13}, \quad Q_{14}, \quad Q_{15}, \quad Q_{16},$$

whose subscripts involve the number 1. This is the first of the six *negative pentagons* of Edge (1954, pp. 275–6); a complete pentagon (or 'pentastigm') whose ten edges are negative tangents.

Similarly, our symbol $a$ is represented by the set of five points

$$Q_{ab}, \quad Q_{ac}, \quad Q_{ad}, \quad Q_{ae}, \quad Q_{af}$$

(the first row of table 1), which form a *positive pentagon*. Accepting Edge's numbering of the negative pentagons, or of their reciprocals, the 'negative pentahedra', we see from table 1 that his 'positive pentahedra'

$$\text{I, \quad II, \quad III, \quad IV, \quad V, \quad VI}$$

correspond to our symbols

$$b, \quad c, \quad d, \quad a, \quad e, \quad f.$$

He comes close to our notation for the points on the quadric when he writes: 'The 10 points of $F$ answer to the 10 separations of the 6 pentahedra into complementary triads' (Edge 1954, p. 278).

He observes (p. 276) that the quadric is invariant for 1440 collineations (of which 720 are 'direct'). The distinction between positive and negative points is preserved by 720 of these collineations, representing the symmetric group $\mathfrak{S}_6$. (Among these, the 360 'direct' collineations represent the alternating group $\mathfrak{A}_6$.) Thus the whole group of order 1440 is the group of automorphisms of $\mathfrak{S}_6$, in agreement with our knowledge of the group of Tutte's 8-cage. The ten points on the quadric provide a representation for this interesting group by permutations of degree ten.

## 7. TWO HEXASTIGMS IN $PG(5, 3)$

The existence of collineations interchanging positive and negative points suggests that the non-ruled quadric, which we derived from the hexastigm *123456*, could have been derived equally well from another hexastigm *abcdef*. In other words, since the symbols *1*, ..., *6*, which denote points, also denote negative pentagons, surely the analogous symbols $a, ..., f$, which denote positive pentagons, should also denote points!

A somewhat similar problem arises two dimensions lower, where a quadrangular set of six points on a line (in a real or complex projective plane) can be derived from a complete quadrangle (or 'tetrastigm') in two distinct ways. Möbius (1828, p. 443) derived the quadrangular set from two quadrangles in distinct planes, and thereby obtained a three-dimensional configuration of eight points which form two quadrangles in four ways and also form two mutually inscribed tetrahedra in four ways.

Analogously, let us embed our *PG*(4, 3) in a *PG*(5, 3) and use two hexastigms, *123456* and *abcdef*, in distinct 4-spaces whose common 3-space contains the non-ruled quadric. The thirty points $Q_{12}, ..., Q_{56}, Q_{ab}, ..., Q_{ef}$, which lie on the edges and transversals of the first hexastigm, may be expected to lie on the transversals and edges of the second. For this purpose, we take an arbitrary point $f$ outside the 4-space *12345*, and use *12345f* as a simplex of reference.

As we saw in §4, the point $Q_{af}$ or (16, 25, 34) is

$$P_{25} - P_{34} = 2 - 3 - 4 + 5.$$

Equating this to $a - f$, and dealing similarly with $Q_{bf}, Q_{cf}, Q_{df}, Q_{ef}$ (which are derived from $Q_{af}$ by cyclically permuting 12345), we define the remaining vertices of the new hexastigm as follows:

$$\left. \begin{aligned} a &= 2-3-4+5+f, \\ b &= 3-4-5+1+f, \\ c &= 4-5-1+2+f, \\ d &= 5-1-2+3+f, \\ e &= 1-2-3+4+f. \end{aligned} \right\} \tag{7.1}$$

The fact that all six points lie in the same 4-space is seen in the consequent syzygy

$$a+b+c+d+e+f = 0$$

(which, like (3.1), is to be interpreted as a congruence mod 3). We can immediately deduce expressions for all the forty points of the 3-space *12345.abcde* in terms of $a, ..., f$ instead of *1, ..., 6*:

$$a-b = 2+3-5-1 = (15, 23, 46) = Q_{ab},$$
$$b-c = 3+4-1-2 = (12, 34, 56) = Q_{bc},$$
$$c-d = 4+5-2-3 = (16, 23, 45) = Q_{cd},$$
$$d-e = 5+1-3-4 = (15, 26, 34) = Q_{de},$$
$$e-a = 1+2-4-5 = (12, 36, 45) = Q_{ae},$$

$$a-c = 1+4-3-5 = (14, 26, 35) = Q_{ac},$$
$$b-d = 2+5-4-1 = (14, 25, 36) = Q_{bd},$$
$$c-e = 3+1-5-2 = (13, 25, 46) = Q_{ce},$$
$$d-a = 4+2-1-3 = (13, 24, 56) = Q_{ad},$$
$$e-b = 5+3-2-4 = (16, 24, 35) = Q_{be},$$

$$Q_{12} = \quad 1-2 \qquad\quad = (1+4-3-5)+(5+3-2-4) = a-c+e-b,$$
$$Q_{13} = \quad 1-3 \qquad\quad = (4+5-2-3)+(1+2-4-5) = c-d+e-a,$$
$$Q_{16} = -1+2+3+4+5 = (1+2-4-5)+(3-4-5+1) = e-a+b-f,$$
$$a+b+c = -2-4-5 = 1+3+6 = P_{136},$$

and so on. We see, in particular, that the points $Q_{ab}$, ..., $Q_{ac}$, ... do indeed lie on edges of the new hexastigm *abcdef*, and the points

$$Q_{12} = (ae, bc, df), \quad ..., \quad Q_{13} = (ad, bf, ce), \quad ...$$

on its transversals.

### 8. SIXTY-SIX PAIRS OF HEXASTIGMS

We soon find that the twelve points

$$1, \quad 2, \quad 3, \quad 4, \quad 5, \quad 6, \quad a, \quad b, \quad c, \quad d, \quad e, \quad f, \tag{8.1}$$

in $PG(5, 3)$, form a configuration which is far more symmetrical than our construction for it would lead us to expect. In fact, our designation of the points by six numbers and six letters is merely one of many equally valid partitions into two sets of six. The first step in this direction is the observation that the above syzygies, and any others that we may obtain by combining them, all involve six (or more) of the twelve points. This means that no five of the points lie in a 3-space. The complete story unfolds when we write out all the syzygies that involve just six points, as in table 2. Every such syzygy indicates that six points lie in a 4-space, i.e. that the 4-space spanned by any five of the six contains the remaining one also. After the first row in the table, the syzygies are conveniently arranged in sets of five, related by the permutation (3.1). (The implied special role of *6* and *f* is easily seen to be illusory.) Each row contains two 'complementary' syzygies, to exhibit the sixty-six ways in which the twelve points lie in two hyperplanes, six in each.

TABLE 2.   132 CONGRUENCES (MOD 3) FORMING A STEINER SYSTEM $\mathfrak{S}(5, 6, 12)$

| | |
|---|---|
| $1+2+3+4+5+6 = 0$ | $a+b+c+d+e+f = 0$ |
| | |
| $1+2+3+a+c+f = 0$ | $4+5+6+b+d+e = 0$ |
| $2+3+4+b+d+f = 0$ | $5+1+6+c+e+a = 0$ |
| $3+4+5+c+e+f = 0$ | $1+2+6+d+a+b = 0$ |
| $4+5+1+d+a+f = 0$ | $2+3+6+e+b+c = 0$ |
| $5+1+2+e+b+f = 0$ | $3+4+6+a+c+d = 0$ |
| | |
| $1+2+4+c+d+e = 0$ | $3+5+6+a+b+f = 0$ |
| $2+3+5+d+e+a = 0$ | $4+1+6+b+c+f = 0$ |
| $3+4+1+e+a+b = 0$ | $5+2+6+c+d+f = 0$ |
| $4+5+2+a+b+c = 0$ | $1+3+6+d+e+f = 0$ |
| $5+1+3+b+c+d = 0$ | $2+4+6+e+a+f = 0$ |
| | |
| $1+2+3 = b+d+e$ | $4+5+6 = a+c+f$ |
| $2+3+4 = c+e+a$ | $5+1+6 = b+d+f$ |
| $3+4+5 = d+a+b$ | $1+2+6 = c+e+f$ |
| $4+5+1 = e+b+c$ | $2+3+6 = d+a+f$ |
| $5+1+2 = a+c+d$ | $3+4+6 = e+b+f$ |
| | |
| $1+2+4 = a+b+f$ | $3+5+6 = c+d+e$ |
| $2+3+5 = b+c+f$ | $4+1+6 = d+e+a$ |
| $3+4+1 = c+d+f$ | $5+2+6 = e+a+b$ |
| $4+5+2 = d+e+f$ | $1+3+6 = a+b+c$ |
| $5+1+3 = e+a+f$ | $2+4+6 = b+c+d$ |

TABLE 2 *(cont.)*

$$3+4+a = 5+2+f$$
$$4+5+b = 1+3+f$$
$$5+1+c = 2+4+f$$
$$1+2+d = 3+5+f$$
$$2+3+e = 4+1+f$$

$$5+2+a = 1+6+f$$
$$1+3+b = 2+6+f$$
$$2+4+c = 3+6+f$$
$$3+5+d = 4+6+f$$
$$4+1+e = 5+6+f$$

$$1+6+a = 3+4+f$$
$$2+6+b = 4+5+f$$
$$3+6+c = 5+1+f$$
$$4+6+d = 1+2+f$$
$$5+6+e = 2+3+f$$

$$5+1+a = 2+3+b$$
$$1+2+b = 3+4+c$$
$$2+3+c = 4+5+d$$
$$3+4+d = 5+1+e$$
$$4+5+e = 1+2+a$$

$$2+3+a = 4+6+b$$
$$3+4+b = 5+6+c$$
$$4+5+c = 1+6+d$$
$$5+1+d = 2+6+e$$
$$1+2+e = 3+6+a$$

$$4+6+a = 5+1+b$$
$$5+6+b = 1+2+c$$
$$1+6+c = 2+3+d$$
$$2+6+d = 3+4+e$$
$$3+6+e = 4+5+a$$

$$4+1+a = 2+6+c$$
$$5+2+b = 3+6+d$$
$$1+3+c = 4+6+e$$
$$2+4+d = 5+6+a$$
$$3+5+e = 1+6+b$$

$$2+6+a = 3+5+c$$
$$3+6+b = 4+1+d$$
$$4+6+c = 5+2+e$$
$$5+6+d = 1+3+a$$
$$1+6+e = 2+4+b$$

$$3+5+a = 4+1+c$$
$$4+1+b = 5+2+d$$
$$5+2+c = 1+3+e$$
$$1+3+d = 2+4+a$$
$$2+4+e = 3+5+b$$

$$6+c+d = 1+e+b$$
$$6+d+e = 2+a+c$$
$$6+e+a = 3+b+d$$
$$6+a+b = 4+c+e$$
$$6+b+c = 5+d+a$$

$$3+d+e = 4+b+c$$
$$4+e+a = 5+c+d$$
$$5+a+b = 1+d+e$$
$$1+b+c = 2+e+a$$
$$2+c+d = 3+a+b$$

$$2+b+d = 5+c+e$$
$$3+c+e = 1+d+a$$
$$4+d+a = 2+e+b$$
$$5+e+b = 3+a+c$$
$$1+a+c = 4+b+d$$

$$6+c+e = 4+d+f$$
$$6+d+a = 5+e+f$$
$$6+e+b = 1+a+f$$
$$6+a+c = 2+b+f$$
$$6+b+d = 3+c+f$$

$$1+c+f = 5+d+e$$
$$2+d+f = 1+e+a$$
$$3+e+f = 2+a+b$$
$$4+a+f = 3+b+c$$
$$5+b+f = 4+c+d$$

$$2+e+f = 3+c+d$$
$$3+a+f = 4+d+e$$
$$4+b+f = 5+e+a$$
$$5+c+f = 1+a+b$$
$$1+d+f = 2+b+c$$

$$3+e+b = 5+d+f$$
$$4+a+c = 1+e+f$$
$$5+b+d = 2+a+f$$
$$1+c+e = 3+b+f$$
$$2+d+a = 4+c+f$$

$$4+e+f = 1+b+d$$
$$5+a+f = 2+c+e$$
$$1+b+f = 3+d+a$$
$$2+c+f = 4+e+b$$
$$3+d+f = 5+a+c$$

$$6+b+f = 2+d+e$$
$$6+c+f = 3+e+a$$
$$6+d+f = 4+a+b$$
$$6+e+f = 5+b+c$$
$$6+a+f = 1+c+d$$

Since the 66 pairs of hexastigms determine 66 non-ruled quadrics (and 8-cages) lying in distinct 3-spaces, the twelve points form a configuration whose group is of order

$$66.1440 = 95040 = 12.11.10.9.8.$$

Since every five of the points appear together in just one of the hyperplanes, the configuration is a Steiner system $\mathfrak{S}(5, 6, 12)$ (Witt 1938$a$). Since there is only one Steiner system $\mathfrak{S}(5, 6, 12)$ (Witt 1938$b$), it follows that this group is the quintuply transitive group $M_{12}$ (Mathieu 1861, p. 274), which is the only simple group of order 95040 (Stanton 1951).

We have thus found a geometrical justification for a remark of Carmichael (1937, p. 432): 'that the Mathieu group of degree 12 contains a subgroup of order 10.9.8 each element of which leaves fixed a given one of the 132 sextuples. This subgroup is intransitive, having two transitive constituents each of degree 6. It thus sets up a simple isomorphism of the symmetric group of degree 6 with itself.... This outer isomorphism is therefore an essential element in the structure of the Mathieu group of degree 12.'

It appears, incidentally, that Carmichael anticipated Witt's discovery of the connexion between $M_{12}$ and the Steiner system (though he did not use that name for it).

## 9. TWO MUTUALLY INSCRIBED SIMPLEXES

The syzygies (7·1) and (3·1) show that each vertex of the simplex $abcde6$ lies in a hyperplane (or 'face') of the simplex $12345f$. The complementary syzygies

$$1 = -b + c + d - e + 6,$$
$$2 = -c + d + e - a + 6,$$
$$3 = -d + e + a - b + 6,$$
$$4 = -e + a + b - c + 6,$$
$$5 = -a + b + c - d + 6,$$
$$f = -a - b - c - d - e$$

show that each vertex of $12345f$ lies in a hyperplane of $abcde6$. It follows that the two simplexes

$$12345f \quad \text{and} \quad abcde6$$

are mutually inscribed, like Möbius's two tetrahedra (§ 7). However, there is one interesting difference, as we shall see.

The quadrics with respect to which Möbius's tetrahedra are polar (Edge 1936, p. 349; Benneton 1947, p. 30) do not belong to the bundle of quadrics through their vertices (which are eight associated points). But our two mutually inscribed simplexes admit a quadric with respect to which each vertex of either simplex is the pole of the incident hyperplane of the other; therefore the quadric passes through the twelve points (cf. Baker 1937, p. 36).

In terms of co-ordinates for which *12345f* is the simplex of reference, the vertices of the other simplex *abcde6* may be extracted from (7·1) as follows:

$$( \ \ 0, \ \ \ 1, -1, -1, \ \ \ 1, \ \ 1),$$
$$( \ \ 1, \ \ \ 0, \ \ \ 1, -1, -1, \ \ 1),$$
$$(-1, \ \ \ 1, \ \ \ 0, \ \ \ 1, -1, \ \ 1),$$
$$(-1, -1, \ \ \ 1, \ \ \ 0, \ \ \ 1, \ \ 1),$$
$$( \ \ 1, -1, -1, \ \ \ 1, \ \ \ 0, \ \ 1),$$
$$( \ \ 1, \ \ \ 1, \ \ \ 1, \ \ \ 1, \ \ \ 1, \ \ 0).$$

Since this, like the simplex of reference, is a self-polar simplex with respect to the quadric

$$x_1^2 + x_2^2 + x_3^2 + x_4^2 + x_5^2 + x_6^2 = 0$$

(so that the six vertices, regarded as vectors, are mutually orthogonal), the same scheme provides tangential co-ordinates for the six hyperplanes, and also provides the matrix

$$\begin{pmatrix} 0 & 1 & -1 & -1 & 1 & 1 \\ 1 & 0 & 1 & -1 & -1 & 1 \\ -1 & 1 & 0 & 1 & -1 & 1 \\ -1 & -1 & 1 & 0 & 1 & 1 \\ 1 & -1 & -1 & 1 & 0 & 1 \\ 1 & 1 & 1 & 1 & 1 & 0 \end{pmatrix} \qquad (9\cdot1)$$

for the polarity by which these hyperplanes can be derived from the vertices of the simplex of reference. The self-conjugate points in this polarity constitute the quadric

$$x_1x_2 + x_2x_3 + x_3x_4 + x_4x_5 + x_5x_1 - x_1x_3 - x_2x_4 - x_3x_5 - x_4x_1 - x_5x_2$$
$$+ x_1x_6 + x_2x_6 + x_3x_6 + x_4x_6 + x_5x_6 = 0, \quad (9\cdot2)$$

which can be expressed as

$$y_1^2 + y_2^2 + y_3^2 + y_4^2 + y_5^2 + y_6^2 = 0 \qquad (9\cdot3)$$

in terms of new co-ordinates

$$y_1 = \ \ \ x_1 + x_2 - x_3 - x_4 + x_5 + x_6,$$
$$y_2 = \ \ \ x_1 + x_2 + x_3 - x_4 - x_5 + x_6,$$
$$y_3 = -x_1 + x_2 + x_3 + x_4 - x_5 + x_6,$$
$$y_4 = -x_1 - x_2 + x_3 + x_4 + x_5 + x_6,$$
$$y_5 = \ \ \ x_1 - x_2 - x_3 + x_4 + x_5 + x_6,$$
$$y_6 = \ \ \ x_1 + x_2 + x_3 + x_4 + x_5 + x_6.$$

Dickson (1901, p. 158) and Edge (1955, p. 134) showed that *PG*(5, 3) (and, in fact, *PG*(n, 3) for any odd *n*) admits two types of non-degenerate quadric: the number of negative squares in its equation is odd for one type, even for the other. Edge was entirely concerned with the first type: quadrics containing two systems of planes.

But the quadric (9·2) or (9·3), which reciprocates two mutually inscribed simplexes into each other, is of the second type. To see that it contains lines, but no planes, we may examine its section by the tangent hyperplane

$$y_1+y_2+y_3+y_4+y_5+y_6 = 0.$$

This section is evidently a cone, joining the point of contact $(1, 1, 1, 1, 1, 1)$ to the non-ruled quadric

$$y_1^2+y_2^2+y_3^2+y_4^2+y_5^2 = 0$$

in the 3-space

$$y_1+y_2+y_3+y_4+y_5 = y_6 = 0 \quad \text{(see § 5)}.$$

We have observed that every five of the twelve points (8·1) determine a hyperplane which contains just one more point of the set. Each of the remaining six points forms a simplex with the original five. Since also each simplex contains six sets of five, there are altogether

$$\binom{12}{5} = 792$$

simplexes. In other words, the configuration may be regarded in 396 ways as a pair of 'Möbius' simplexes.

For instance, the five points *12345* form a simplex not only with *f* but equally well with *a* or *b* or *c* or *d* or *e*. The simplex *12345a* has the companion *fcebd6*, since the respective hyperplanes

$$2345a, \quad 1345a, \quad 1245a, \quad 1235a, \quad 1234a, \quad 12345$$

(see table 2) contain

$$f, \qquad c, \qquad e, \qquad b, \qquad d, \qquad 6.$$

Since the 132 syzygies occur in complementary pairs, it follows automatically that the vertices of the former simplex lie in the corresponding hyperplanes of the latter.

Of course, the 396 pairs of simplexes are related by 396 quadrics such as (9·2). For instance, the quadric for the pair

$$12345a, \quad fcebd6 \qquad\qquad\qquad (9\cdot4)$$

is $\quad x_1x_2+x_2x_4+x_4x_3+x_3x_5+x_5x_1-x_1x_3-x_2x_5-x_4x_1-x_3x_2-x_5x_4+x_1x_6 = 0.$

## 10. Involutory generators for $M_{12}$

Instead of regarding (9·1) as a polarity, we may equally well regard it as an involutory collineation. This is an automorphism of the configuration, namely the permutation (1·1) of the twelve points.

The matrix itself is of period four: it transforms the 'weighted' point *1* into *a*, *a* into $-1$, $-1$ into $-a$, $-a$ into *1*, and similarly for the remaining symbols. This transformation is clearly visible in table 2, where the syzygy

$$1+2+3+a+c+f = 0$$

yields $a+b+c-1-3-6 = 0$, which appears in the table as

$$1+3+6 = a+b+c.$$

The 396 pairs of simplexes are related by 396 such collineations of period two; e.g. the collineation for the pair (9·4) is

$$(1f) \ (2c) \ (3e) \ (4b) \ (5d) \ (6a),$$

and the collineation for the pair *123abc, d6e4f5* is

$$(1d) \ (26) \ (3e) \ (4a) \ (5c) \ (bf).$$

These 396 permutations generate a normal subgroup of $M_{12}$. But $M_{12}$ is simple. Therefore they generate the whole group. The corresponding matrices (of period 4) generate a group of order 190080 which may reasonably be called a *binary* Mathieu group (by analogy with the binary polyhedral groups, which arise from the ordinary polyhedral groups when their rotations are replaced by quaternions).

Perhaps an energetic reader will select three of the 396 collineations in such a way as to generate $M_{12}$ with an elegant set of defining relations. Since these elements of $M_{12}$ are of period 2, two of them can only generate a dihedral group. The conjecture that three (suitably chosen) will suffice for the whole group is suggested by Carmichael (1937, p. 151, Ex. 12), whose generator $U$ can evidently be identified with one of our involutory collineations.

## REFERENCES

Baker, H. F. 1929 *Principles of geometry*, 1. Cambridge University Press.
Baker, H. F. 1930 *Principles of geometry*, 2. Cambridge University Press.
Baker, H. F. 1937 *Ann. Mat. pura. appl.* (4), **16**, 35.
Baker, H. F. 1940 *Principles of geometry*, 4. Cambridge University Press.
Baker, H. F. 1946 *A locus with 25920 linear self-transformations.* Cambridge University Press.
Benneton, G. 1947 *Ann. École Norm.* (3), **64**, 1.
Carmichael, R. D. 1937 *Groups of finite order.* Boston: Ginn and Company.
Coxeter, H. S. M. 1950 *Bull. Amer. Math. Soc.* **56**, 413.*
Coxeter, H. S. M. 1958 *Canad. J. Math.* **10**, 484.
Dickson, L. E. 1901 *Linear groups, with an exposition of the Galois field theory.* Leipzig: Teubner.
Edge, W. L. 1936 *Proc. Lond. Math. Soc.* (2), **41**, 337.
Edge, W. L. 1954 *Proc. Roy. Soc.* A, **222**, 262.
Edge, W. L. 1955 *Proc. Roy. Soc.* A, **228**, 129.
Mathieu, E. 1861 *J. de Math.* (2), **6**, 241.
Möbius, A. F. 1828 *Gesammelte Werke*, **1** (1886).
Richmond, H. W. 1900 *Quart. J. Math.* **31**, 125.
Stanton, R. G. 1951 *Canad. J. Math.* **3**, 164.
Sylvester, J. J. 1844 *Collected mathematical papers*, **1**. Cambridge University Press (1904).
Sylvester, J. J. 1861 *Collected mathematical papers*, **2**. Cambridge University Press (1908).
Tutte, W. T. 1958 *Canad. J. Math.* **10**, 481.
Whitehead, A. N. 1898 *A treatise on universal algebra*, **1**. Cambridge University Press.
Witt, E. 1938*a* *Abh. math. Sem. Univ. Hamburg* **12**, 256.
Witt, E. 1938*b* *Abh. math. Sem. Univ. Hamburg* **12**, 265.

*Reference "Coxeter 1950" is Chapter 6 of this book.

# 8

# ARRANGEMENTS
# OF EQUAL SPHERES
# IN NON-EUCLIDEAN SPACES

Reprinted, with permission of the publisher,
from *Acta Mathematica Acad. Sci. Hungaricae*,
Tomus V (1954).

# 1. Introduction

We consider two kinds of arrangements of equal spheres: *packings*, where no point is inside more than one sphere, and *coverings*, where no point is outside every sphere. Each arrangement determines a honeycomb whose cells are called *Dirichlet regions*. The Dirichlet region for any particular sphere is a polytope whose interior consists of all points that are nearer to the centre of that sphere than to the centre of any other [13, pp. 216—219]* If all the Dirichlet regions are of the same size, the ratio of the content of the sphere to the content of the Dirichlet region is called the *density* of the arrangement. If they are of various sizes, we use a suitable average. We are interested in cases where the density is as near to 1 as possible: close packings and thin coverings. In such cases the spheres are respectively inscribed in, and circumscribed about, the Dirichlet regions.

The two-dimensional results were obtained by FEJES TÓTH, who proved that the elliptic plane cannot be packed as closely, nor covered as thinly, as the Euclidean plane, and that no arrangements of finite circles in the hyperbolic plane are as "good" as the best arrangement of horocycles. Elliptic 3-space presents a different state of affairs: sixty spheres of radius $\pi$ 10, each touching twelve others, form a packing of density 0,774, and somewhat larger spheres (having the same centres) form a covering of density 1,439, whereas in Euclidean space the maximum packing-density and minimum covering density are almost certainly 0,740 and 1,464 [14, pp. 171, 174].

In hyperbolic 3-space, no superior arrangements of finite spheres are known, but horospheres provide a packing of density 0,853 and a covering of density 1,280.

Finally, hyperbolic 4-space admits remarkable arrangements of finite spheres: a packing of density 0,691 and a covering of density 1,558, whereas the best densities known for Euclidean 4-space are 0,617 and 1,766.

In the course of these investigations we use some results on hyberbolic geometry (§§ 4 and 5) that may also be of some intrinsic interest.

*For references, see page 178.

## 2. Two-dimensional arrangements

Most of the results in this section have already been given in a different notation by SCHLEGEL [20, p. 360] and FEJES TÓTH [15; 16], but are included here for the sake of completeness.

Let $\{p, q\}$ denote the regular tessellation of $p$-gons, $q$ at each vertex, filling a Euclidean or non-Euclidean plane. Consider the characteristic triangle $P_0 P_1 P_2$ formed by a vertex, the mid-point of one of the edges meeting at this vertex, and the centre of either of the faces meeting at this edge. Since the angles at $P_0, P_1, P_2$ are

$$\pi/q, \quad \pi/2, \quad \pi/p$$

the plane must be elliptic, Euclidean, or hyperbolic according as the number

$$2p + 2q - pq = 4 - (p-2)(q-2)$$

is positive, zero, or negative.

Let $\psi$ and $\chi$ denote the in-radius and circum-radius of a face, so that

$$\psi = P_1 P_2, \quad \chi = P_0 P_2.$$

The formulae for a right-angled triangle yield

$$\cos \psi = \operatorname{cosec} \frac{\pi}{p} \cos \frac{\pi}{q}, \quad \cos \chi = \operatorname{ctg} \frac{\pi}{p} \operatorname{ctg} \frac{\pi}{q}$$

in the elliptic case [9, pp. 21, 24], and the same expressions for $\operatorname{ch} \psi$, $\operatorname{ch} \chi$ in the hyperbolic case.

It is known [14, p. 58] that the in- and circum-circles of the faces of the Euclidean tessellation $\{6, 3\}$ form the closest packing and thinnest covering of equal circles in the Euclidean plane. Similarly, the in- and circum-circles of the faces of $\{p, 3\}$ ($p < 6$ or $p > 6$), in the elliptic or hyperbolic plane, form a closer packing and a thinner covering than any other arrangement of circles of respective radii $\psi$ and $\chi$, where

$$\cos \psi \text{ or } \operatorname{ch} \psi = \frac{1}{2} \operatorname{cosec} \frac{\pi}{p}, \quad \cos \chi \text{ or } \operatorname{ch} \chi = 3^{-\frac{1}{2}} \operatorname{ctg} \frac{\pi}{p}.$$

Clearly, the face of $\{p, 3\}$ is a $p$-gon of angle $2\pi/3$. Comparing its area

$$|p - 6| \pi/3$$

with the area

$$2\pi \left| \begin{matrix} \cos \\ \operatorname{ch} \end{matrix} \psi - 1 \right| = \pi \left| \operatorname{cosec} \frac{\pi}{p} - 2 \right|$$

of its in-circle and with the area

$$2\pi \left| \begin{matrix} \cos \\ \operatorname{ch} \end{matrix} \chi - 1 \right| = \frac{2\pi}{\sqrt{3}} \left| \operatorname{ctg} \frac{\pi}{p} - \sqrt{3} \right|$$

of its circum-circle, we obtain the densities

$$d(p) = \frac{3}{p-6} \left( \operatorname{cosec} \frac{\pi}{p} - 2 \right) \quad \text{and} \quad D(p) = \frac{2\sqrt{3}}{p-6} \left( \operatorname{ctg} \frac{\pi}{p} - \sqrt{3} \right)$$

for the packing and covering, respectively. Taking the limits as $p$ tends to 6, we obtain the densities

$$d(6) = \frac{\pi}{2\sqrt{3}} = 0{,}9069\ldots \quad \text{and} \quad D(6) = \frac{2\pi}{3\sqrt{3}} = 1{,}2092\ldots$$

for the closest packing and thinnest covering of the Euclidean plane. Taking instead the limits as $p$ tends to infinity, we obtain the densities

$$d(\infty) = \frac{3}{\pi} = 0{,}9549\ldots \quad \text{and} \quad D(\infty) = \frac{2\sqrt{3}}{\pi} = 1{,}1026\ldots$$

for the packing and covering of the hyperbolic plane by the in- and circumscribed horocycles of the faces of the limiting tessellation $\{\infty, 3\}$, whose rotation group is the modular group $S^3 = T^2 = 1$ [12, p. 425; 15, p. 107; 16, p. 113].

Since $d(p)$ is an increasing function, every packing of equal circles in the elliptic plane (or of equal small circles on a sphere) has density less than $d(6)$, and every packing of equal circles in the hyberbolic plane [15, p. 106] has density less than $d(\infty)$.

Since $D(p)$ is a decreasing function, every covering of the elliptic plane by equal circles (or of a sphere by equal small circles) has density greater than $D(6)$, and every covering of the hyperbolic plane by equal circles [16, p. 112] has density greater than $D(\infty)$.

## 3. Three-dimensional arrangements

The symbol $\{p, q\}$, which we have used for a spherical tessellation when $(p-2)(q-2) < 4$, is naturally used also for the corresponding regular polyhedron (or Platonic solid) in Euclidean or non-Euclidean space. Repetitions of this solid, of such a size that $r$ of them can be fitted round a common edge, form a regular honeycomb $\{p, q, r\}$ [9, p. 138].

Consider the *characteristic tetrahedron* $P_0 P_1 P_2 P_3$ formed by a vertex, the mid-point of an incident edge, the centre of an incident plane face, and the centre of an incident solid cell. Since the dihedral angles at the edges $P_2 P_3$, $P_0 P_3$, $P_0 P_1$ are

$$\pi/p, \quad \pi/q, \quad \pi/r,$$

while the remaining three are right angles, the space must be elliptic, Euclidean, or hyperbolic according as

$$\sin\frac{\pi}{p} \sin\frac{\pi}{r} - \cos\frac{\pi}{q}$$

is positive, zero, or negative [9, p. 135].

The in-radius and circum-radius of a cell are evidently

$$\psi = P_2 P_3, \quad \chi = P_0 P_3$$

[9, p. 139]. When the space is hyperbolic, we might allow the vertices to lie on the absolute quadric, so that $\chi$ is infinite; e. g., the tetrahedron, octahedron and hexahedron described in an earlier paper [6, p. 29] are cells of $\{3, 3, 6\}$, $\{3, 4, 4\}$ and $\{4, 3, 6\}$, respectively.

The problems of close packing and thin covering in Euclidean space have not yet been completely solved. The difficulty may be attributed to the fact that, since the equation

$$\sin\frac{\pi}{p}\ \sin\frac{\pi}{3}\ -\cos\frac{\pi}{3}=0$$

yields a value of $p$ between 5 and 6, there is no Euclidean honeycomb of type $\{p, 3, 3\}$. (There is an obviously closest arrangement for four solid spheres, but this cannot be regularly continued.)

This difficulty disappears in the non-Euclidean spaces, where we have the elliptic (or spherical) $\{5, 3, 3\}$ and the hyperbolic $\{6, 3, 3\}$ [12, pp. 426, 428]. The former consists of 60 regular dodecahedra $\{5, 3\}$ filling the elliptic space (or 120 filling the spherical space) [9, pp. 138, 176, 273; 10, p. 116; 11, p. 471; 12, p. 425]. The in- and circum-spheres of these dodecahedra provide a packing of spheres of radius

$$\psi = \pi/10$$

[9, p. 293] and a covering by spheres of radius

$$\chi = \frac{1}{3}\pi - \eta$$

where

$$\eta = \frac{1}{2}\ \text{arc cos}\ \frac{1}{4}\ .$$

Dividing the volumes

$$\pi(2\psi - \sin 2\psi)\quad \text{and}\quad \pi(2\chi - \sin 2\chi)$$

of these spheres by the volume

$$\pi^2/60$$

of the dodecahedral cell, we obtain the density

$$\frac{60}{\pi}(2\psi - \sin 2\psi) = \frac{60}{\pi}\left(\frac{\pi}{5} - \sin\frac{\pi}{5}\right) = 12\left(1 - \frac{5}{\pi}\ \sin\frac{\pi}{5}\right) = 0{,}774\ldots$$

for the packing, and

$$\frac{60}{\pi}(2\chi - \sin 2\chi) = \frac{60}{\pi}\left(\frac{2\pi}{3} - 2\eta - \frac{\sqrt{3}}{4}\ \tau\right) = 40 - \frac{15}{\pi}(8\eta + \sqrt{3}\tau) = 1{,}439\ldots$$

$\left(\text{where}\ \tau = \dfrac{\sqrt{5}+1}{2}\right)$ for the covering.

Since the cells are dodecahedra, each sphere in the packing touches twelve others [14, p. 177, Fig. 121]: the same number as in the familiar close-packings of Euclidean space [14, pp. 171—172, Figs. 116, 118], whose density

$$\frac{\pi}{3\sqrt{2}} = 0,74048\ldots$$

is probably the greatest that can be maintained throughout the whole space, although greater densities are possible in a tubular region [2, pp. 308—310]. Fejes Tóth [14, pp. 175—176] has shown that the density of the closest possible packing is almost certainly less than

$$\frac{\pi}{14} 5^{-1/4} \tau^{7/2} = 0,7545\ldots,$$

an amount still far short of the above elliptic density, 0,774.... 

Similarly, the elliptic covering density, 1,439..., compares favourably with

$$\frac{5\sqrt{5}\pi}{24} = 1,464\ldots,$$

which is the density of the covering of Euclidean space by the circumspheres of the cells (truncated octahedra) of the honeycomb $t_{1,2}\delta_4$ [7, p. 403; 14, p. 174]. Bambah [1, p. 26] has proved that this covering, in which each sphere intersects fourteen others, is the thinnest covering by spheres whose centres form a lattice (in this case, the body-centred cubic lattice); but the possibility of thinner irregular coverings remains open.

### 4. A digression on the hyperbolic sine and tangent

Before investigating $\{6, 3, 3\}$, we need two lemmas about hyperbolic geometry. The first, which was proved quite differently by Sommerville [21, pp. 62, 77], may be regarded as the hyperbolic counterpart of the Euclidean

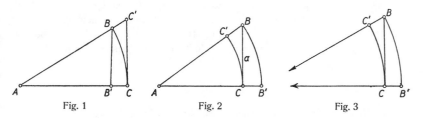

| Fig. 1 | Fig. 2 | Fig. 3 |

definition for the sine and tangent of an arc of a unit circle: if lines $BB'$ and $CC'$ are drawn through the ends of an arc $BC$, perpendicular to the radius $AC$, so that $B'$ lies on $AC$, and $C'$ on $AB$ produced (as in Fig. 1), then

$$BB' = \sin BC, \quad CC' = \operatorname{tg} BC.$$

In the hyperbolic plane, let arcs $BB'$ and $CC'$ of two circles with centre $A$ be drawn through the vertices $B$ and $C$ of a right triangle $ABC$, so that $B'$ lies on $AC$ produced, and $C'$ on $AB$ (as in Fig. 2). The sides $a = BC$, $b = CA$, $c = AB$ are connected by the classical formulae [8, p. 238]

$$\text{sh } a = \sin A \text{ sh } c, \quad \text{th } a = \text{tg } A \text{ sh } b,$$

while the two arcs (of radii $c$ and $b$) are

(4. 1) $$BB' = A \text{ sh } c, \quad CC' = A \text{ sh } b.$$

Hence

$$BB' = \frac{A}{\sin A} \text{ sh } a, \quad CC' = \frac{A}{\text{tg } A} \text{ th } a.$$

Taking the limit as $A$ tends to zero, so that the triangle becomes singly-asymptotic (as in Fig. 3), we obtain the desired lemma:

*If $BB'$ and $C'C$ are corresponding arcs of two concentric horocycles, so situated that the tangent at $C$ passes through $B$, then*

$$BB' = \text{sh } BC, \quad CC' = \text{th } BC.$$

## 5. The volume of a sector of a horosphere

Our second lemma is the solid analogue of another theorem of plane hyperbolic geometry: the area of a horocyclic sector is equal to its arc [8, p. 250].

Here it is to be understood that we are measuring length and area in terms of the natural units of LOBATSCHEFSKY: the unit arc of a horocycle is such that the tangent at one end is parallel to the diameter through the other end (produced outwards), and the unit area is bounded by this arc and the (parallel) diameters at its two ends.

Consider the "rectangle" bounded by two corresponding arcs of concentric horocycles and the intercepted segments of the diameters through their ends (as in Fig. 4). If the straight sides are both $x$, while the curved sides are $s$ and $s_x$ ($s > s_x$), we have from (4. 1)

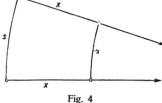

Fig. 4

(5. 1) $$\frac{s_x}{s} = \lim_{c \to \infty} \frac{\text{sh}(c-x)}{\text{sh } c} = e^{-x}$$

[3, p. 95]. It follows that the area of the horocyclic sector is

$$\int_0^\infty s_x \, dx = s \int_0^\infty e^{-x} \, dx = s.$$

Passing now to hyperbolic space, we observe that the surface of a horosphere is developable on the Euclidean plane [4, p. 205; 8, pp. 220, 251].

Accordingly, the natural unit of area on a horosphere is the "square" formed by four unit horocyclic arcs. Consider the "solid sector" bounded by this "square" and portions of the diametral planes through its sides. The section of this "solid sector" by a concentric horosphere, distant $x$ from the bounding horosphere, is a "square" of side $e^{-x}$ (by (5.1) with $s = 1$). Hence the volume of the "solid sector" is

$$\int_0^\infty (e^{-x})^2 \, dx = \int_0^\infty e^{-2x} \, dx = \frac{1}{2}.$$

Since any area can be expressed as a numerical multiple of the unit square, we deduce the desired lemma:

*The volume of any "solid sector" of a horosphere is equal to one-half the area of its horospherical boundary.*

This result can evidently be generalized to $n$ dimensions, with $1/(n-1)$ in place of $\frac{1}{2}$.

## 6. Three-dimensional arrangements, continued

Since $\{6, 3\}$ is a Euclidean tessellation, the cell of the hyperbolic honeycomb $\{6, 3, 3\}$ is an infinite polyhedron having an inscribed horosphere and a concentric circumscribed horosphere. Analogy with the above discussion of $\{\infty, 3\}$ would suggest that every packing of equal spheres is less dense than the arrangement of horospheres inscribed in the cells of $\{6, 3, 3\}$, and that every covering of hyperbolic space by equal spheres is denser than the arrangement of horospheres circumscribed about these same cells.

Since the cells are infinite, the densities are most conveniently computed by comparing the singly-asymptotic characteristic tetrahedron $P_0 P_1 P_2 P_3$ ($P_3$ at infinity) with the corresponding solid sectors of the respective horospheres.

The inscribed horosphere, touching a hexagonal face of the cell at its centre $P_2$, cuts out from the tetrahedron $P_0 P_1 P_2 P_3$ a horospherical triangle $P_0' P_1' P_2$, whose angles are $\pi/3$, $\pi/2$, $\pi/6$. The plane triangle $P_0 P_1 P_2$ has likewise a right angle at $P_1$ and $\pi/6$ at $P_2$, while its side $P_0 P_1$ is

$$\varphi = \frac{1}{2} \log 2$$

[12, p. 427], so that th $\varphi = \frac{1}{3}$. Setting $a = P_0 P_1$, $b = P_1 P_2$, $A = \pi/6$ in the formula

$$\text{sh } b = \frac{\text{th } a}{\text{tg } A}$$

for a right-angled triangle $ABC$, we obtain

$$\text{sh } P_1 P_2 = \sqrt{\frac{1}{3}}.$$

Using the lemma in § 4, we see that this side $P_1 P_2$ is the tangent corresponding to the horocyclic arc

$$P_1' P_2 = \text{th } P_1 P_2 = \frac{1}{2}.$$

By Euclidean mensuration, the area of the horocyclic triangle $P_0' P_1' P_2$ is

$$\frac{1}{8\sqrt{3}}.$$

By the lemma in § 5, the volume of the corresponding solid sector is

$$\frac{1}{16\sqrt{3}}.$$

Similarly we have, on the circumscribed horosphere through $P_0$, a horospherical triangle $P_0 P_1'' P_2''$ whose angles are again $\pi/3$, $\pi/2$, $\pi/6$. The side $P_0 P_1 = \varphi$ of the corresponding plane triangle is the "sine" of the horocyclic arc

$$P_0 P_1'' = \text{sh } \varphi = \frac{1}{2\sqrt{2}}.$$

Hence the area of $P_0 P_1'' P_2''$ is $\sqrt{3}/16$, and the volume of the corresponding solid sector is

$$\frac{\sqrt{3}}{32}.$$

Using an observation of SCHLÄFLI [19, p. 162; see also 6, pp. 23, 27, 29], we obtain the volume of the singly-asymptotic tetrahedron $P_0 P_1 P_2 P_3$ in the form

$$\frac{i}{4} S\left(\frac{\pi}{3}, \frac{\pi}{3}, \frac{\pi}{6}\right) = \frac{i}{16} S\left(\frac{\pi}{6}, \frac{\pi}{6}, \frac{\pi}{6}\right) = \frac{i}{24} S\left(\frac{\pi}{3}, \frac{\pi}{3}, \frac{\pi}{3}\right) =$$

$$= \frac{\sqrt{3}}{32}\left(1 - \frac{1}{2^2} + \frac{1}{4^2} - \frac{1}{5^2} + \frac{1}{7^2} - \frac{1}{8^2} + \cdots\right) =$$

$$= \frac{1}{16\sqrt{3}}\left(1 + \frac{1}{2^2} - \frac{1}{4^2} - \frac{1}{5^2} + \frac{1}{7^2} + \frac{1}{8^2} - \cdots\right).$$

Hence the density of the packing is

$$\left(1 + \frac{1}{2^2} - \frac{1}{4^2} - \frac{1}{5^2} + \frac{1}{7^2} + \frac{1}{8^2} - \cdots\right)^{-1} = 0{,}853\cdots,$$

and the density of the covering is

$$\left(1 - \frac{1}{2^2} + \frac{1}{4^2} - \frac{1}{5^2} + \frac{1}{7^2} - \frac{1}{8^2} + \cdots\right)^{-1} = 1{,}280\cdots.$$

When P. SCHERK saw the ratio of these two series, he proved a generalization in which the squares are replaced by $s$th powers. Setting

$$f(s) = 1 - \frac{1}{2^s} + \frac{1}{4^s} - \frac{1}{5^s} + \frac{1}{7^s} - \frac{1}{8^s} + \cdots,$$

$$g(s) = 1 + \frac{1}{2^s} - \frac{1}{4^s} - \frac{1}{5^s} + \frac{1}{7^s} + \frac{1}{8^s} - \cdots,$$

he observed that, if $s > 1$,

$$g(s) - f(s) = 2\left(\frac{1}{2^s} - \frac{1}{4^s} + \frac{1}{8^s} - \frac{1}{10^s} + \frac{1}{14^s} - \frac{1}{16^s} + \cdots\right) = 2^{1-s} f(s),$$

whence

$$g(s) = (1 + 2^{1-s}) f(s)$$

and, in particular,

$$g(2) = \frac{3}{2} f(2).$$

## 7. Arrangements in hyperbolic 4-space

We saw in § 3 that, when

$$\sin \frac{\pi}{p} \sin \frac{\pi}{r} > \cos \frac{\pi}{q},$$

the honeycomb $\{p, q, r\}$ fills elliptic or spherical 3-space. Taking the latter interpretation, we use the symbol $g_{p,q,r}$ to denote the number of repetitions of the characteristic tetrahedron $P_0 P_1 P_2 P_3$ that will fill the whole spherical space, so that the volume of each is $2\pi^2/g_{p,q,r}$. This is a rather complicated function of $p, q, r$. For our present purpose it will suffice to note that

$$g_{3,3,3} = 120, \quad g_{5,3,3} = 120^2 = 14\,400$$

[9, pp. 133, 153].

The "Schläfli symbol" $\{p, q, r\}$ is naturally used also for the corresponding regular polytope in Euclidean or non-Euclidean 4-space. Repetitions of this polytope, of such a size that $s$ of them can be fitted round a common plane face, form a regular honeycomb $\{p, q, r, s\}$, whose *characteristic simplex* $P_0 P_1 P_2 P_3 P_4$ is formed by a vertex, the mid-point of an incident edge, and the centres of other incident elements of dimension 2, 3, 4 [9, p. 136]. Since the condition for the space to be hyperbolic is

$$\frac{\cos^2 \pi/q}{\sin^2 \pi/p} + \frac{\cos^2 \pi/r}{\sin^2 \pi/s} > 1,$$

the only instance having finite cells, with $s = 3$, is

$$\{5, 3, 3, 3\}.$$

Accordingly it is a natural conjecture that the closest packing and thinnest covering of hyperbolic 4-space are provided by the in- and circum-spheres of the cells of this honeycomb. Since $\{5, 3, 3\}$ is bounded by 120 dodeca-hedra, each sphere has 120 neighbours.

It was shown by SCHLÄFLI [19, p. 283; cf. 9, p. 233] that the content of the characteristic simplex for $\{p, q, r, s\}$ is

$$\frac{\pi^2}{6}\left(\frac{8}{g_{p,q,r}} + \frac{8}{g_{q,r,s}} + \frac{2}{ps} - \frac{1}{p} - \frac{1}{q} - \frac{1}{r} - \frac{1}{s} + 1\right).$$

In the present case this becomes

$$\frac{\pi^2}{6}\left(\frac{8}{14\,400} + \frac{8}{120} + \frac{2}{15} - \frac{1}{5}\right) = \frac{\pi^2}{10\,800}.$$

Multiplying by $g_{5,3,3} = 14\,400$, we deduce that the content of a cell $\{5, 3, 3\}$ of the honeycomb $\{5, 3, 3, 3\}$ is

$$\frac{4}{3}\,\pi^2.$$

This is to be compared with the contents of spheres (or rather, 4-dimensional hyperspheres), whose radii,

$$\psi = P_3 P_4 \quad \text{and} \quad \chi = P_0 P_4,$$

are given by the formulae

$$\text{sech}^2\,\psi = (3, 5)\,(-1, 4)/(-1, 3), \qquad \text{sech}^2\,\chi = (0, 5)\,(-1, 4)$$

[9, pp. 161, 162], where, in the case of $\{5, 3, 3, 3\}$,

$$(j, k) = k - j \qquad (0 \leqq j \leqq k)$$

and

$$(-1, k) = 1 - k\tau^{-3} \qquad \left(\tau = \frac{\sqrt{5}+1}{2},\ \tau^{-3} = \sqrt{5} - 2\right),$$

so that

$$(3, 5) = 2, \quad (-1, 4) = \tau^{-6}, \quad (-1, 3) = 2\tau^{-4}, \quad (0, 5) = 5,$$

$$\text{ch}\,\psi = \tau, \qquad \text{ch}\,\chi = 5^{-\frac{1}{2}}\,\tau^3.$$

A sphere of radius $\varrho$ in spherical 4-space has the same 3-dimensional content as one of radius $\sin\varrho$ in Euclidean 4-space, namely $2\pi^2 \sin^3\varrho$. Therefore its 4-dimensional content is

$$\int_0^\varrho 2\pi^2 \sin^3\theta\, d\theta = \frac{2}{3}\,\pi^2(2 + \cos\varrho)\,(1 - \cos\varrho)^2.$$

It follows that the 4-dimensional content of a sphere of radius $\varrho$ in hyper-bolic 4-space is

$$\frac{2}{3}\,\pi^2(2 + \text{ch}\,\varrho)\,(1 - \text{ch}\,\varrho)^2.$$

Setting $\varrho = \psi$ and $\chi$ in turn, we obtain

$$\frac{2}{3}\,\pi^2(2+\tau)\,(1-\tau)^2 = \frac{2}{3}\,\pi^2\sqrt{5}\,\tau^{-1}$$

and

$$\frac{2}{3}\,\pi^2\!\left(2+5^{-\frac{1}{2}}\tau^3\right)\!\left(1-5^{-\frac{1}{2}}\tau^3\right)^2 = \frac{8}{3}\,\pi^2\frac{2+3\sqrt{5}}{5\sqrt{5}}$$

for the contents of the in- and circum-spheres.

Dividing by $\frac{4}{3}\,\pi^2$, we obtain the packing density

$$\frac{1}{2}\,\sqrt{5}\,\tau^{-1} = \frac{5-\sqrt{5}}{4} = 0{,}69098\ldots$$

and the covering density

$$\frac{2(2+3\sqrt{5})}{5\sqrt{5}} = 1{,}55777\ldots.$$

It is interesting to compare these with the corresponding results in Euclidean 4-space, where the greatest known packing density

$$\frac{\pi^2}{16} = 0{,}61685\ldots$$

is attained by the in-spheres of the cells $\{3, 4, 3\}$ of the regular honeycomb $\{3, 4, 3, 3\}$ [9, pp. 153—154, 156, 296; 18, p. 42], while the smallest known covering density

$$\frac{2\,\pi^2}{5\sqrt{5}} = 1{,}76553\ldots$$

is attained by the circum-spheres of the cells $t_{0,1,2,3}\alpha_4$ of HINTON's honeycomb [5, p. 334].

## 8. Postscript

After the above sections were written, Professor FEJES TÓTH kindly sent me the galley proof of his own work on the same subject [17]. This shows that he anticipated my use of the in-spheres of the cells of $\{5, 3, 3\}$ for a packing in elliptic 3-space, and of the inscribed horospheres of the cells of $\{6, 3, 3\}$ for a packing in hyperbolic 3-space. He and I obtained the same expressions for the density in both cases.

## References

[1] R. P. Bambah, On lattice coverings by spheres, *Proc. Nat. Inst. Sci. India*, **20** (1954), pp. 25—52.

[2] A. H. Boerdijk, Some remarks concerning close-packing of equal spheres, *Philips Research Reports*, **7** (1952), pp. 303—313.

[3] H. S. Carslaw, *The elements of non-Euclidean plane geometry and trigonometry* (London, 1916).

[4] J. L. Coolidge, *The elements of non-Euclidean geometry* (Oxford, 1909).

[5] H. S. M. Coxeter, Wythoff's construction for uniform polytopes, *Proc. London Math. Soc. (2)*, **38** (1935), pp. 327—339.*

[6] H. S. M. Coxeter, The functions of Schläfli and Lobatschefsky, *Quart. J. Math.*, **6** (1935), pp. 13—29.*

[7] H. S. M. Coxeter, Regular and semi-regular polytopes I, *Math. Z.*, **46** (1940), pp. 380 - 407.

[8] H. S. M. Coxeter, *Non-Euclidean geometry* (Toronto, 1947).

[9] H. S. M. Coxeter, *Regular polytopes* (New York, 1963).

[10] H. S. M. Coxeter, Interlocked rings of spheres, *Scripta Mathematica*, **18** (1952), pp. 113—121.

[11] H. S. M. Coxeter, Regular honeycombs in elliptic space, *Proc. London Math. Soc. (3)*, **4** (1954), pp. 471—501.

[12] H. S. M. Coxeter and G. J. Whitrow, World structure and non-Euclidean honeycombs, *Proc. Roy. Soc. A*, **201** (1950), pp. 417—437.

[13] G. L. Dirichlet, Über die Reduction der positiven quadratischen Formen mit drei unbestimmten ganzen Zahlen, *J. reine angew. Math.*, **40** (1850), pp. 209—227.

[14] L. Fejes Tóth, *Lagerungen in der Ebene, auf der Kugel und im Raum* (Berlin, 1953).

[15] L. Fejes Tóth, Kreisausfüllungen der hyperbolischen Ebene, *Acta Math. Acad. Sci. Hung.*, **4** (1953), pp. 103—110.

[16] L. Fejes Tóth, Kreisüberdeckungen der hyperbolischen Ebene, *Acta Math. Acad. Sci. Hung.*, **4** (1953), pp. 111—114.

[17] L. Fejes Tóth, On close-packings of spheres in spaces of constant curvature, *Publicationes Mathematicae (Debrecen)*, **3** (1953), pp. 158—167.

[18] D. Hilbert und S. Cohn-Vossen, *Anschauliche Geometrie* (Berlin, 1932).

[19] L. Schläfli, *Gesammelte mathematische Abhandlungen*, **2** (Basel, 1953).

[20] V. Schlegel, Theorie der homogenen zusammengesetzten Raumgebilde, *Verh. K. Leopold.-Carol. D. Akad. Naturf.*, **44** (1883), pp. 343—459.

[21] D. M. Y. Sommerville, *The elements of non-Euclidean geometry* (London, 1914).

*References 5 and 6 are Chapters 3 and 1 of this book.

**9**

# AN UPPER BOUND
# FOR THE NUMBER OF EQUAL
# NONOVERLAPPING SPHERES
# THAT CAN TOUCH
# ANOTHER OF THE SAME SIZE

Reprinted, with permission of the publisher,
from *Proceedings of Symposia in Pure Mathematics*
Vol. VII, pp. 53–72. Copyright © 1963,
The American Mathematical Society.

**1. Introduction.** The problem indicated in the title is regarded as the case $\phi = \pi/6$ of the $n$-dimensional analog of the problem of packing small circles of angular radius $\phi$ on a sphere in ordinary space (Coxeter 1962).* Reasons are given for believing that the number of such $(n-2)$-spheres that can be packed on an $(n-1)$-sphere has the upper bound

$$2F_{n-1}(\alpha)/F_n(\alpha) \, ,$$

where $\alpha$ is given by $\sec 2\alpha = \sec 2\phi + n - 2$ and the function $F$ is defined recursively by

$$F_{n+1}(\alpha) = \frac{2}{\pi} \int_{(\arcsec n)/2}^{\alpha} F_{n-1}(\beta)d\theta \, , \qquad \sec 2\beta = \sec 2\theta - 2 \, ,$$

with the initial conditions $F_0(\alpha) = F_1(\alpha) = 1$. This upper bound is attained when the $(n-2)$-spheres are inscribed in the cells of a regular polytope

$$\{p, 3, \cdots, 3\} \, ,$$

in which case $\alpha = \pi/p$.

**2. Kepler, Gregory, and Hales.** Kepler (1611) discovered the *cubic close-packing* of equal spheres in Euclidean 3-space. Their centers, forming the face-centered cubic lattice, may be taken to have integral Cartesian coordinates with an even sum; then the common radius of the spheres is $1/\sqrt{2}$. Each sphere touches 12 others; for instance, the one with center $(0, 0, 0)$ touches those with centers $(0, \pm 1, \pm 1), (\pm 1, 0, \pm 1), (\pm 1, \pm 1, 0)$. These 12 centers are the vertices of a quasi-regular polyhedron, the cuboctahedron, which was described by Plato. The existence of the packing is closely associated with the fact that the circumradius of this polyhedron is equal to its edge-length.

Among the unpublished papers of David Gregory, H. W. Turnbull found notes of a conversation with Newton in 1694 about the distribution of stars of various magnitudes. The question arose: Can a rigid material sphere be brought into contact with 13 other such spheres of the same size? Gregory said "Yes", and Newton said "No"; but 180 years were to elapse before a conclusive answer was given.

Stephen Hales (1727) made a random close-packing of dried peas, filled the interstices with water, and kept the vessel closed. In his own words,

I compressed several fresh parcels of Pease in the same Pot, with a force equal to 1600, 800, and 400 pounds, in which Experiments, tho' the Pease dilated, yet they did

*For references, see pages 197–198.

not raise the lever, because what they increased in bulk was, by the great incumbent weight, pressed into the interstices of the Pease, which they adequately filled up, being thereby formed into pretty regular Dodecahedrons.

Hales presumably reached his conclusion by observing some pentagonal faces on his dilated peas. Modern versions of his experiment have confirmed the prevalence of pentagonal faces. But the solid cells could not all be regular dodecahedra. For, since the dihedral angle of this Platonic solid is less than 120°, three specimens with a common edge will leave an angular gap (about 10°19′). In other words, twelve unit spheres with their centers at the vertices of a regular icosahedron of circumradius 2 (and edge $2 \sec 18° = 2.102924\cdots$) will all touch one central unit sphere, but will not touch one another. Gregory may have imagined that, by letting the twelve spheres roll on the central one until the gaps are all concentrated in one direction, there might somehow be enough space for one more, so that the central sphere would touch thirteen others. It may reasonably be argued that there *is* enough space (actually enough for 13.397 spheres, as we shall see in § 9). The problem is to arrange the twelve spheres so as to concentrate the free space and make room for the thirteenth. The impossibility of such a redistribution was eventually proved by R. Hoppe (see Bender 1874), thus settling the Gregory-Newton controversy in favor of Newton. Simpler proofs were given by Günter (1875), Schütte and van der Waerden (1953), and Leech (1956).

**3. Dirichlet and Barlow.** To be precise, an arrangement of equal spheres is called a *packing* if no point of space is inside more than one sphere. In any packing, we may associate with each sphere a *Dirichlet region* (or *Voronoi polyhedron*) consisting of all the points that are as near to the center of that sphere as to the center of any other. Such regions, each surrounding a sphere, fit together to fill the whole space without interstices (Dirichlet 1850; Voronoi 1908). The *density* of the packing may be defined as the average of the ratio of the volume of a sphere to the volume of the Dirichlet region that surrounds it. For instance, the density of Kepler's cubic close-packing is equal to the ratio of the volume of a sphere to the volume of a circumscribed rhombic dodecahedron (the reciprocal of the cuboctahedron), namely,

$$\frac{\pi}{3\sqrt{2}} = 0.74048\cdots$$

(see, e.g., Coxeter 1961, pp. 407–408). W. Barlow (1883) described an equally dense packing in which each sphere still touches twelve others, although the centers do not form a lattice. For this *hexagonal close-packing* the Dirichlet region is a trapezo-rhombic dodecahedron, which may be constructed by cutting a rhombic dodecahedron into equal halves by a plane perpendicular to six parallel edges and then sticking the two halves together again after turning one of them through 60° (Fejes Tóth 1953, p. 173).

**4. Schläfli.** Schläfli (1855) studied polytopes in Euclidean $n$-space and in spherical $(n-1)$-space, extending spherical trigonometry from $n = 3$ to $n > 3$. He found that an $(n-1)$-sphere of radius $R$ in Euclidean $n$-space has $(n-1)$-dimensional content (or "surface") $S_n R^{n-1}$ and $n$-dimensional content (or "volume") $J_n R^n$, where

$$S_n = \frac{2\pi^{n/2}}{\Gamma\left(\dfrac{n}{2}\right)} , \qquad J_n = \frac{\pi^{n/2}}{\Gamma\left(\dfrac{n}{2} + 1\right)} ,$$

so that

$$S_n = n J_n = 2\pi J_{n-2} ,$$

$$S_1 = 2, \ S_2 = 2\pi, \ S_3 = 4\pi, \ S_4 = 2\pi^2, \ S_5 = \frac{8}{3}\pi^2, \ S_6 = \pi^3, \cdots ,$$

$$J_1 = 2, \ J_2 = \pi, \ J_3 = \frac{4}{3}\pi, \ J_4 = \frac{\pi^2}{2}, \ J_5 = \frac{8}{15}\pi^2, \ J_6 = \frac{\pi^3}{6}, \cdots .$$

He defined (on p. 177) a remarkable function $F_n(\alpha)$, in terms of which a regular simplex of dihedral angle $2\alpha$ in such a spherical space (with $R = 1$) has $(n - 1)$-dimensional content

(4.1) $$2^{-n} n! \, S_n F_n(\alpha) .$$

When $\alpha = \pi/4$, the bounding hyperplanes of the simplex are mutually orthogonal, so that the content is $2^{-n} S_n$; therefore

(4.2) $$F_n\left(\frac{\pi}{4}\right) = 1/n! .$$

When $\alpha = \pi/3$, the spherical simplex corresponds by central projection to one of the $n + 1$ cells of a regular Euclidean simplex; therefore its content is $S_n/(n + 1)$, and

(4.3) $$F_n\left(\frac{\pi}{3}\right) = 2^n/(n + 1)! .$$

The boundary of the spherical simplex decomposes the spherical space into two regions: an "inside" whose dihedral angles are $2\alpha$ and an "outside" whose dihedral angles are $2\pi - 2\alpha$; therefore

$$2^{-n} n! \, F_n(\alpha) + 2^{-n} n! \, F_n(\pi - \alpha) = 1 ,$$

that is,

(4.4) $$F_n(\alpha) + F_n(\pi - \alpha) = 2^n/n! .$$

In particular,

(4.41) $$F_n\left(\frac{\pi}{2}\right) = \frac{2^{n-1}}{n!} .$$

When $n = 3$, the "simplex" is a spherical triangle with angles $2\alpha, 2\alpha, 2\alpha$ and area $6\alpha - \pi$; therefore, since $S_3 = 4\pi$,

(4.5) $$F_3(\alpha) = \frac{2\alpha}{\pi} - \frac{1}{3} .$$

When $n = 2$, we merely have a circular arc of length $2\alpha$; therefore, since

$S_2 = 2\pi$,

(4.6) $$F_2(\alpha) = \frac{2\alpha}{\pi} .$$

Since a regular Euclidean simplex has dihedral angle arcsec $n$ in $n$ dimensions, or arcsec $(n-1)$ in $n-1$ dimensions, and since an infinitesimal spherical simplex is Euclidean, we deduce

(4.7) $$F_n(\alpha) = 0 \quad \text{for} \quad \alpha = \frac{1}{2} \operatorname{arcsec}(n-1) .$$

One of his most brilliant discoveries (Schläfli 1855, pp. 167, 168) is that

$$dF_n(\alpha) = F_{n-2}(\beta)dF_2(\alpha) ,$$

where $\beta$ is given by

$$\sec 2\beta = \sec 2\alpha - 2 .$$

Thus the Schläfli function can be defined recursively by the formula

(4.8) $$F_n(\alpha) = \frac{2}{\pi} \int_{(\operatorname{arcsec}(n-1))/2}^{\alpha} F_{n-2}(\beta)d\theta , \qquad \sec 2\beta = \sec 2\theta - 2 ,$$

with the initial conditions

(4.9) $$F_0(\alpha) = F_1(\alpha) = 1 .$$

This function has been studied from another standpoint by Ruben (1961, p. 262).

5. **Höhn, Peschl, and Guinand.** Various formulae connecting the angles of the general $n$-dimensional simplex have been discovered by Poincaré, Sommerville, Höhn, and Peschl. Their equivalence has been established by Guinand (1959), who observes that, in the symbolic or umbral notation (with $a^k$ standing for $a_k$), Höhn's formulae become

$$a^r = (1 - a)^r \qquad (r = 1, 2, 3, \cdots) ,$$

where $a_k$ (symbolically $a^k$) is the average angle at an $(n-k)$-cell, expressed as a fraction of the whole angle at an $(n-k)$-flat. When the simplex is regular, we have

$$a_k = 2^{-k}k! \, F_k(\alpha) ,$$

so that the formula

$$a_r = \sum_{k=0}^{r} (-1)^k \binom{r}{k} a_k$$

becomes

$$F_r(\alpha) = \sum_{k=0}^{r} \frac{(-1)^k 2^{r-k}}{(r-k)!} F_k(\alpha) = (-1)^r \sum_{j=0}^{r} \frac{(-2)^j}{j!} F_{r-j}(\alpha) .$$

When $r$ is even, $F_r(\alpha)$ cancels, leaving

$$\sum_{j=1}^{r} \frac{(-2)^j}{j!} F_{r-j}(\alpha) = 0 \qquad\qquad (r \text{ even}).$$

But when $r$ is odd, we obtain an expression for $F_r(\alpha)$ in terms of simpler functions:

$$(5.1) \qquad\qquad F_r(\alpha) = \sum_{j=1}^{r} \frac{(-2)^{j-1}}{j!} F_{r-j}(\alpha) \qquad\qquad (r \text{ odd}).$$

From such equations with $r = 1, 3, 5, \cdots, n$, where $n$ is odd, we can eliminate $F_1(\alpha), F_3(\alpha), \cdots, F_{n-2}(\alpha)$, obtaining

$$(5.2) \qquad\qquad F_n(\alpha) = F_{n-1}(\alpha) - \frac{1}{3} F_{n-3}(\alpha) + \frac{2}{15} F_{n-5}(\alpha) - \cdots \qquad (n \text{ odd}),$$

where the series ends with the term in $F_0(\alpha)$ and the coefficients are the same as in the expansion

$$\tanh x = x - \frac{1}{3} x^3 + \frac{2}{15} x^5 - \cdots$$

(Schläfli 1855, p. 178).

Alternatively, we can deduce (5.2) from Peschl's formula

$$(2B + a)^r = (B + a)^r \qquad\qquad (r \text{ even})$$

(Guinand 1959, p. 59) or

$$\sum_{j=0}^{r} (2^j - 1)\binom{r}{j} B_j a_{r-j} = 0$$

$$\left(B_0 = 1, \ B_1 = -\frac{1}{2}, \ B_2 = \frac{1}{6}, \ B_3 = B_5 = B_7 = \cdots = 0, \ B_4 = -\frac{1}{30}, \cdots\right),$$

which yields

$$\sum_{j=1}^{r} \frac{2^j(2^j - 1)}{j!} B_j F_{r-j}(\alpha) = 0$$

or

$$F_{r-1}(\alpha) = \sum_{j=2}^{r} \frac{2^j(2^j - 1)}{j!} B_j F_{r-j}(\alpha) \qquad\qquad (r \text{ even}).$$

Since $F_{n+1}(\alpha) = 0$ when $\alpha = (\text{arcsec } n)/2$, we have, for this value of $\alpha$ and $n$ even,

$$F_n(\alpha) = \frac{1}{3} F_{n-2}(\alpha) - \frac{2}{15} F_{n-4}(\alpha) + \frac{17}{315} F_{n-6}(\alpha) - \cdots$$

$$(5.3) \qquad\qquad\qquad (\alpha = (\text{arcsec } n)/2, \ n \text{ even}).$$

For instance,

$$F_4\left(\frac{1}{2} \text{ arcsec } 4\right) = \frac{1}{3} F_2\left(\frac{1}{2} \text{ arcsec } 4\right) - \frac{2}{15}$$

$$(5.4) \qquad\qquad\qquad = \frac{1}{3}\left(\frac{\text{arcsec } 4}{\pi} - \frac{2}{5}\right)$$

(Schläfli 1855, p. 182). Other special values that are known explicitly are

(5.5) $$F_4\left(\frac{\pi}{5}\right) = \frac{1}{900}, \quad F_4\left(\frac{2\pi}{5}\right) = \frac{191}{900}$$

(Schläfli 1855, p. 181; Coxeter 1935, pp. 18, 19). Taking (4.2), (4.3), (4.4), and (4.7) into consideration, we can now assert that $F_4(\alpha)$ is elementary for at

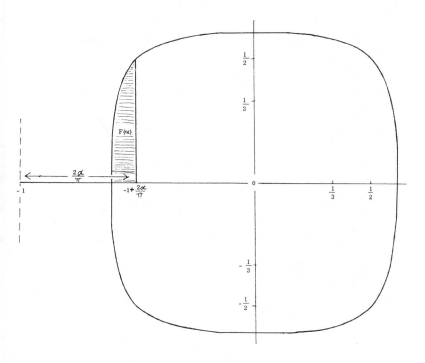

FIGURE 1

least 13 particular values of $\alpha$. These results provide a convenient check when numerical computation is applied to (4.8) (with $n = 4$) or, equivalently, to

(5.6) $$F_4(\alpha) = \int_{-1+(\operatorname{arcsec} 3)/\pi}^{-1+2\alpha/\pi} y\,dx \;,$$

where $y$ is related to $x$ by the symmetrical equation

(5.61) $$\sec \pi x + \sec \pi y + 2 = 0 \;.$$

Figure 1 is a sketch of the curve (5.61) with $F(\alpha)$ indicated as an area in a typical case, actually the case $F(\pi/4) = 1/24$. Figure 2 is a sketch of the curve $y = F_4(x)$, with the thirteen "elementary" points marked, namely (with the temporary abbreviation $s_k = \text{arcsec } k$):

| $\alpha$ | $\frac{1}{2}s_3$ | $\frac{1}{5}\pi$ | $\frac{1}{2}s_4$ | $\frac{1}{4}\pi$ | $\frac{1}{3}\pi$ | $\frac{2}{5}\pi$ | $\frac{1}{2}\pi$ | $\frac{3}{5}\pi$ | $\frac{2}{3}\pi$ | $\frac{3}{4}\pi$ | $\pi - \frac{1}{2}s_4$ | $\frac{4}{5}\pi$ | $\pi - \frac{1}{2}s_3$ |
|---|---|---|---|---|---|---|---|---|---|---|---|---|---|
| $F_4(\alpha)$ | $0$ | $\frac{1}{900}$ | $\frac{1}{3}\left(\frac{s_4}{\pi} - \frac{2}{5}\right)$ | $\frac{1}{24}$ | $\frac{2}{15}$ | $\frac{191}{900}$ | $\frac{1}{3}$ | $\frac{409}{900}$ | $\frac{8}{15}$ | $\frac{5}{8}$ | $\frac{1}{3}\left(\frac{12}{5} - \frac{s_4}{\pi}\right)$ | $\frac{599}{900}$ | $\frac{2}{3}$ |

## 6. Minkowski and Blichfeldt. We shall return to this function, but first

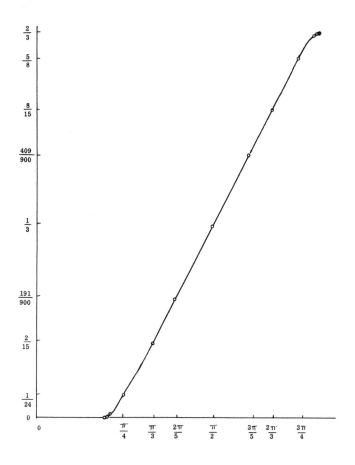

FIGURE 2

we must mention one more historic event: Minkowski (1905, p. 247) proved that the densest packing of equal spheres whose centers form a lattice (in Euclidean $n$-space) consists of spheres of radius $1/\sqrt{2}$ whose centers have integral Cartesian coordinates with an even sum, not only when $n = 3$ but also when $n = 4$ or 5. Thus the sphere with its center at the origin touches $2n(n - 1)$ others, whose centers are given by the permutations of

$$(\pm 1, \pm 1, 0, 0) \qquad\qquad (n = 4)$$

or

$$(\pm 1, \pm 1, 0, 0, 0) \qquad\qquad (n = 5) .$$

In other words, the centers of all the spheres are the vertices of the honeycomb $h\delta_{n+1}$ whose cells are cross polytopes $\beta_n$ and "half measure polytopes" $h\gamma_n$ (Coxeter 1930, p. 365; 1948, p. 155). When $n = 4$, this uniform honeycomb is regular, since $h\gamma_4 = \beta_4$ and $h\delta_5 = \{3, 3, 4, 3\}$; the spheres are inscribed in the Dirichlet regions of the lattice, that is, in the cells $\{3, 4, 3\}$ of the dual honeycomb $\{3, 4, 3, 3\}$.

Such a packing is rigid for all values of $n$, but is the densest lattice packing only when $n = 3, 4$, or 5. When $n = 6, 7$, or 8 (Blichfeldt 1935) the closest packings are such that each sphere touches 72, 126, or 240 others, respectively. The centers of the spheres are the vertices of the uniform honeycombs

$$2_{22} , \qquad 3_{31} , \qquad 5_{21}$$

(Coxeter 1930, pp. 379, 393-397; 1948, pp. 201-205). In other words, for spheres of radius $3/\sqrt{2}$, convenient coordinates for the centers are nine integers, mutually congruent modulo 3 and satisfying the equations

$$x_1 + x_2 + x_3 = x_4 + x_5 + x_6 = x_7 + x_8 + x_9 = 0 \qquad (n = 6) ,$$
$$x_1 + x_2 + x_3 + x_4 + x_5 + x_6 = x_7 + x_8 + x_9 = 0 \qquad (n = 7) ,$$
$$x_1 + x_2 + x_3 + x_4 + x_5 + x_6 + x_7 + x_8 + x_9 = 0 \qquad (n = 8) .$$

Since Barlow's hexagonal close-packing (in Euclidean 3-space) has the same density as cubic close-packing, namely,

$$\frac{\pi}{3\sqrt{2}} = 0.74048 \cdots ,$$

it is natural to ask whether some other non-lattice packing may have a greater density. According to Rogers (1958, p. 610), "many mathematicians believe, and all physicists know" that no such denser packing exists in 3 dimensions. But this has never been rigorously proved, and dense non-lattice packings may reasonably be expected when the number of dimensions is sufficiently great.

Leech has observed that in five dimensions, as in three, there is a non-lattice packing that has the same density as the closest lattice packing. In fact, the three-dimensional honeycomb $h\delta_4$, formed by the centers in cubic close-packing, may be broken up into a sequence of layers, each consisting of tetrahedra $h\gamma_3 = \alpha_3$ and octahedra $\beta_3$ sandwiched between two tessellations

$\{3, 6\}$. Hexagonal close-packing is derived by shifting the layers until each triangle $\{3\}$ of $\{3, 6\}$ belongs to two tetrahedra or two two octahedra, instead of belonging to one tetrahedron and one octahedron. Somewhat similarly, the five-dimensional honeycomb $h\delta_6$ may be broken up into a sequence of layers, each consisting of $h\gamma_5$'s and $\beta_4$-pyramids sandwiched between two four-dimensional honeycombs $h\delta_5$. Leech (1967) derived three non-lattice packings by shifting the layers until each cell $\beta_4$ of $h\delta_5$ that belongs to a $\beta_4$-pyramid, instead of belonging to another $\beta_4$-pyramid (so as to complete a $\beta_5$), belongs to a $h\gamma_5$.

When the lattice requirement has been abandoned, the packing problem in Euclidean space does not differ greatly from the packing problem in spherical space. The latter begins with the problem of packing small circles on the surface of an ordinary sphere. Meschkowski (1960) states this vividly, as follows:

> On a planet, say, ten inimical dictators govern. How must the residences of these gentlemen be located in order to get as far as possible from one another?

**7. Rankin.** Consider a regular polytope inscribed in an $(n - 1)$-sphere of radius 1 in Euclidean $n$-space. Let $2\phi$ denote the angle subtended at the center by an edge (Coxeter 1948, p. 133). Clearly, the $N_0$ vertices are the centers of $N_0$ nonoverlapping $(n - 2)$-spheres of angular radius $\phi$, packed on the $(n - 1)$-sphere (e.g., when $n = 3$, spherical caps on an ordinary sphere).

If the polytope is a $k$-dimensional regular simplex $\alpha_k$ $(1 \leqq k \leqq n)$, we have $2\phi = \pi - \mathrm{arcsec}\, k$ and $N_0 = k + 1$ (Coxeter 1948, p. 295). Rankin (1955) proved that this arrangement is the closest packing of $k + 1$ $(n - 2)$-spheres $(1 \leqq k \leqq n)$. In other words, letting $N(\phi)$ denote the maximum number of such "caps," of angular radius $\phi$, that can be packed on the unit $(n - 1)$-sphere, we have

$$(7.1) \qquad N(\phi) = [\sec(\pi - 2\phi)] + 1 \qquad \text{for} \quad \pi - \mathrm{arcsec}\, n \leqq 2\phi \leqq \pi \,.$$

Davenport and Hajós (1951) proved that $N(\phi)$ cannot take a value between $n + 1$ and $2n$, the latter value appearing when the polytope is the cross-polytope or "$n$-dimensional octahedron" $\beta_n$ (Coxeter 1948, p. 121). In other words, although

$$(7.2) \qquad\qquad N\left(\frac{\pi}{4}\right) = 2n \,,$$

we have

$$(7.3) \qquad\qquad N(\phi) = n + 1 \qquad \text{for} \quad \frac{\pi}{2} < 2\phi \leqq \pi - \mathrm{arcsec}\, n \,.$$

Thus $N(\phi)$ is known precisely whenever $\phi \geqq \pi/4$. For smaller values of $\phi$, Rankin was content to prove an inequality which yields, for instance, $N(\pi/10) < 258$ when $n = 4$.

**8. The proposed new bound.** It is intuitively obvious that $n$ equal $(n - 2)$-spheres are packed as closely as possible when they all touch one another, so that their centers are the $n$ vertices of a regular simplex of (angular) edge

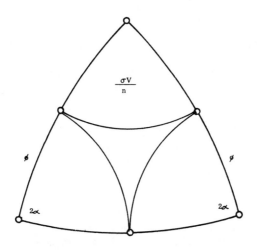

$2\phi$.  In other words, we may reasonably expect that the density of a packing of at least $n$ "caps" of radius $\phi$ can never exceed

$$\frac{\sigma V}{\Sigma} ,$$

where $\sigma$ is the sum of the vertex angles of the simplex, expressed as a fraction of the total angle at a point in $n - 1$ dimensions, $V$ is the $(n - 1)$-dimensional content of such an $(n - 2)$-sphere, namely,

$$V = S_{n-1}\int_0^\phi \sin^{n-2}\rho\,d\rho ,$$

and $\Sigma$ is the $(n - 1)$-dimensional content of the simplex.  The analogous conjecture for a packing in Euclidean space has been justified by Rogers (1958), with the conclusion that the density cannot exceed

$$2^{-3n/2}(n!)^2(n + 1)^{1/2}J_nF_n\left(\frac{1}{2}\operatorname{arcsec} n\right) .$$

The truth of the original "intuitively obvious" conjecture has been established for $n = 3$ (see Figure 3) by Fejes Tóth (1959, p. 311; see also Coxeter 1962). If it holds universally, we can deduce a bound for $N(\phi)$ by observing that, since the $(n - 1)$-sphere is packed with $N(\phi)$ $(n - 2)$-spheres of content $V$, the density is

$$N(\phi)\frac{V}{S_n} ,$$

and therefore

$$N(\phi) \leqq \frac{\sigma S_n}{\Sigma} .$$

Here $\Sigma$ is the content of a regular $(n-1)$-dimensional spherical simplex whose dihedral angle $2\alpha$ is a certain function of its edge $2\phi$; and $\sigma$ is the sum of the $n$ vertex angles, each equal to the content of a regular $(n-2)$-dimensional spherical simplex having the same dihedral angle $2\alpha$. By (4.1),

$$\Sigma = 2^{-n} n! \, S_n F_n(\alpha)$$

and

$$\sigma = n \cdot 2^{-(n-1)} (n-1)! \, F_{n-1}(\alpha) ,$$

so that

$$\frac{\sigma S_n}{\Sigma} = \frac{2 F_{n-1}(\alpha)}{F_n(\alpha)} .$$

To express $\alpha$ in terms of $\phi$, a simple procedure has been suggested by Rogers. Let the vertices of the simplex have coordinates

$$(c + a, a, \cdots, a), \ (a, c + a, \cdots, a), \ \cdots, \ (a, a, \cdots, c + a) .$$

The angle $2\phi$ subtended by two of these points at the origin is given by

$$\cos 2\phi = \frac{2a(c + a) + (n - 2)a^2}{(c + a)^2 + (n - 1)a^2} = \frac{2ac + na^2}{c^2 + 2ac + na^2} ,$$

whence

$$\sec 2\phi = \frac{c^2}{2ac + na^2} + 1 .$$

The internal angle $2\alpha$ between two bounding hyperplanes

$$\{c + (n - 1)a\} x_1 - a(x_2 + \cdots + x_n) = 0$$

and

$$a(x_1 + \cdots + x_{n-1}) - \{c + (n - 1)a\} x_n = 0$$

is given by

$$\cos 2\alpha = \frac{2\{ac + (n - 1)a^2\} - (n - 2)a^2}{\{c + (n - 1)a\}^2 + (n - 1)a^2} = \frac{2ac + na^2}{c^2 + (n - 1)(2ac + na^2)} ,$$

whence

$$\sec 2\alpha = \frac{c^2}{2ac + na^2} + n - 1$$
$$= \sec 2\phi + n - 2 .$$

Thus our final result is

(8.1) $$N(\phi) \leqq \frac{2F_{n-1}(\alpha)}{F_n(\alpha)}, \qquad \sec 2\alpha = \sec 2\phi + n - 2 .$$

This upper bound is attained whenever the pattern of $n$ spheres touching one another can be continued over the whole spherical space, that is, whenever the centers of the spheres coincide with the vertices of a polytope

$$\{3, 3, \cdots, 3, p\} ,$$

where $p = \pi/\alpha$. In the notation of Coxeter (1948, pp. 160–162; see also Böhm 1959, 1960) we may conveniently take $(n - 2, n) = (\sec^2 \alpha)/2$,

$$(j, k) = k - j \qquad\qquad (j \leqq k \leqq n - 1) ,$$

and

$$(j, n) = \frac{(j, n - 1)(j + 1, n) - 1}{(j + 1, n - 1)} = \frac{(n - 1 - j)(j + 1, n) - 1}{n - 2 - j} ,$$

whence

$$(j, n) = \frac{1}{2}(n - 1 - j) \sec^2 \alpha - (n - 2 - j) ,$$

$$\tan^2 \phi = \frac{(-1, n)}{(1, n)} = \frac{\frac{1}{2} n \sec^2 \alpha - (n - 1)}{\frac{1}{2}(n - 2) \sec^2 \alpha - (n - 3)} ,$$

$$\cos 2\phi = \frac{\cos 2\alpha}{1 - (n - 2) \cos 2\alpha} ,$$

$$\sec 2\phi = \sec 2\alpha - (n - 2) ,$$

as before.

For instance, in four dimensions, the regular 600-cell $\{3, 3, 5\}$ has $\alpha = \pi/5$ and $\phi = \pi/10$, so that

$$N\left(\frac{\pi}{10}\right) = \frac{2F_3\left(\frac{\pi}{5}\right)}{F_4\left(\frac{\pi}{5}\right)} = \frac{2\left(\frac{2}{5} - \frac{1}{3}\right)}{\frac{1}{900}} = 120 .$$

The remaining four-dimensional results are epitomized in **Figure 4**, which shows the smooth curve obtained by plotting the upper bound

$$\tilde{N}(\phi) = \frac{2F_3(\alpha)}{F_4(\alpha)} , \qquad \sec 2\alpha = \sec 2\phi + 2$$

for $\phi \leqq (\pi - \text{arcsec } 3)/2$, and also the "step function" which expresses the exact value of $N(\phi)$ for every $\phi \geqq \pi/4$. For convenience, the accompanying table gives the values of $\phi$ and $\alpha$ in degrees instead of radians, so that (4.5) yields $2F_3(\alpha) = (\alpha - 30)/45$:

| $\phi$ | 18 | 45 | $52\frac{1}{4}$ | 54 | $54\frac{3}{4}$ | 54 | $52\frac{1}{4}$ | 45 | 18 |
|---|---|---|---|---|---|---|---|---|---|
| $\alpha$ | 36 | 45 | 60 | 72 | 90 | 108 | 120 | 135 | 144 |
| $2F_3(\alpha)$ | $\dfrac{2}{15}$ | $\dfrac{1}{3}$ | $\dfrac{2}{3}$ | $\dfrac{14}{15}$ | $\dfrac{4}{3}$ | $\dfrac{26}{15}$ | 2 | $\dfrac{7}{3}$ | $\dfrac{38}{15}$ |
| $F_4(\alpha)$ | $\dfrac{1}{900}$ | $\dfrac{1}{24}$ | $\dfrac{2}{15}$ | $\dfrac{191}{900}$ | $\dfrac{1}{3}$ | $\dfrac{409}{900}$ | $\dfrac{8}{15}$ | $\dfrac{5}{8}$ | $\dfrac{599}{900}$ |
| $\tilde{N}(\phi)$ | 120 | 8 | 5 | $\dfrac{840}{191}$ | 4 | $\dfrac{1560}{409}$ | $\dfrac{15}{4}$ | $\dfrac{56}{15}$ | $\dfrac{2280}{599}$ |

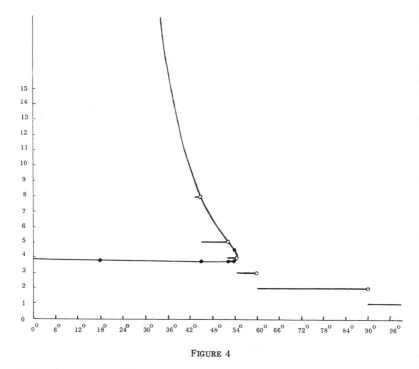

FIGURE 4

With only two exceptions (one repeated), every entry in this table is *exact*. The exceptions are $54\frac{3}{4}$ and $52\frac{1}{4}$, which are the approximate numbers of degrees in $(\text{arcsec}\,(-3))/2$ and $(\text{arcsec}\,(-4))/2$, respectively.

Since $F_4(\alpha) = 0$ when $\phi = 0$, $\tilde{N}(0) = \infty$: the curve for $\tilde{N}(\phi)$ has the vertical axis as an asymptote. The point (18, 120) is too high to be conveniently shown. The "white" marked points are

$$N(45) = 8, \ N(52\tfrac{1}{4}) = 5, \ N(54\tfrac{3}{4}) = 4, \ N(60) = 3, \ N(90) = 2 \ .$$

The first three of these, along with $N(18) = 120$, are the only places where $N(\phi)$ and $\tilde{N}(\phi)$ exactly agree. The "black" marked points help us to plot the curve for $\tilde{N}(\phi)$ because, although they have no geometric significance, they indicate further known cases in which the expression $2F_3(\alpha)/F_4(\alpha)$ takes rational values. The lower part of this curve arises because the relation

$$\sec 2\alpha = \sec 2\phi + 2$$

makes each value of $\phi$ yield two supplementary values for $\alpha$, and hence two values for $2F_3(\alpha)/F_4(\alpha)$, the larger of which serves as a bound for $N(\phi)$ while the smaller remains in the interval from 3.733 to 3.815 for all $\phi < 54°$. Since

$$\frac{d}{d\phi}\tilde{N}(\phi) = \frac{4}{\pi F_4(\alpha)}\left(1 - \frac{\phi}{\pi}\tilde{N}(\phi)\right)\frac{\sec 2\phi \tan 2\phi}{\sec 2\alpha \tan 2\alpha}$$

is zero when $\phi = \pi/\tilde{N}(\phi)$, the minimum on this nearly straight portion of the curve occurs where $\phi = \pi/3.733$, which is about 48°.

**9. The case when $\phi = \pi/6$.** In any number of dimensions, the centers of three mutually tangent equal spheres form an equilateral triangle. Hence the problem indicated in the title of this paper amounts to finding an upper bound for $N(\pi/6)$. By (8.1), such an upper bound (attained only when $n = 1$ or 2) is

(9.1)
$$\frac{2F_{n-1}\left(\dfrac{1}{2}\arcsec n\right)}{F_n\left(\dfrac{1}{2}\arcsec n\right)} \ .$$

It is interesting to see how this works out for small values of $n$.

When $n = 2$, the value

$$\frac{2F_1\left(\dfrac{\pi}{6}\right)}{F_2\left(\dfrac{\pi}{6}\right)} = \frac{2}{\dfrac{1}{3}} = 6$$

is attained by the six circles that can obviously touch one of the same size.

When $n = 3$, the bound

$$\frac{2F_2\left(\dfrac{1}{2}\arcsec 3\right)}{F_3\left(\dfrac{1}{2}\arcsec 3\right)} = \frac{6}{3 - \pi/\arcsec 3} = 13.39733257 \cdots$$

(Coxeter 1958, p. 757), like Rankin's

$$N_3^* = 6\sqrt{2} + 8 = 16.48528 \cdots ,$$

is not strict enough to disprove Gregory's conjecture that 13 spheres in ordinary space could touch one equal sphere.

When $n = 4$, we use (5.3) to compute

$$F_4\left(\frac{1}{2}\operatorname{arcsec} 4\right) = \frac{1}{3}F_2\left(\frac{1}{2}\operatorname{arcsec} 4\right) - \frac{2}{15} = \frac{1}{3}\left(\frac{\operatorname{arcsec} 4}{\pi} - \frac{2}{5}\right).$$

Since, by (4.5),

$$F_3\left(\frac{1}{2}\operatorname{arcsec} 4\right) = \frac{\operatorname{arcsec} 4}{\pi} - \frac{1}{3},$$

it follows that

$$\frac{2F_3\left(\frac{1}{2}\operatorname{arcsec} 4\right)}{F_4\left(\frac{1}{2}\operatorname{arcsec} 4\right)} = 5\left(1 + \frac{1}{5 - 2\pi/\operatorname{arcsec} 4}\right) = 26.44009910\cdots.$$

Though this number is less than Rankin's

$$N_4^* = \frac{6\pi}{10 - 3\pi} = 32.769\cdots,$$

it leaves open the question whether a 3-sphere in Euclidean 4-space can touch only 24 others (as in the closest lattice packing) or 25 or 26.

**10. Leech's computation.** When $n > 4$, it is convenient to write $f_n(\sec 2\alpha)$ for $F_n(\alpha)$, so that

$$f_n(x) = F_n\left(\frac{1}{2}\operatorname{arcsec} x\right).$$

In this notation, the formulae (4.6), (4.7), (4.8), (5.2), (5.3) become

$$f_2(x) = \frac{\operatorname{arcsec} x}{\pi}, \qquad\qquad\qquad f_n(n-1) = 0,$$

$$f_n(x) = \frac{1}{\pi}\int_{n-1}^{x}\frac{f_{n-2}(x-2)}{x\sqrt{x^2-1}}\,dx = f_n(n) + \frac{1}{\pi}\int_{n}^{x}\frac{f_{n-2}(x-2)}{x\sqrt{x^2-1}}\,dx,$$

$$f_n(x) = f_{n-1}(x) - \frac{1}{3}f_{n-3}(x) + \frac{2}{15}f_{n-5}(x) - \frac{17}{315}f_{n-7}(x) + \cdots \qquad (n\text{ odd}),$$

$$f_n(n) = \frac{1}{3}f_{n-2}(n) - \frac{2}{15}f_{n-4}(n) + \frac{17}{315}f_{n-6}(n) - \cdots \qquad (n\text{ even}),$$

and the function (9.1) is simply

$$2f_{n-1}(n)/f_n(n).$$

Leech helpfully undertook to evaluate this function for $n \leq 8$. He began by tabulating

$$\frac{\operatorname{arcsec}(x-2)}{\pi^2 x\sqrt{x^2-1}}$$

to twelve decimal places for $x = 3.5(0.1)8.2$, and integrated by the trapezium rule with central difference correction at the ends of the ranges. From $f_4(x)$ he deduced

$$f_5(x) = f_4(x) - \frac{1}{3}f_2(x) + \frac{2}{15}$$

and

$$f_6(6) = \frac{1}{3}f_4(6) - \frac{2}{15}f_2(6) + \frac{17}{315}.$$

He obtained $f_6(x)$ by a similar integration of

$$\frac{f_4(x-2)}{\pi x \sqrt{x^2 - 1}}.$$

Finally, he deduced

$$f_7(x) = f_6(x) - \frac{1}{3}f_4(x) + \frac{2}{15}f_2(x) - \frac{17}{315}$$

and

$$f_8(8) = \frac{1}{3}f_6(8) - \frac{2}{15}f_4(8) + \frac{17}{315}f_2(8) - \frac{62}{2835}.$$

The results are as follows:

$f_4(4) = 0.00652\ 31255\ 82$ ,  $f_6(6) = 0.00003\ 02840\ 12$ ,

$f_4(5) = 0.01248\ 11393\ 80$ ,  $f_6(7) = 0.00010\ 31285\ 78$ ,

$f_4(6) = 0.01686\ 59385\ 17$ ,  $f_6(8) = 0.00018\ 73637\ 63$ ,

$f_4(7) = 0.02014\ 88228\ 26$ ,  $f_7(7) = 0.00000\ 14072\ 22$ ,

$f_4(8) = 0.02268\ 05970\ 96$ ,  $f_7(8) = 0.00000\ 64990\ 72$ ,

$f_5(5) = 0.00051\ 25449\ 97$ ,  $f_8(8) = 0.00000\ 00531\ 355$ ,

$f_5(6) = 0.00129\ 93981\ 97$ ,

$$2\, f_3(4)/f_4(4) = \ \ 26.440 \cdots ,$$
$$2\, f_4(5)/f_5(5) = \ \ 48.702 \cdots ,$$
$$2\, f_5(6)/f_6(6) = \ \ 85.814 \cdots ,$$
$$2\, f_6(7)/f_7(7) = 146.570 \cdots ,$$
$$2\, f_7(8)/f_8(8) = 244.622 \cdots .$$

Thus the maximum number of spheres that can touch another of the same size (in Euclidean $n$-space) is $N_n$, where

$$N_1 = 2 , \qquad N_2 = 6 , \qquad N_3 = 12 ,$$
$$24 \leqq N_4 \leqq \ \ 26 ,$$
$$40 \leqq N_5 \leqq \ \ 48 ,$$
$$72 \leqq N_6 \leqq \ \ 85 ,$$
$$126 \leqq N_7 \leqq 146 ,$$
$$240 \leqq N_8 \leqq 244 .$$

Notice the remarkably small difference between the upper and lower bounds for $N_8$. This closeness seems to be a manifestation of the extraordinary "near-regularity" of the honeycomb $5_{21}$, whose vertices are the centers of the spheres in the lattice packing. Any simplicial cell of the honeycomb indicates a set of nine spheres, perfectly packed. Of every 137 cells, 128 are simplexes and only 9 are cross polytopes (Coxeter 1948, p. 204). The same honeycomb appears again in the theory of Cayley numbers (Coxeter 1946, p. 571) and in the related theory of the simple Lie group $E_8$ (Coxeter 1951, pp. 412-414, 420-426).

**11. Spheres in non-Euclidean spaces.** The problem that plays the title role in this investigation remains significant when the equal spheres are packed in a non-Euclidean space (spherical, elliptic, or hyperbolic). The 3-dimensional case has already been considered by Fejes Tóth (1954), who uses spheres of radius 1 in a space of curvature $K$. The number of such unit spheres that can touch another unit sphere is evidently $N(\phi)$, where $\phi$ is half the angle of the equilateral triangle of side 2 whose vertices are the centers of three mutually tangent spheres. By the familiar non-Euclidean formula

$$\sin A = \frac{\sin a \sqrt{K}}{\sin c \sqrt{K}}$$

for the angle $A$ of a right-angled triangle $ABC$, we have

$$\sin \phi = \frac{\sin \sqrt{K}}{\sin 2\sqrt{K}} = \frac{1}{2 \cos \sqrt{K}} \, ,$$

whence

$$\sec 2\phi = \sec 2\sqrt{K} + 1 \, .$$

By (8.1), an upper bound for the number of spheres that can touch one is

$$(11.1) \qquad \tilde{N} = \frac{2F_{n-1}(\alpha)}{F_n(\alpha)} \, ,$$

where $\alpha$ is given by

$$(11.2) \qquad \sec 2\alpha = \sec 2\sqrt{K} + n - 1 \, .$$

In the special case when $n = 3$, we see from (4.5) and (4.6) that

$$\tilde{N} = \frac{12\alpha}{6\alpha - \pi} \, .$$

By (11.2), we have $\sec 2\alpha - 1 = \sec 2\sqrt{K} + 1$, whence

$$\frac{1 + \tan^2 \alpha}{1 - \tan^2 \alpha} - 1 = \frac{1 + \tan^2 \sqrt{K}}{1 - \tan^2 \sqrt{K}} + 1, \quad \frac{\tan^2 \alpha}{1 - \tan^2 \alpha} = \frac{1}{1 - \tan^2 \sqrt{K}} \, ,$$

$$\cot^2 \alpha + \tan^2 \sqrt{K} = 2 \, .$$

Since

$$\alpha = \frac{\tilde{N}}{\tilde{N} - 2} \frac{\pi}{6},$$

this is equivalent to the formula of Fejes Tóth (1954, p. 162).

**12. An asymptotic bound.** How fast does the number of spheres increase when the number of dimensions tends to infinity? According to Rogers (1961), when the number

$$b^{-1} = \sec 2\alpha - n + 1$$

(cf. (11.2)) is bounded,

$$F_n(\alpha) \sim \sqrt{\frac{1 + nb}{2}} \frac{1}{n! \, e^{1/b}} \left( \frac{2e}{\pi nb} \right)^{n/2}.$$

When we apply this asymptotic formula to (8.1), we have $b^{-1} = \sec 2\phi - 1$ in the denominator, and $b^{-1} = \sec 2\phi$ in the numerator. Thus the asymptotic upper bound for $N(\phi)$ is

(12.1) $$\frac{2^{1-n/2}\sqrt{\pi \cos 2\phi}}{\sin^{n-1}\phi} \left( \frac{n}{n-1} \right)^{(n-1)/2} \left( \frac{n}{e} \right)^{3/2} \sim \frac{2^{1-n/2}\sqrt{\pi \cos 2\phi} \, n^{3/2}}{e \sin^{n-1}\phi}.$$

This excels, by the factor $2/e$, the bound

$$\frac{2^{-n/2}\sqrt{\pi \cos 2\phi} \, n^{3/2}}{\sin^{n-1}\phi}$$

of Rankin (1955, p. 139; our $\phi$ is his $\alpha$).

Setting $\phi = \pi/6$ in (10.1), we deduce, for the bound named in the title of this paper, the asymptotic expression

$$\frac{2^{(n-1)/2}\sqrt{\pi}}{e} n^{3/2}.$$

**REFERENCES**

W. Barlow 1883. *Probable nature of the internal symmetry of crystals*, Nature **29**, 186-188.

C. Bender 1874. *Bestimmung der grössten Anzahl gleich grosser Kugeln, welche sich auf eine Kugel von demselben Radius, wie die übrigen, auflegen lassen*, Grunert Arch. **56**, 302-313.

H. F. Blichfeldt 1935. *The minimum values of positive quadratic forms in six, seven and eight variables*, Math. Z. **39**, 1-15.

J. Böhm 1959. *Simplexinhalt in Räumen konstanter Krümmung beliebiger Dimension*, J. Reine Angew. Math. **202**, 16-51.

———— 1960. *Inhaltsmessung im R₅ konstanter Krümmung*, Arch. Math. **11**, 298-309.

H. S. M. Coxeter 1930. *The polytopes with regular-prismatic vertex figures* (I), Philos. Trans. Roy. Soc. London, Ser. A **229**, 329-425.

———— 1935. *The functions of Schläfli and Lobatschefsky*, Quart. J. Math. Oxford Ser. **6**, 13-29.*

———— 1946. *Integral Cayley numbers*, Duke Math. J. **13**, 561-578.*

———— 1948. *Regular polytopes*, 2nd ed., Macmillan, New York.

———— 1951. *Extreme forms*, Canad. J. Math. **3**, 391-441.

———— 1954. *Arrangements of equal spheres in non-Euclidean spaces*, Acta Math. Acad. Sci. Hungar. **5**, 263-274∗

———— 1958. *Close-packing and froth*, Illinois J. Math. **2**, 746-758.

———— 1961. *Introduction to geometry*, Wiley, New York.

———— 1962. *The problem of packing a number of equal nonoverlapping circles on a sphere*, Trans. New York Acad. Sci. (2) **24**, 320-331.

H. Davenport and G. Hajós 1951. *Aufgabe* 35, Mat. Lapok **2**, 68.

G. L. Dirichlet 1850. *Über die Reduktion der positiven quadratischen Formen mit drei unbestimmten ganzen Zahlen*, J. Reine Angew. Math. **40**, 209-227.

L. Fejes Tóth 1953. *Lagerungen in der Ebene, auf der Kugel und im Raum*, Springer, Berlin.

———— 1954. *On close-packings of spheres in spaces of constant curvature*, Publ. Math. Debrecen **3**, 158-167.

———— 1959. *Kugelunterdeckungen und Kugelüberdeckungen in Räumen konstanter Krummung*, Arch. Math. **10**, 303-313.

S. Günter 1875. *Ein stereometrisches Problem*, Grunert Arch. **57**, 209-215.

A. P. Guinand 1959. *A note on the angles in an n-dimensional simplex*, Proc. Glasgow Math. Assoc. **4**, 58-61.

S. Hales 1727. *Vegetable staticks*, London.

J. Kepler 1611. *The six-cornered snowflake*, Oxford, London, 1966.

J. Leech 1956. *The problem of the thirteen spheres*, Math. Gaz. **40**, 22-23.

———— 1967. *Five-dimensional non-lattice sphere packings*. Canad. Math. Bull. **10**, 387–393.

H. Meschkowski 1960. *Ungelöste und unlösbare Probleme der Geometrie*, Vieweg, Braunschweig.

H. Minkowski 1905. *Diskontinuitätsbereich für arithmetische Äquivalenz*, J. Reine Angew. Math. **129**, 220-274.

R. A. Rankin 1955. *The closest packing of spherical caps in n dimensions*, Proc. Glasgow Math. Assoc. **2**, 139-144.

C. A. Rogers 1958. *The packing of equal spheres*, Proc. London Math. Soc. (3) **8**, 609-620.

———— 1961. *An asymptotic expansion for certain Schläfli functions*, J. London Math. Soc. **36**, 78-80.

H. Ruben 1961. *A multidimensional generalization of the inverse sine function*, Quart. J. Math. Oxford Ser. (2) **12**, 257-264.

L. Schläfli 1855. *Réduction d'une intégrale multiple, qui comprend l'arc de cercle et l'aire du triangle sphérique comme cas particuliers*, Gesammelte mathematische Abhandlungen, Bd. 2, pp. 164-190. (See also pp. 219-258.)

K. Schütte and B. L. van der Waerden 1953. *Das Problem der dreizehn Kugeln*, Math. Ann. **125**, 325-334.

G. Voronoi 1908. *Recherches sur les paralléloèdres primitives*, J. Reine Angew. Math. **134**, 198-287.

∗References "Coxeter 1935, 1946, 1954" are Chapters 1, 2, 8 of this book.

# 10

# REGULAR HONEYCOMBS IN HYPERBOLIC SPACE

Reprinted, with permission,
from the *Proceedings of the International
Congress of Mathematicians,
Amsterdam*, 1954 (Vol. III).

1. *Introduction.* Schlegel (1883,* pp. 444, 454) made a study of honeycombs whose cells are equal regular polytopes in spaces of positive, zero, and negative curvature. The spherical and Euclidean honeycombs had already been described by Schläfli (1855), but the only earlier mention of the hyperbolic honeycombs was when Stringham (1880, pp. 7, 12, and *errata*) discarded them as "imaginary figures", or, for the two-dimensional case, when Klein (1879) used them in his work on automorphic functions. Interest in them was revived by Sommerville (1923), who investigated their metrical properties.

The honeycombs considered by the above authors have finite cells and finite vertex figures. It seems desirable to make a slight extension so as to allow infinite cells, and infinitely many cells at a vertex, because of applications to indefinite quadratic forms (Coxeter and Whitrow 1950, pp. 424, 428) and to the close packing of spheres (Fejes Tóth 1953, p. 159). However, we shall restrict consideration to cases where the fundamental region of the symmetry group has a finite content, like that of a space group in crystallography. This extension increases the number of three-dimensional honeycombs from four to fifteen, the number of four-dimensional honeycombs from five to seven, and the number of five-dimensional honeycombs from zero to five.

A further extension allows the cell or vertex figure to be a *star*-polytope, so that the honeycomb covers the space several times. Some progress in this direction was made in two earlier papers: one (Coxeter 1933), not insisting on finite fundamental regions, was somewhat lacking in rigour; the other (Coxeter 1946) was restricted to two dimensions. The present treatment is analogous to § 14.8 of *Regular Polytopes* (Coxeter 1948, p. 283). We shall find that there are four regular star-honeycombs in hyperbolic 4-space, as well as two infinite families of them in the hyperbolic plane.

2. *Two-dimensional honeycombs.* In the Euclidean plane, the angle of a regular $p$-gon, $\{p\}$, is $(1 - 2/p)\pi$. In the hyperbolic plane it is smaller, gradually decreasing to zero when the side increases from 0 to $\infty$. Hence, if $p$ and $q$ are positive integers satsifying

2.1 $$(p - 2)(q - 2) > 4,$$

we can adjust the size of the polygon so as to make the angle $2\pi/q$. Then $q$

*For references, see page 214.

such $\{p\}$'s will fit together round a common vertex, and we can add further $\{p\}$'s indefinitely. In this manner we construct a two-dimensional honeycomb or *tessellation* $\{p, q\}$, which is an infinite collection of regular $p$-gons, $q$ at each vertex, filling the whole hyperbolic plane just once (Schlegel 1883, p. 360). We call $\{p\}$ the *face*, $\{q\}$ the *vertex figure*. The centres of the faces of $\{p, q\}$ are the vertices of the *reciprocal* (or dual) tesselation $\{q, p\}$, whose edges cross those of $\{p, q\}$. The simplest instances, $\{7, 3\}$ and $\{3, 7\}$, are shown conformally in Figs. 1 and 2.

As limiting cases, we admit $\{\infty, p\}$, whose faces $\{\infty\}$ are inscribed in horocycles instead of finite circles, and its reciprocal $\{p, \infty\}$, whose vertices are all at infinity (i.e., on the absolute conic).

The lines of symmetry, in which $\{p, q\}$ reflects into itself, are its edges (produced) and the lines of symmetry of its faces. They form a network of congruent triangles whose angles are $\pi/q$ (at a vertex of $\{p, q\}$), $\pi/2$ (at the mid-point of an edge), and $\pi/p$ (at the centre of a face). The symmetry group is generated by reflections in the sides of such a *characteristic* triangle $P_0P_1P_2$. Alternatively, we may begin with the triangle and derive $\{p, q\}$ by Wythoff's construction (Coxeter 1948, p. 87) as indicated by the symbol

$$\underset{p \qquad q}{\odot\!\!-\!\!-\!\!\bullet\!\!-\!\!-\!\!\bullet}$$

This means that the vertices of $\{p, q\}$ are the images of the vertex $P_0$ (where the angle is $\pi/q$) in the kaleidoscope formed by mirrors along the three sides of the triangle. (The nodes of the graph represent mirrors, those not directly joined being at right angles.)

For some instances of the network of characteristic triangles, see Klein (1879, p. 448), Fricke (1892, p. 458), and Coxeter (1939, pp. 126, 127).

The classical formulae for a right-angled triangle (Coxeter 1947, p. 238) enable us to compute the sides

$$\varphi = P_0P_1, \qquad \chi = P_0P_2, \qquad \psi = P_1P_2$$

in the form

$$\cosh \varphi = \cos \frac{\pi}{p} \,/ \sin \frac{\pi}{q}, \quad \cosh \chi = \cot \frac{\pi}{p} \cot \frac{\pi}{q}, \quad \cosh \psi = \cos \frac{\pi}{q} \,/ \sin \frac{\pi}{p}$$

(Sommerville 1923, p. 86; cf. Coxeter 1948, pp. 21, 64). Then we may describe $\{p, q\}$ as a tesselation of edge $2\varphi$, whose faces are $p$-gons of circum-radius $\chi$ and in-radius $\psi$.

3. *Three-dimensional honeycombs.* When

$$(p - 2)(q - 2) < 4,$$

the symbol $\{p, q\}$ denotes a Platonic solid (e.g., $\{4, 3\}$ is a regular hexahedron) which exists not only in Euclidean space but equally well in hyperbolic space. When the edge increases from 0 to $\infty$, the dihedral angle decreases from its Euclidean value

$$2 \arcsin \left( \cos \frac{\pi}{q} \Big/ \sin \frac{\pi}{p} \right)$$

to $(1 - 2/q)\pi$. For, the solid angle at a vertex resembles a Euclidean solid angle in that its section by a sphere is a spherical $q$-gon, whose angle must exceed $(1 - 2/q)\pi$.

Thus the dihedral angle can take the value $2\pi/r$ whenever $p$, $q$, $r$ are integers, greater than 2, satisfying

3.1 $$\sin \frac{\pi}{p} \sin \frac{\pi}{r} < \cos \frac{\pi}{q}$$

and $(q - 2)(r - 2) < 4$. Then $r$ such $\{p, q\}$'s will fit together round a common edge, and we can add further $\{p, q\}$'s indefinitely. In this manner we construct a three-dimensional honeycomb $\{p, q, r\}$, which is an infinite collection of $\{p, q\}$'s, $r$ at each edge, filling the whole hyperbolic space just once. The arrangement of cells round a vertex is like the arrangement of faces of the polyhedron $\{q, r\}$, which is called the *vertex figure*. The centres of the cells of $\{p, q, r\}$ are the vertices of the *reciprocal* honeycomb $\{r, q, p\}$, whose edges cross the $\{p\}$'s of $\{p, q, r\}$.

The actual instances are

$$\{3, 5, 3\}, \qquad \{4, 3, 5\}, \qquad \{5, 3, 4\}, \qquad \{5, 3, 5\},$$

in which the cells are respectively: an icosahedron of angle $2\pi/3$, a hexahedron of angle $2\pi/5$, and dodecahedra of angles $\pi/2$ and $2\pi/5$ (Schlegel 1883, p. 444). The existence of the "hyperbolic dodecahedron space" (Weber and Seifert 1933, pp. 241—243) reveals an interesting property of the self-reciprocal honeycomb $\{5, 3, 5\}$: its symmetry group has a subgroup (of index 120) which is transitive on the vertices and has the whole cell for a fundamental region. This resembles the translation group of the Euclidean honeycomb of cubes, $\{4, 3, 4\}$, only now instead of translations we have screws.

Returning to the general discussion, we allow the dihedral angle of the Platonic solid $\{p, q\}$ to take its minimum value $(1 - 2/q)\pi$, so as to obtain the further honeycombs

$$\{3, 4, 4\}, \qquad \{3, 3, 6\}, \qquad \{4, 3, 6\}, \qquad \{5, 3, 6\},$$

whose vertices are all at infinity (i.e., on the absolute quadric). We naturally consider also the respective reciprocals

$$\{4, 4, 3\}, \qquad \{6, 3, 3\}, \qquad \{6, 3, 4\}, \qquad \{6, 3, 5\},$$

whose cells are inscribed in horospheres instead of finite spheres (Coxeter and Whitrow 1950, p. 426), as well as the three self-reciprocal honeycombs

$$\{6, 3, 6\}, \qquad \{4, 4, 4\}, \qquad \{3, 6, 3\},$$

which suffer from both peculiarities at once.

In other words, necessary and sufficient conditions for the existence of a hyperbolic honeycomb $\{p, q, r\}$ are 3.1 and

3.2 $$(p - 2)(q - 2) \leqq 4, \qquad (q - 2)(r - 2) \leqq 4.$$

The honeycomb can be derived by Wythoff's construction from its characteristic simplex, which is the quadrirectangular hyperbolic tetrahedron

(Coxeter 1948, p. 139). The inequalities 3.2 ensure that this tetrahedron has a finite volume, being entirely accessible except that it may have one or two vertices at infinity. From relations between the edges and angles of the tetrahedron, we find the edge-length of $\{p, q, r\}$ to be $2\varphi$ while its cell $\{p, q\}$ has circum-radius $\chi$ and in-radius $\psi$, where

$$\cosh \varphi = \cos \frac{\pi}{p} \sin \frac{\pi}{r} \Big/ \sin \frac{\pi}{h_{q,r}}, \quad \cosh \psi = \sin \frac{\pi}{p} \cos \frac{\pi}{r} \Big/ \sin \frac{\pi}{h_{p,q}},$$

$$\cosh \chi = \cos \frac{\pi}{p} \cos \frac{\pi}{q} \cos \frac{\pi}{r} \Big/ \sin \frac{\pi}{h_{p,q}} \sin \frac{\pi}{h_{q,r}},$$

$h_{p,q}$ being given by

$$\cos^2 \frac{\pi}{h_{p,q}} = \cos^2 \frac{\pi}{p} + \cos^2 \frac{\pi}{q}$$

(Coxeter 1948, p. 19). For the particular cases, see Table III, the first part of which was given earlier by Sommerville (1923, p. 96), whose $e$, $R$, $r$ are our $2\varphi$, $\chi$, $\psi$.

4. *Four-dimensional honeycombs*. When

4.1 $$\sin \frac{\pi}{p} \sin \frac{\pi}{r} > \cos \frac{\pi}{q},$$

the symbol $\{p, q, r\}$ denotes a regular four-dimensional polytope (Coxeter 1948, p. 135), which has smaller angles in hyperbolic space than in Euclidean. A discussion analogous to that of § 3 shows that infinitely many such cells $\{p, q, r\}$ can be fitted together, $s$ round each plane face, to make a four-dimensional honeycomb

$$\{p, q, r, s\},$$

whenever both $\{p, q, r\}$ and $\{q, r, s\}$ are finite polytopes or Euclidean honeycombs such that

4.2
$$\frac{\cos^2 \pi/q}{\sin^2 \pi/p} + \frac{\cos^2 \pi/r}{\sin^2 \pi/s} > 1$$

(cf. Coxeter 1948, p. 136). In this manner, the six polytopes

$$\{3, 3, 3\}, \quad \{3, 3, 4\}, \quad \{4, 3, 3\}, \quad \{3, 4, 3\}, \quad \{3, 3, 5\}, \quad (5, 3, 3),$$

and the cubic honeycomb $\{4, 3, 4\}$, yield the seven hyperbolic honeycombs

$$\{3, 3, 3, 5\}, \quad \{4, 3, 3, 5\}, \quad \{5, 3, 3, 5\}, \quad \{5, 3, 3, 4\}, \quad \{5, 3, 3, 3\},$$
$$\{3, 4, 3, 4\}, \qquad \{4, 3, 4, 3\}.$$

Each of these can be derived from the appropriate characteristic simplex

by Wythoff's construction (Coxeter 1948, p. 199).

The edge-length $2\varphi$, and the circum- and in-radii of a cell, $\chi$ and $\psi$, are given by the formulae

$$\text{sech}^2 \varphi = \frac{(-1, 1)(0, 5)}{(1, 5)}, \qquad \text{sech}^2 \psi = \frac{(-1, 4)(3, 5)}{(-1, 3)},$$
$$\text{sech}^2 \chi = (-1, 4)(0, 5)$$

(Coxeter 1948, p. 161), where the symbols $(j, k)$ are derived by the recurrence formula

$$(j, k) = \frac{(j, k-1)(j+1, k) - 1}{(j+1, k-1)}, \quad (j, j+1) = 1,$$

from any convenient sequence of numbers

$$(-1, 1), \qquad (0, 2), \qquad (1, 3), \qquad (2, 4), \qquad (3, 5)$$

satisfying

$$(-1, 1)(0, 2) = \sec^2 \frac{\pi}{p}, \quad (0, 2)(1, 3) = \sec^2 \frac{\pi}{q},$$
$$(1, 3)(2, 4) = \sec^2 \frac{\pi}{r}, \quad (2, 4)(3, 5) = \sec^2 \frac{\pi}{s}.$$

In any particular case, the values are most easily computed by arranging the $(j, k)$'s in a triangular table:

$$(-1, 1) \qquad (0, 2) \qquad (1, 3) \qquad (2, 4) \qquad (3, 5)$$
$$(-1, 2) \qquad (0, 3) \qquad (1, 4) \qquad (2, 5)$$
$$(-1, 3) \qquad (0, 4) \qquad (1, 5)$$
$$(-1, 4) \qquad (0, 5)$$
$$(-1, 5)$$

e.g., the respective tables for $\{5, 3, 3, 5\}$ and $\{4, 3, 4, 3\}$ are

| | | | | | | | | | | |
|---|---|---|---|---|---|---|---|---|---|---|
| $2\tau^{-2}$ | 2 | 2 | 2 | $2\tau^{-2}$ | | 1 | 2 | 2 | 1 | 4 |
| $\sqrt{5}\tau^{-3}$ | 3 | 3 | $\sqrt{5}\tau^{-3}$ | | | 1 | 3 | 1 | 3 | |
| $2\tau^{-4}$ | 4 | $2\tau^{-4}$ | | | | 1 | 1 | 2 | | |
| $\tau^{-6}$ | $\tau^{-6}$ | | | | | 0 | 1 | | | |
| $-2\sqrt{5}\tau^{-6}$ | | | | | | $-1$ | | | | |

where $\tau = \tfrac{1}{2}(\sqrt{5} + 1)$, so that $\tau^{-1} = \tfrac{1}{2}(\sqrt{5} - 1)$. The negative value of $(-1, 5)$ provides a verification that the honeycomb is hyperbolic, and the zero for $(-1, 4)$ indicates that the cell of $\{4, 3, 4, 3\}$ is infinite. (For the results of this computation, see Table IV.)

5. *Five-dimensional honeycombs.* Similarly, the cell $\{p, q, r, s\}$ and vertex figure $\{q, r, s, t\}$ of a five-dimensional honeycomb $\{p, q, r, s, t\}$ must occur among the finite polytopes

$$\{3, 3, 3, 3\}, \qquad \{3, 3, 3, 4\}, \qquad \{4, 3, 3, 3\}$$

or among the Euclidean honeycombs

$$\{3, 3, 4, 3\}, \qquad \{3, 4, 3, 3\}, \qquad \{4, 3, 3, 4\}$$

(Coxeter 1948, p. 136). Since

$$\{3, 3, 3, 3, 3\}, \qquad \{3, 3, 3, 3, 4\}, \qquad \{4, 3, 3, 3, 3\}$$

are finite, while $\{4, 3, 3, 3, 4\}$ is Euclidean, the only hyperbolic honeycombs $\{p, q, r, s, t\}$ are

$$\{3, 3, 3, 4, 3\}, \quad \{4, 3, 3, 4, 3\}, \quad \{3, 3, 4, 3, 3\}, \quad \{3, 4, 3, 3, 4\}, \quad \{3, 4, 3, 3, 3\}$$

all of which have either infinite cells or all their vertices at infinity.

The edge-length and radii are now given by the formulae

$$\text{sech}^2\,\varphi = \frac{(-1, 1)(0, 6)}{(1, 6)}, \quad \text{sech}^2\,\psi = \frac{(-1, 5)(4, 6)}{(-1, 4)},$$

$$\text{sech}^2\,\chi = (-1, 5)(0, 6),$$

with $(3, 5)(4, 6) = \sec^2 \pi/t$. The triangular tables for $\{3, 3, 3, 4, 3\}$ and $\{3, 3, 4, 3, 3\}$ are

| | | | | | | | | | | | | |
|---|---|---|---|---|---|---|---|---|---|---|---|---|
| 2 | 2 | 2 | 2 | 1 | 4 | | 2 | 2 | 2 | 1 | 4 | 1 |
| 3 | 3 | 3 | 1 | 3 | | | 3 | 3 | 1 | 3 | 3 | |
| 4 | 4 | 1 | 2 | | | | 4 | 1 | 2 | 2 | | |
| 5 | 1 | 1 | | | | | 1 | 1 | 1 | | | |
| 1 | 0 | | | | | | 0 | 0 | | | | |
| $-1$ | | | | | | | $-1$ | | | | | |

yielding $\varphi = \chi = \infty$, $\psi = \log \tau$ for the former, and $\varphi = \chi = \psi = \infty$ for the latter. (See Table V.)

This is the end of the story, so far as honeycombs of density 1 are concerned. For, if $n > 5$, the only finite polytopes and Euclidean honeycombs that might serve as cells and vertex figures are

$$\alpha_n = \{3, 3, \ldots, 3, 3\}, \qquad \beta_n = \{3, 3, \ldots, 3, 4\},$$
$$\gamma_n = \{4, 3, \ldots, 3, 3\}, \qquad \delta_n = \{4, 3, \ldots, 3, 4\};$$

and these yield only $\alpha_{n+1}$, $\beta_{n+1}$, $\gamma_{n+1}$, and $\delta_{n+1}$. Hence (Schlegel 1883, p. 455) *There are no regular honeycombs in hyperbolic space of six or more dimensions.*

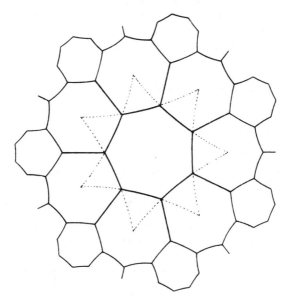

Figure 1

6. *Two-dimensional star-honeycombs.* If $n/d$ is a fraction in its lowest terms, whose value is greater than 2, the symbol $\{n/d\}$ denotes a regular star-polygon whose $n$ sides surround its centre $d$ times: briefly, a regular $n$-gon of density $d$, such as the pentagram $\{\frac{5}{2}\}$. It is natural to ask whether the symbol $\{p, q\}$ for a hyperbolic tessellation remains valid when $p$ or $q$ is fractional. We can obviously *begin* to construct such a star-tessellation whenever 2.1 is satisfied. The question is whether it will cover the plane a finite number of times, i.e., whether its *density* is finite.

The symmetry group of such a tessellation $\{p, q\}$ is still generated by reflections in the sides of its characteristic triangle, whose angles are $\pi/q$, $\pi/2$, and $\pi/p$ (Coxeter 1948, p. 109). If $p$ or $q$ is fractional, this triangle is dissected into smaller triangles by "virtual mirrors" (Coxeter 1948, pp. 75, 76), and the process of subdivision will continue until we come to a triangle all of whose angles are submultiples of $\pi$ (or possibly zero). There might conceivably be

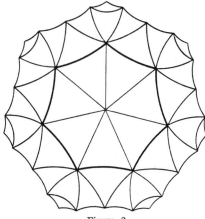

Figure 2

several different triangles of this kind, but the smallest of them will serve as a fundamental region for the group. We denote this smallest triangle by $(l \, m \, n)$ to indicate that its angles are $\pi/l$, $\pi/m$, $\pi/n$. The number of repetitions of it that fill the characteristic triangle $(2 \, p \, q)$ is an integer $D > 1$, which is equal to the density of the tessellation (Coxeter 1948, p. 110).

Since each angle of $(2 \, p \, q)$ must be a multiple of one of the angles of $(l \, m \, n)$, a zero angle of the former implies a zero angle of the latter. But since the subdivision of an asymptotic triangle $(2 \, p \, \infty)$ yields at least one finite piece, this is impossible; both $p$ and $q$ must be finite.

Since the triangle $(2 \, p \, q)$ is filled with $D$ triangles $(l \, m \, n)$, of area $(1 - l^{-1} - m^{-1} - n^{-1})\pi$, we have

$$\tfrac{1}{2} - p^{-1} - q^{-1} = D(1 - l^{-1} - m^{-1} - n^{-1}),$$

where the numerators of the rational numbers $p$ and $q$ are divisors of one or two of the integers $l$, $m$, $n$.

The numerators of $p$ and $q$ cannot divide two *different* integers among $l, m, n$, say $m$ and $n$; for then we would have

$$\tfrac{1}{2} - p^{-1} - q^{-1} \leqq \tfrac{1}{2} - m^{-1} - n^{-1} \leqq 1 - l^{-1} - m^{-1} - n^{-1},$$

implying $D \leqq 1$. Thus we may assume that these numerators both divide $m$, so that $p = m/x$ and $q = m/y$, where $x$ and $y$ are positive integers. Since $p$ and $q$ are not both integers,

$$x + y \geqq 3.$$

Moreover, $D \geqq 3$, since the triangle $(2\ p\ q)$ obviously cannot be *bisected* by a line through one of its acute vertices. Hence

$$\frac{1}{2} = \frac{x}{m} + \frac{y}{m} + D\left(1 - \frac{1}{l} - \frac{1}{m} - \frac{1}{n}\right)$$

$$\geqq \frac{3}{m} + 3\left(1 - \frac{1}{l} - \frac{1}{m} - \frac{1}{n}\right) = 3\left(1 - \frac{1}{l} - \frac{1}{n}\right),$$

i.e.,

$$\frac{1}{l} + \frac{1}{n} \geqq \frac{5}{6}.$$

Since also

$$\frac{1}{l} + \frac{1}{n} < \frac{1}{l} + \frac{1}{m} + \frac{1}{n} < 1,$$

the integers $l$ and $n$ must be 2 and 3, and

$$x + y = D = 3.$$

Thus the only possibility is the triangle $(2\ m\ m/2)$ dissected into three triangles $(2\ m\ 3)$. In other words,

*The only regular star-tessellations in the hyperbolic plane are*

$$\left\{\frac{m}{2}, m\right\} \text{ and } \left\{m, \frac{m}{2}\right\},$$

*of density* 3, *where m is any odd number greater than* 5.

Such star-tessellations exist also when $m = 5$, but then they are not hyperbolic but spherical. In fact, they are essentially the small stellated dodecahedron of Kepler and the great dodecahedron of Poinsot (Coxeter 1948, p. 95). But the other two Kepler-Poinsot polyhedra have no hyperbolic analogues.

We may describe the faces of $\{m/2, m\}$ as the stellated faces of $\{m, 3\}$. (One face of $\{\frac{7}{2}, 7\}$ is indicated by broken lines in Fig. 1.) Dually $\{m, m/2\}$ is derived from $\{3, m\}$ by regarding the same vertices and edges as forming $m$-gons (such as the heptagon of $\{7, \frac{7}{2}\}$ which is emphasized in Fig. 2) instead of triangles. The density 3 can be observed in the fact that each triangle of $\{3, m\}$ lies within three $m$-gons of $\{m, m/2\}$. In ascribing the same density to $(m/2, m)$, we regard the middle part of each $\{m/2\}$ as being covered twice over. The relationship of $(m/2, m)$ and $\{m, m/2\}$ is such that any face of either has the same vertices as a corresponding face of the other. Both have the same vertices as $\{3, m\}$, and the same face-centres as $\{m, 3\}$. Thus

$$\chi(m, m/2) = 2\varphi(m, m/2) = 2\varphi(3, m),$$

in agreement with Table II.

7. *Four-dimensional star-honeycombs.* One might expect to find a three-dimensional star-honeycomb $\{p, q, r\}$ whose cell $\{p, q\}$ or vertex figure $(q, r)$ is one of the Kepler-Poinsot polyhedra

$$\{\tfrac{5}{2}, 5\}, \qquad \{5, \tfrac{5}{2}\}, \qquad \{\tfrac{5}{2}, 3\}, \qquad \{3, \tfrac{5}{2}\}.$$

However, in every instance (Coxeter 1948, p. 264) the values of $p, q, r$ satisfy 4.1, not 3.1. Hence

*There are no regular star-honeycombs in hyperbolic 3-space.*

Hoping for a more positive result in four dimensions, we seek a honeycomb $\{p, q, r, s\}$ in which both $\{p, q, r\}$ and $\{q, r, s\}$ occur among the sixteen regular polytopes (Coxeter 1948, pp. 293—294) while at least one of $p, q, r. s$ has the fractional value $\tfrac{5}{2}$. A list of the twenty-three possibilities reveals that only four satisfy 4.2:

$$\{3, 3, 5, \tfrac{5}{2}\}, \qquad \{3, 5, \tfrac{5}{2}, 5\}, \qquad \{5, \tfrac{5}{2}, 5, 3\}, \qquad \{\tfrac{5}{2}, 5, 3, 3\}$$

(Coxeter 1948, p. 264: 14.15). That these four are genuine hyperbolic honeycombs, of finite density, may be seen by verifying that the reflections in the bounding hyperplanes of their characteristic simplexes

generate discrete groups. In Table I (cf. Coxeter 1948, p. 283) these simplexes appear as $X$ and $Z$, and we see how they are dissected, by virtual mirrors (bisecting their dihedral angles $2\pi/5$, which are indicated by the mark $\tfrac{5}{2}$ in the graphical symbols), into smaller simplexes $T$ and $W$, $W$ and $Y$. Similarly $W$ and $Y$, having angles $2\pi/3$, are further subdivided, until the process ends with the "quantum" $T$, which is thus seen to be the fundamental region for both these groups, as it is also for the symmetry group of the ordinary honeycombs $\{3, 3, 3, 5\}$ and $\{5, 3, 3, 3\}$.

The accuracy of Table I can be checked by observing that, in each graph, the branch with a fractional mark forms, with any third node, the symbol for a spherical triangle whose dissection is obvious; e.g., the dissection $Y = 2V$ embodies

$$(5\ 5\ \tfrac{3}{2}) = (5\ 2\ 3) + (2\ 5\ 3) \text{ and } (3\ 2\ \tfrac{3}{2}) = (3\ \tfrac{3}{2}\ 3) + (3\ 2\ 3)$$

(Coxeter 1948, p. 113). Since $X = 5T$ and $Z = 10T$, the densities of the star-honeycombs are 5 and 10.

Thus the six hyperbolic honeycombs

$$\{5, 3, 3, 3\}, \quad \{\tfrac{5}{2}, 5, 3, 3\}, \quad \{5, \tfrac{5}{2}, 5, 3\}, \quad \{3, 5, \tfrac{5}{2}, 5\}, \quad \{3, 3, 5, \tfrac{5}{2}\}, \quad \{3, 3, 3, 5\}$$

all have the same symmetry group, and their densities are the binomial coefficients

$$1, \quad 5, \quad 10, \quad 10, \quad 5, \quad 1.$$

It is interesting to compare them with the five spherical honeycombs (or Euclidean polytopes)

$$\{5, 3, 3\}, \quad \{\tfrac{5}{2}, 5, 3\}, \quad \{5, \tfrac{5}{2}, 5\}, \quad \{3, 5, \tfrac{5}{2}\}, \quad \{3, 3, 5\},$$

whose densitities are 1, 4, 6, 4, 1, and with the four spherical tessellations (or Euclidean polyhedra)

$$\{5, 3\}, \quad \{\tfrac{5}{2}, 5\}, \quad \{5, \tfrac{5}{2}\}, \quad \{3, 5\},$$

whose densities are 1, 3, 3, 1. We can summarize these results by saying that, when the Schläfli symbol for an $n$-dimensional honeycomb has $\tfrac{5}{2}$ in the $r^{th}$ place with 5 before and after and 3 everywhere else, the density is $\binom{n}{r}$.

The dissection of characteristic simplexes can be translated into a direct derivation of the honeycombs from one another. The first stage is to derive $\{\tfrac{5}{2}, 5, 3, 3\}$ from $\{5, 3, 3, 3\}$ by stellating each cell $\{5, 3, 3\}$ to form a $\{\tfrac{5}{2}, 5, 3\}$ (Coxeter 1948, p. 264). The second stage is to replace each $\{\tfrac{5}{2}\}$ of $\{\tfrac{5}{2}, 5, 3, 3\}$ by the pentagon that has the same vertices, and consequently each $\{\tfrac{5}{2}, 5\}$ by a $\{5, \tfrac{5}{2}\}$ and each $\{\tfrac{5}{2}, 5, 3\}$ by a $\{5, \tfrac{5}{2}, 5\}$. In this process the vertex figure gets stellated from $\{5, 3, 3\}$ to $\{\tfrac{5}{2}, 5, 3\}$, and the result is $\{5, \tfrac{5}{2}, 5, 3\}$. The third stage is to replace each $\{5, \tfrac{5}{2}\}$ by the icosahedron that has the same edges, and consequently each $\{5, \tfrac{5}{2}, 5\}$ by a $\{3, 5, \tfrac{5}{2}\}$. The vertex figure $\{\tfrac{5}{2}, 5, 3\}$ is changed into $\{5, \tfrac{5}{2}, 5\}$, and the result is $\{3, 5, \tfrac{5}{2}, 5\}$. The fourth stage is to replace each $\{3, 5, \tfrac{5}{2}\}$ by the $\{3, 3, 5\}$ that has the same faces, so that the vertex figure $\{5, \tfrac{5}{2}, 5\}$ is changed into $\{3, 5, \tfrac{5}{2}\}$ (which has the same edges), and we obtain $\{3, 3, 5, \tfrac{5}{2}\}$. The fifth and last stage is to replace the cell $\{3, 3, 5\}$ by a cluster of 600 regular simplexes $\{3, 3, 3\}$ surrounding a common vertex, so as to obtain $\{3, 3, 3, 5\}$. Each simplex belongs to five of the clusters, as we would expect from the fact that $\{3, 3, 5, \tfrac{5}{2}\}$ has density 5.

Taking the same stages in the reverse order, we may say that the vertices, edges, faces and cells of $\{3, 3, 3, 5\}$ belong also to $\{3, 3, 5, \tfrac{5}{2}\}$; the vertices, edges and faces belong also to $\{3, 5, \tfrac{5}{2}, 5\}$; the vertices and edges belong also to $\{5, \tfrac{5}{2}, 5, 3\}$; and the vertices belong also to $\{\tfrac{5}{2}, 5, 3, 3\}$, which is the "stellated" $\{5, 3, 3, 3\}$.

The formulae for $\varphi$, $\chi$, $\psi$ (§ 4) apply to star-honeycombs without any alteration. We see from Table IV that, since $\tau^3 = 2\tau^2 - 1$, the value of $\chi$ for each of them is equal to the value of $2\varphi$ for $\{3, 3, 3, 5\}$.

Finally, since there are no regular star-polytopes in five or more dimensions (Coxeter 1948, p. 278) to serve as cell or vertex figure,

*There are no regular star-honeycombs in hyperbolic space of five or more dimensions.*

*Table I. The dissection of characteristic simplexes in hyperbolic 4-space.*

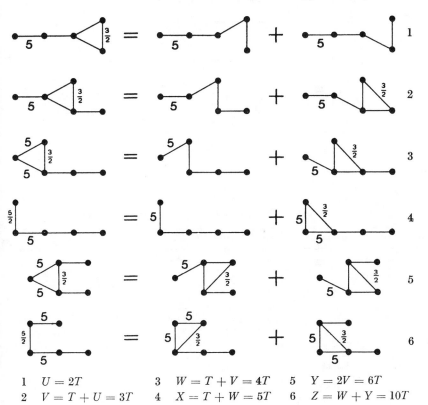

| 1 | $U = 2T$ | 3 | $W = T + V = 4T$ | 5 | $Y = 2V = 6T$ |
| 2 | $V = T + U = 3T$ | 4 | $X = T + W = 5T$ | 6 | $Z = W + Y = 10T$ |

Table II. Regular honeycombs in the hyperbolic plane

| Tessellation | Density | $\cosh \varphi$ | $\cosh \chi$ | $\cosh \psi$ |
|---|---|---|---|---|
| $\{p, q\}$ (integers) | 1 | $\cos \dfrac{\pi}{p} \operatorname{cosec} \dfrac{\pi}{q}$ | $\cot \dfrac{\pi}{p} \cot \dfrac{\pi}{q}$ | $\operatorname{cosec} \dfrac{\pi}{p} \cos \dfrac{\pi}{q}$ |
| $\left\{\dfrac{m}{2}, m\right\}$ | | $\cos \dfrac{2\pi}{m} \operatorname{cosec} \dfrac{\pi}{m}$ | $\cot \dfrac{\pi}{m} \cot \dfrac{2\pi}{m}$ | $\dfrac{1}{2} \operatorname{cosec} \dfrac{\pi}{m}$ |
| ($m$ odd) $\left\{m, \dfrac{m}{2}\right\}$ | 3 | $\dfrac{1}{2} \operatorname{cosec} \dfrac{\pi}{m}$ | | $\cos \dfrac{2\pi}{m} \operatorname{cosec} \dfrac{\pi}{m}$ |

Table III. The fifteen regular honeycombs in hyperbolic 3-space

| Honeycomb | Density | $\cosh^2\varphi$ | $\cosh^2 \chi$ | $\cosh^2 \psi$ | Honeycomb | Density | $\cosh^2 \varphi$ | $\cosh^2 \chi$ | $\cosh^2 \psi$ |
|---|---|---|---|---|---|---|---|---|---|
| $\{3, 5, 3\}$ | 1 | $\frac{3}{4}\tau^2$ | $\frac{1}{4}\tau^6$ | $\frac{3}{4}\tau^2$ | $\{3, 6, 3\}$ | 1 | $\infty$ | $\infty$ | $\infty$ |
| $\{4, 3, 5\}$ | 1 | $\frac{1}{2}\sqrt{5}\,\tau$ | $\frac{1}{2}\tau^4$ | $\frac{1}{2}\tau^2$ | $\{4, 3, 6\}$ | 1 | $\infty$ | $\infty$ | $\frac{3}{2}$ |
| $\{5, 3, 4\}$ | | $\frac{1}{2}\tau^2$ | | $\frac{1}{2}\sqrt{5}\,\tau$ | $\{6, 3, 4\}$ | | $\frac{3}{2}$ | | $\infty$ |
| $\{5, 3, 5\}$ | 1 | $\frac{1}{4}\sqrt{5}\,\tau^3$ | $\frac{1}{4}\tau^8$ | $\frac{1}{4}\sqrt{5}\,\tau^3$ | $\{4, 4, 4\}$ | 1 | $\infty$ | $\infty$ | $\infty$ |
| $\{3, 3, 6\}$ | 1 | $\infty$ | $\infty$ | $\frac{9}{8}$ | $\{5, 3, 6\}$ | 1 | $\infty$ | $\infty$ | $\frac{3}{4}\sqrt{5}\,\tau$ |
| $\{6, 3, 3\}$ | | $\frac{9}{8}$ | | $\infty$ | $\{6, 3, 5\}$ | | $\frac{3}{4}\sqrt{5}\,\tau$ | | $\infty$ |
| $\{3, 4, 4\}$ | 1 | $\infty$ | $\infty$ | $\frac{3}{2}$ | $\{6, 3, 6\}$ | 1 | $\infty$ | $\infty$ | $\infty$ |
| $\{4, 4, 3\}$ | | $\frac{3}{2}$ | | $\infty$ | | | | | |

$$\left(\tau = \frac{\sqrt{5} + 1}{2}\right)$$

Table IV. The eleven regular honeycombs in hyperbolic 4-space

| Honeycomb | Density | $\cosh\varphi$ | $\cosh\chi$ | $\cosh\psi$ |
|---|---|---|---|---|
| $\{3, 3, 3, 5\}$ | 1 | $\tau$ | $\sqrt{\tfrac{1}{5}}\,\tau^3$ | $\sqrt{\tfrac{2}{5}}\,\tau$ |
| $\{5, 3, 3, 3\}$ | | $\sqrt{\tfrac{2}{5}}\,\tau$ | | $\tau$ |
| $\{4, 3, 3, 5\}$ | 1 | $\sqrt{2}\,\tau$ | $\tau^3$ | $\sqrt{\tfrac{1}{2}}\,\tau$ |
| $\{5, 3, 3, 4\}$ | | $\sqrt{\tfrac{1}{2}}\,\tau$ | | $\sqrt{2}\,\tau$ |
| $\{5, 3, 3, 5\}$ | 1 | $\tau^2$ | $\tau^6$ | $\tau^2$ |

| Honeycomb | Density | $\cosh\varphi$ | $\cosh\chi$ | $\cosh\psi$ |
|---|---|---|---|---|
| $\{3, 4, 3, 4\}$ | 1 | $\infty$ | $\infty$ | $\sqrt{2}$ |
| $\{4, 3, 4, 3\}$ | | $\sqrt{2}$ | | $\infty$ |
| $\{\tfrac{5}{2}, 5, 3, 3\}$ | 5 | $\sqrt{2}\,\tau$ | $\tau^3$ | $\tau$ |
| $\{3, 3, 5, \tfrac{5}{2}\}$ | | $\tau$ | | $\sqrt{2}\,\tau$ |
| $\{3, 5, \tfrac{5}{2}, 5\}$ | 10 | $\tau$ | $\tau^3$ | $\tau$ |
| $\{5, \tfrac{5}{2}, 5, 3\}$ | | | | |

Table V. The five regular honeycombs in hyperbolic 5-space

| Honeycomb | Density | $\varphi$ | $\chi$ | $\psi$ |
|---|---|---|---|---|
| $\{3, 3, 3, 4, 3\}$ | 1 | $\infty$ | $\infty$ | $\log\tau$ |
| $\{3, 4, 3, 3, 3\}$ | | $\log\tau$ | | $\infty$ |
| $\{3, 3, 4, 3, 3\}$ | 1 | $\infty$ | $\infty$ | $\infty$ |
| $\{3, 4, 3, 3, 4\}$ | 1 | $\infty$ | $\infty$ | $\infty$ |
| $\{4, 3, 3, 4, 3\}$ | | | | |

REFERENCES

H. S. M. COXETER 1933. The densities of the regular polytopes, Part 3, Proc. Camb. Phil. Soc. **29**, 1—22.

H. S. M. COXETER 1939. The abstract groups $G^{m, n, p}$, Trans. Amer. Math. Soc. **45**, 73—150.

H. S. M. COXETER 1946. The nine regular solids, Proc. 1st Canadian Math. Congress, 252—264.

H. S. M. COXETER 1947. Non-Euclidean geometry, Toronto (University Press).

H. S. M. COXETER 1948. Regular polytopes, 2nd ed., New York (Macmillan, 1963).

H. S. M. COXETER and G. J. WHITROW 1950. World-structure and non-Euclidean honeycombs, Proc. Roy. Soc. A, **201**, 417—437.

L. FEJES TÓTH 1953. On close-packings of spheres in spaces of constant curvature, Publ. Math. Debrecen, **3**, 158—167.

R. FRICKE 1892. Ueber den arithmetischen Charakter der zu den Verzweigungen (2, 3, 7) und (2, 4, 7) gehörenden Dreiecksfunctionen, Math. Ann. **41**, 443—468.

F. KLEIN 1879. Ges. Math. Abh. **3** (Berlin, 1923), 90—136.

L. SCHLÄFLI 1855. Ges. Math. Abh. **2** (Basel, 1953), 164—190.

V. SCHLEGEL 1883. Theorie der homogen zusammengesetzten Raumgebilde, Nova Acta Leop. Carol. **44**, 343—459.

D. M. Y. SOMMERVILLE 1923. Division of space by congruent triangles and tetrahedra, Proc. Roy. Soc. Edinburgh, **43**, 85—116.

W. I. STRINGHAM 1880. Regular figures in $n$-dimensional space, Amer. J. Math. **3**, 1—14.

C. WEBER and H. SEIFERT 1933. Die beiden Dodekaederräume, Math. Z, **37**, 237—253.

# 11

# REFLECTED LIGHT SIGNALS

**Abstract:** In a generalized Michelson-Morley experiment, two un-accelerated observers emit light signals alternately, each emitting a signal at the instant when he receives one from the other, and the first observer records the sequence of proper times for these events. If both observers measure time from the events when their space-like separation $\phi$ is perpendicular to both world lines (which are, in general, skew lines), if their relative rapidity at this origin of time is $\psi$, and if the world is Minkowskian, the proper time for the first observer's jth event is found to be

$$t_{2j} = t_1 U_{2j-1}(\cosh \psi) - t_0 U_{2j-2}(\cosh \psi),$$

where $t_1 = t_0 \cosh \psi + (t_0 \sinh^2\psi + \phi^2)^{1/2}$ and $U_m(x)$ is a Chebyshev polynomial of the second kind. The analogous formulae in de Sitter's world[14] are

$$\tanh \frac{t_n + t_{n-2}}{2} = k \tanh t_{n-1}, \quad k = \frac{\cosh \psi}{\cos \phi},$$

$$\cosh t_1 \cosh t_0 \cos \phi - \sinh t_1 \sinh t_0 \cosh \psi = 1.$$

## 1. Introduction

The space-time of relativity theory is a geometrical diagram consisting of a four-dimensional space whose points represent events (idealized so that an event has no size and no duration).* The locus of events in the history of a particle is a directed curve in the 4-space. Following Minkowski, we call this curve the world line of the particle. Each event O is the vertex of a light cone whose generators, called null lines, are the world lines of photons that could be emitted from O or received at O. † The world line of any other kind of particle, beginning at O, is a curve whose tangent at O is a timelike line, interior to the cone. In particular, the world line of an unaccelerated particle is a timelike line itself. A third kind of line, called spacelike, joins two events that are essentially incapable of influencing each other.

In the special theory of relativity, the four-dimensional space is Minkowskian, that is, real affine 4-space (embedded, for convenience, in a real projective 4-space) with a pseudo-Euclidean metric determined by a non-ruled quadric in the hyperplane at infinity (which is a real projective 3-space). The light cone at any event O joins O to all the points of this quadric.

---

* See reference 10, pp. 5 and 6. (For references, see page 228.)

† op. cit., p. 41

It is convenient to use a system of affine coordinates $x_1, \ldots, x_4$ based on projective coordinates $x_1, \ldots, x_4, x_5$ in such a way that the hyperplane at infinity is $x_5 = 0$ and the absolute quadric is

$$x_1^2 + x_2^2 + x_3^2 - x_4^2 = 0, \ x_5 = 0.$$

Since the x's are homogeneous, any point for which $x_5 \neq 0$ may be expressed in the form $(x_1, x_2, x_3, x_4, 1)$, which we abbreviate to $(x_1, x_2, x_3, x_4)$ or simply $(x)$. The light cone at any point, or event, $(a)$ is

$$(x_1 - a_1)^2 + (x_2 - a_2)^2 + (x_3 - a_3)^2 - (x_4 - a_4)^2 = 0.$$

The line joining two events, $A = (a)$ and $B = (b)$, is timelike, null, or spacelike in accordance with whether the number

$$(a_1 - b_1)^2 + (a_2 - b_2)^2 + (a_3 - b_3)^2 - (a_4 - b_4)^2$$

is negative, zero, or positive. The interval, or <u>separation</u>, AB between these two events is defined to be the square root of the absolute magnitude of this number. If A and B are two events in the history of an unaccelerated particle, and we take the velocity of light to be 1, the separation AB is an interval of <u>proper time</u> for the particle.

If two unaccelerated particles part from a common event with relative velocity v, their world lines a and b form an angle whose measure depends on v. More precisely, if $a'$ and $b'$ are perpendicular to a and b, respectively, in their plane, the cross ratio $\{aa', bb'\}$ is equal to $v^2$.[6] Unit vectors $(a)$ and $(b)$ along these world lines satisfy

$$-a_1^2 - a_2^2 - a_3^2 + a_4^2 = -b_1^2 - b_2^2 - b_3^2 + b_4^2 = 1$$

(see reference 3, p. 281) and form a hyperbolic angle $\psi$ which is given by

$$\cosh \psi = (a) \cdot (b) = -a_1 b_1 - a_2 b_2 - a_3 b_3 + a_4 b_4.$$

Robb[9] (p. 406) calls $\psi$ the relative <u>rapidity</u> of the two particles.* By choosing the $x_3$-axis along the bisector of this "angle," we can express $(a)$ and $(b)$ in the symmetrical form

$$(0, 0, \pm\sinh \tfrac{1}{2} \psi, \cosh \tfrac{1}{2} \psi), \tag{1}$$

which gives their inner product the desired value

$$\sinh^2 \tfrac{1}{2} \psi + \cosh^2 \tfrac{1}{2} \psi = \cosh \psi.$$

## 2. Summary of Results

Figure 1 duplicates one of the figures of Robb† representing a zigzag $A_0 B_0 A_1 B_1 A_2 \ldots$ formed by segments of null lines joining points $A_j$ on one

---

* See also reference 8, p. 22, and reference 10, p. 126.

† See reference 9, p. 108.

timelike line a to points $B_j$ and $B_{j-1}$ on another timelike line b, so that $B_{j-1}A_j$ and $A_jB_j$ are generators of the light cone at $A_j$, and $A_jB_j$ and $B_jA_{j+1}$ are generators of the light cone at $B_j$. Physically, we can regard a and b as the world lines of two unaccelerated observers who emit light signals alternately, each emitting a signal at the instant when he receives one from the other (as if he could reflect it in exactly the right direction).[14] The first observer records his proper time $t_{2j}$ for the event $A_j$, and the second records his proper time $t_{2j+1}$ for $B_j$. The purpose of this note is to describe the sequence of times $t_0, t_2, t_4, \ldots$ (or $t_1, t_3, t_5, \ldots$). In section 3, we shall find that, in Minkowski's world of special relativity theory, if the observers measure time from the events when their spacelike separation $\phi$ is minimax (in a sense that will be made precise later), and if their relative rapidity at this origin of time is $\psi$, then

$$t_1 = t_0 \cosh \psi + (t_0^2 \sinh^2 \psi + \phi^2)^{1/2}$$

and

$$t_{2j} = \frac{t_1 \sinh 2j\psi - t_0 \sinh (2j-1)\psi}{\sinh \psi}.$$

In section 5, we shall improve an earlier solution of the analogous problem in de Sitter's world, where the absolute quadric surface in the hyperplane at infinity is replaced by a nondegenerate quadric hypersurface, and the light cone at any event O is the enveloping cone to this quadric hypersurface from O. Projective 4-space was used also by Veblen and Hoffmann[11] for more sophisticated developments in relativity theory.

### 3. Minkowski's World

The simplest case is that of the celebrated Michelson-Morley experiment, where the two observers are at rest (relatively), with a constant spacelike separation $\phi$, so that we have the arithmetic progression

$$t_{2j} = t_0 + 2j\phi.$$

Another possibility is that the observers were together, say at time zero, and are parting with relative velocity v, so that we have the geometric progression

$$t_{2j} = t_0V^j,$$

where V is a suitable function of v. The same formula (with $t_0$ negative and $V < 1$) covers the case when the observers are approaching each other. The only really interesting case is when the observers are unable ever to meet because their world lines are skew.

If a and b (fig. 1) are skew lines, they span a three-dimensional subspace which we can take to be $x_1 = 0$. In this subspace, we can draw a plane through each line parallel to the other, and thus place the two lines in parallel planes, say a in $x_2 = \frac{1}{2}\phi$ and b in $x_2 = -\frac{1}{2}\phi$. Then, using the Minkowskian definition for orthogonality, we can project each line ortho-

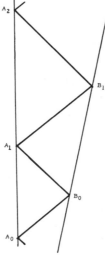

Figure 1

gonally on the plane through the other and so construct a common perpendicular, which we take to be the $x_2$-axis. The events at which the spacelike separation attains its <u>minimax</u>, $\phi$, are

$$(0, \pm \tfrac{1}{2} \phi, 0, 0),$$

and we may assume these events to be at time zero for both observers.

The above use of the word minimax was suggested by Wigner, who explains it as follows: for each point A on a, there is a corresponding point B on b, which is the foot of the perpendicular from A to b; i.e., the intersection of b with the plane through A perpendicular to b. This is the point on b that is <u>farthest</u> from A. (As we saw in figure 1, there are two points on b at zero distance from A.) The common perpendicular yields the positions of A and B for which this maximal distance AB is minimal. These are the events from which both observers agree to measure their proper time.

Applying the vectors in eq. (1), we deduce that their events at time t are

$$(0, \pm \tfrac{1}{2} \phi, \pm t \sinh \tfrac{1}{2} \psi, t \cosh \tfrac{1}{2} \psi).$$

Thus, the events $A_j$ and $B_j$ are $(0, \tfrac{1}{2} \phi, t_{2j} \sinh \tfrac{1}{2} \psi, t_{2j} \cosh \tfrac{1}{2} \psi)$ and $(0, -\tfrac{1}{2} \phi, -t_{2j+1} \sinh \tfrac{1}{2} \psi, t_{2j+1} \cosh \tfrac{1}{2} \psi)$.

Since the separation between two events (x) and (y) is the square root of

$$|(x_1 - y_1)^2 + (x_2 - y_2)^2 + (x_3 - y_3)^2 - (x_4 - y_4)^2|,$$

the null separations $A_jB_j$ and $B_{j-1}A_j$ yield

$$\phi^2 + (t_{2j} + t_{2j\pm 1})^2 \sinh^2 \tfrac{1}{2} \psi - (t_{2j} - t_{2j\pm 1})^2 \cosh^2 \tfrac{1}{2} \psi = 0.$$

Hence, for all positive integers n,

$$\phi^2 = t_n^2 + t_{n-1}^2 - 2t_n t_{n-1} \cosh \psi. \tag{2}$$

Subtracting the analogous equation with n replaced by $n-1$, and dividing by $t_n - t_{n-2}$, we obtain

$$t_n + t_{n-2} = 2t_{n-1} \cosh \psi = (e^\psi + e^{-\psi})t_{n-1}, \tag{3}$$

whence

$$t_n - e^{-\psi} t_{n-1} = e^\psi (t_{n-1} - e^{-\psi} t_{n-2}) = \dots \tag{4}$$
$$= e^{(n-1)\psi} (t_1 - e^{-\psi} t_0).$$

Adding together n equations such as

$$e^{n\psi}t_n - e^{(n-1)\psi}t_{n-1} = e^{(2n-1)\psi}(t_1 - e^{-\psi}t_0),$$

we deduce

$$e^{n\psi}t_n - t_0 = \frac{e^{2n\psi} - 1}{e^{2\psi} - 1}e^\psi (t_1 - e^{-\psi}t_0)$$
$$= \frac{e^{n\psi} - e^{-n\psi}}{e^\psi - e^{-\psi}} e^{n\psi}t_1 - \frac{e^{(2n-1)\psi} - e^{-\psi}}{e^\psi - e^{-\psi}} t_0$$
$$= \frac{e^{n\psi} - e^{-n\psi}}{e^\psi - e^{-\psi}} e^{n\psi}t_1 - \left(1 + \frac{e^{(n-1)\psi} - e^{-(n-1)\psi}}{e^\psi - e^{-\psi}} e^{n\psi}\right)$$

whence

$$t_n = \frac{\sinh n\psi}{\sinh \psi} t_1 - \frac{\sinh (n-1)\psi}{\sinh \psi} t_0$$
$$= U_{n-1}(\cosh \psi)t_1 - U_{n-2}(\cosh \psi)t_0, \tag{5}$$

where $U_m(X)$ is a Chebyshev polynomial of the second kind, namely

$$U_m(X) = \sum_{r=0}^{[m/2]} (-1)^r \binom{m-r}{r}(2X)^{m-2r}$$

$$= \begin{vmatrix} 2X & 1 & 0 & \dots & 0 & 0 \\ 1 & 2X & 1 & \dots & 0 & 0 \\ & \dots & & & \dots & \\ 0 & 0 & 0 & \dots & 1 & 2X \end{vmatrix}.$$

To express $t_1$ in terms of $t_0$, we set $n = 1$ in eq. (2), obtaining

$$t_1 = t_0 \cosh \psi + (t_0^2 \sinh^2 \psi + \phi^2)^{1/2}. \tag{6}$$

Thus, the desired sequence of proper times $t_{2j}$ is given by eqs. (5) and (6) (with 2j for n).

## 4. Special Cases

If $t_0 = 0$, so that the experiment begins when the observers are at their closest, we have $t_1 = \phi$ and

$$t_n = \frac{\phi \sinh n\psi}{\sinh \psi} = \phi \ U_{n-1}(\cosh \psi). \tag{7}$$

The case when the two observers are at rest (relatively) is given by making $\psi$ tend to zero, so that

$$t_n = n\phi. \tag{8}$$

The other simple case is when $\phi = 0$ (and $t_0 \neq 0$), so that the two world lines intersect. If $t_0 < 0$, so that the observers are approaching each other, eq. (6) yields

$$t_1 = t_0 \cosh \psi + |t_0| \sinh \psi = t_0 (\cosh \psi - \sinh \psi) = e^{-\psi}t_0,$$

whence, by eq. (4),

$$t_n - e^{-\psi}t_{n-1} = 0$$

and

$$t_n = e^{-n\psi}t_0. \tag{9}$$

If $t_0 > 0$, so that the observers are receding from each other, we have

$$t_1 = t_0 (\cosh \psi + \sinh \psi) = e^{\psi}t_0.$$

Then the relation (3) can more usefully be expressed as

$$t_n - e^{\psi}t_{n-1} = e^{-\psi}(t_{n-1} - e^{\psi}t_{n-2}) = e^{-(n-1)\psi}(t_1 - e^{\psi}t_0) = 0,$$

whence

$$t_n = e^{n\psi}t_0. \tag{10}$$

## 5. De Sitter's World

Analogous results have been given elsewhere[4,12,13] for de Sitter's world, which is the part of real projective 4-space that lies outside a nonruled quadric hypersurface $\Omega$.[1,7] The light cone at any given point (event) A is the enveloping cone to $\Omega$ from A. In other words, the null lines are tangents, the timelike lines are secants, and the spacelike lines are nonsecants.

Let A and B be two distinct points (representing events), and let $\alpha$ and $\beta$

be their polar hyperplanes with respect to $\Omega$. Taking sections by the plane ABM, where M is a point of general position on $\Omega$, we obtain for $\Omega$ a conic, and for $\alpha$ and $\beta$, lines a and b which are the polars of A and B with respect to this conic. From the theory of conics,* we know that the line AB meets a and b in points A′ and B′, conjugate to A and B, respectively. If AB is not a tangent, A′ and B′ are distinct, and the involution of conjugate points on this line is adequately described by the symbol (AA′)(BB′). The line AB is a nonsecant (spacelike), or a secant (timelike) line according to whether this involution is elliptic or hyperbolic,† that is, according to whether the point pairs AA′ and BB′ do, or do not, separate each other. In other words, AB is spacelike or timelike according to whether the cross ratio $\{AA′, BB′\}$ is negative or positive‡ that is, according to whether the cross ratio

$$\{AB, \beta\alpha\} = \{AB, B′A′\} = \frac{1}{1 - \{AA′, BB′\}}$$

does, or does not, lie in the interval from 0 to 1.

Since A and B represent events, they are both exterior to the conic. Hence, if AB is a secant, A′ and B′ are interior; if AB is a tangent, A′ and B′ coincide (at the point of contact); if AB is a nonsecant, A and A′ separate B and B′. Thus, it can never happen that A and B separate A′ and B′, or that $\{AB, B′A′\} < 0$. We conclude that AB is timelike, null, or spacelike according to whether

$$\{AB, \beta\alpha\}^{1/2}$$

is greater than 1, equal to 1, or less than 1.

Using the abbreviation

$$(xy) = x_1y_1 + x_2y_2 + x_3y_3 - x_4y_4 + x_5y_5,$$

let us write the equation for $\Omega$ in the form $(xx) = 0$. Since $(xx) = 1$ for the exterior point $(0, 0, 0, 0, 1)$, we can be sure (by considerations of continuity) that $(xx)$ is positive for every exterior point $(x)$. Then, by dividing the coordinates of each point $(x)$ by $(xx)^{1/2}$, we can make $(xx) = 1$ for every exterior point $(x)$. Let A and B be $(x)$ and $(y)$, with their coordinates normalized in this manner. It follows that[6]

$$\{AB, \beta\alpha\} = \frac{(xy)^2}{(xx)(yy)} = (xy)^2,$$

that the line AB is timelike, null, or spacelike according to whether $(xy)^2 > 1, (xy)^2 = 1$, or $(xy)^2 < 1$, and that the separation AB between these two events, $(x)$ and $(y)$, is given by

$$\cosh AB = |(xy)|, \quad AB = 0, \quad \text{or} \quad \cos AB = |(xy)|,$$

respectively.[6]

---

* See reference 5, pp. 60, 62, and 72.

† See reference 2, p. 53 (4. 65).

‡ op. cit., p. 192 (12. 4).

For the discussion of two skew timelike lines a and b, we may restrict consideration to the 3-space spanned by these lines. Taking this 3-space to be the hyperplane $x_5 = 0$ we have, instead of $\Omega$, its section by the 3-space, namely, the ordinary quadric surface

$$x_1^2 + x_2^2 + x_3^2 - x_4^2 = 0. \qquad (11)$$

The coordinates are still homogeneous, but since the fifth coordinate remains zero, we now let (x) stand for

$$(x_1, x_2, x_3, x_4),$$

meaning $(x_1, x_2, x_3, x_4, 0)$, normalized so that

$$x_1^2 + x_2^2 + x_3^2 - x_4^2 = 1.$$

Two such points, (x) and (y), determine a line which is timelike, null, or spacelike, according to whether

$$(xy)^2 > 1, \quad (xy)^2 = 1, \quad \text{or} \quad (xy)^2 < 1,$$

where now $(xy) = x_1y_1 + x_2y_2 + x_3y_3 - x_4y_4$; and the separation between these events is

$$\text{arc cosh } |(xy)|, \quad 0, \quad \text{or} \quad \text{arc cos } |(xy)|,$$

respectively.

The two common perpendiculars of a and b, according to the hyperbolic metric determined by the quadric eq. (11), are the two transversals of these lines and their polar lines. Since these transversals themselves are polar lines, we may choose our self-polar tetrahedron of reference so that they are the two opposite edges

$$x_3 = x_4 = 0, \quad x_1 = x_2 = 0$$

(one a non-secant, the other a secant), and so that the points where a and b meet these edges are

$$(\cos \tfrac{1}{2} \phi, \pm\sin \tfrac{1}{2} \phi, 0, 0), \quad (0, 0, \pm\sinh \tfrac{1}{2} \psi, \cosh \tfrac{1}{2} \psi).$$

Since all events are exterior to the absolute quadric, the events for which the two observers have the minimax spacelike separation ($\phi$) between them, are

$$(\cos \tfrac{1}{2} \phi, \pm\sin \tfrac{1}{2} \phi, 0, 0),$$

and we may assume these events to be at time zero for both observers. The interior points

$$(0, 0, \pm\sinh \tfrac{1}{2} \psi, \cosh \tfrac{1}{2} \psi),$$

on the opposite edge of the tetrahedron of reference, are not events but indicators of velocity, showing that the relative rapidity is $\psi$, as before. Since the line a joins $(\cos \tfrac{1}{2}\phi, \sin \tfrac{1}{2}\phi, 0, 0)$ to $(0, 0, \sinh \tfrac{1}{2}\psi, \cosh \tfrac{1}{2}\psi)$, while b joins $(\cos \tfrac{1}{2}\phi, -\sin \tfrac{1}{2}\phi, 0, 0)$ to $(0, 0, -\sinh \tfrac{1}{2}\psi, \cosh \tfrac{1}{2}\psi)$, the events $A_j$ and $B_j$ (fig. 1) now have the coordinates

$$(\cosh t_{2j} \cos \tfrac{1}{2}\phi, \cosh t_{2j} \sin \tfrac{1}{2}\phi, \sinh t_{2j} \sinh \tfrac{1}{2}\psi,$$
$$\sinh t_{2j} \cosh \tfrac{1}{2}\psi)$$

and

$$(\cosh t_{2j+1} \cos \tfrac{1}{2}\phi, -\cosh t_{2j+1} \sin \tfrac{1}{2}\phi, -\sinh t_{2j+1} \sinh \tfrac{1}{2}\psi,$$
$$\sinh t_{2j+1} \cosh \tfrac{1}{2}\psi).$$

The null separations $A_j B_j$ and $B_{j-1} A_j$ yield

$$\cosh t_{2j} \cosh t_{2j\pm 1} \cos \phi - \sinh t_{2j} \sinh t_{2j\pm 1} \cosh \psi = 1.$$

Hence, for all positive integers n,

$$\cosh t_n \cosh t_{n-1} \cos \phi - \sinh t_n \sinh t_{n-1} \cosh \psi = 1. \tag{12}$$

Subtracting eq. (12) from the analogous equation with n replaced by $n + 1$, and dividing by

$$\sinh \frac{t_{n+1} - t_{n-1}}{2} \cosh \frac{t_{n+1} + t_{n-1}}{2} \cosh t_n,$$

we obtain

$$\tanh \frac{t_{n+1} + t_{n-1}}{2} = k \tanh t_n, \tag{13}$$

where $k = \sec \phi \cosh \psi > 1$. This shows clearly that the increasing sequence of numbers $t_n$ cannot tend either to infinity or to a finite limit. In other words, the experiment must come to an end after a finite number of steps. The first observer's last signal will never be answered, either because it failed to reach the second observer, or because the second observer's reply failed to reach the first observer. The same conclusion follows directly from the geometric fact that de Sitter's world is nonconvex. (See figure 2, where the last null line touches the absolute quadric before it meets the world line of the other observer.) No new theory is needed to explain why a distant galaxy may disappear as if it were moving away from the earth faster than the speed of light.

For computation,[*] it is perhaps desirable to replace eq. (13) with a formula that expresses $\exp t_{n+1}$ as a rational function of $\exp t_n$ and $\exp t_{n-1}$. Two such formulae, easily deduced from eq. (12), are:

---

[*] The discussion in reference 4, p. 442, after eq (6. 1) is marred by an error: the denominators of the three complicated expressions for $Tn$ ($= e^{tn}$) should be reversed in sign.

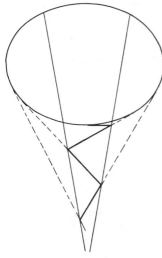

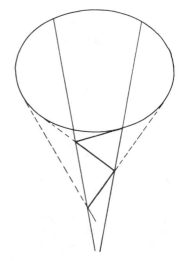

Figure 2                                     Figure 3

$(\cosh t_n \cos \phi - \sinh t_n \cosh \psi)(e^{t_{n-1}} + e^{t_{n+1}}) = 2,$

$(\cosh t_n \cos \phi + \sinh t_n \cosh \psi)(e^{-t_{n-1}} + e^{-t_{n+1}}) = 2.$

Figure 3 illustrates the extreme case in which the final signal only just fails to reach its destination, so that $t_{n+1} = \infty$ and, by eq. (13),

$$\tanh t_n = k^{-1} = \cos \phi \ \text{sech} \ \psi.$$

This $t_n$ is the earliest possible time for the final event.

### 6. Special Cases

If $\phi = 0$, so that the two world lines intersect instead of being skew, equation (12)(divided by $\sinh t_n \sinh t_{n-1}$) yields

$$\cosh \psi = \coth t_n \coth t_{n-1} - \text{csch} \ t_n \ \text{csch} \ t_{n-1}.$$

In terms of a new parameter s, related to t by $\sinh s \sinh t = 1$, we deduce

$$\cosh \psi = \cosh s_n \cosh s_{n-1} - \sinh s_n \sinh s_{n-1}$$
$$= \cosh |s_n - s_{n-1}|, \quad s_n = s_0 - n\psi.$$

If $t_0 < 0$, so that the observers are approaching each other,

$$t_n = \log(\cosh t_n + \sinh t_n) = \log(-\coth s_n + \text{csch} \ s_n)$$
$$= \log\left(-\tanh \frac{s_n}{2}\right) = \log \tanh \frac{n\psi - s_0}{2} = -\log \coth \frac{n\psi - s_0}{2},$$

where $s_0 = -\log \coth (-t_0/2)$. The infinite sequence of times $t_n$ converges to zero through negative values.*

If $t_0 > 0$, so that the observers are parting,

$$t_n = \log (\cosh t_n + \sinh t_n) = \log (\coth s_n + \operatorname{csch} s_n)$$
$$= \log \coth \frac{s_n}{2} = \log \coth \frac{s_0 - n\psi}{2},$$

and the exchange of signals continues till $n\psi$ exceeds

$$s_0 = \log \coth (t_0/2)$$

and then comes to an end.† In other words, the total number of signals received is $[s_0/\psi]$.

If, on the other hand, $\psi = 0$, so that the two world lines are coplanar but ultraparallel, and the observers have a minimax space-like separation $\phi$ occurring at time zero, equation (12) (divided by $\cosh t_n \cosh t_{n-1}$) yields

$$\cos \phi = \operatorname{sech} t_n \operatorname{sech} t_{n-1} + \tanh t_n \tanh t_{n-1}.$$

In terms of a new parameter $\sigma$, related to $t$ by $\sin \sigma \cosh t = 1$, we deduce

$$\cos \phi = \sin \sigma_n \sin \sigma_{n-1} + \cos \sigma_n \cos \sigma_{n-1} = \cos |\sigma_n - \sigma_{n-1}|,$$
$$\sigma_n = \sigma_0 - n\phi,$$

and

$$t_n = \log (\cosh t_n + \sinh t_n) = \log (\csc \sigma_n + \cot \sigma_n)$$
$$= \log \cot \frac{\sigma_n}{2} = \log \cot \frac{\sigma_0 - n\phi}{2}.$$

The exchange of signals continues till $n\phi$ exceeds‡

$$\sigma_0 = 2 \operatorname{arc} \cot e^{t_0}.$$

In other words, the total number of signals received is $[\sigma_0/\phi]$.

There remains one peculiar case that cannot be derived from eq. (12) by any limiting process: the two world lines may be parallel (in the sense of hyperbolic geometry) so that they converge to a common "point at

---

* See reference 4, figure 3.

† See reference 4, figure 2.

‡ op. cit., figure 1.

infinity" on the absolute quadric. (We may perhaps regard this as the de Sitter version of the Michelson-Morley experiment.) It is natural to take the world lines themselves as two sides of the triangle of reference, while the third side joins their remaining points of intersection with the absolute quadric.*

The section of the quadric by this plane may be expressed in the form $x_2x_3 + x_3x_1 + x_1x_2 = 0$, so that

$$(xy) = \tfrac{1}{2} (x_2y_3 + x_3y_2 + x_3y_1 + x_1y_3 + x_1y_2 + x_2y_1),$$

and we can normalize the coordinates so that $(xx) = -1$ for every event $(x)$. The hyperbolic cosine of a timelike separation is now $-(xy)$; for instance, the separation from $(-1, 0, 1)$ to $(-e^{-t}, 0, e^t)$ is t. The null separations $A_jB_j$ and $B_{j-1}A_j$, among the events

$$A_j = (-e^{-t_{2j}}, \; 0, \; e^{t_{2j}}), \quad B_j = (0, \; -e^{-t_{2j+1}}, e^{t_{2j+1}}),$$

yield

$$-\tfrac{1}{2} (-e^{t_n - t_{n-1}} - e^{t_{n-1} - t_n} + e^{-t_n - t_{n-1}}) = 1,$$

whence

$$(e^{t_n} - e^{t_{n-1}})^2 = 1.$$

Since the t's are increasing, we deduce

$$e^{t_n} = 1 + e^{t_{n-1}} = n + e^{t_0}.$$

Thus

$$t_n = \log (n + e^{t_0}).$$

In particular, if $t_0 = 0$,

$$t_n = \log (n + 1).$$

The spacelike separation $\phi$ between the two observers at corresponding times t, that is, between

$$(-e^{-t}, \; 0, \; e^t) \text{ and } (0, \; -e^{-t}, e^t),$$

is given by

$$\cos \phi = -(xy) = -\tfrac{1}{2} (-2 + e^{-2t}) = 1 - \tfrac{1}{2} e^{-2t},$$

whence

$$\sin \frac{\phi}{2} = \frac{e^{-t}}{2} .$$

---

* See reference 4, p. 440.

If $t = t_n = \log (n + 1)$, this becomes

$$\sin \frac{\phi_n}{2} = \frac{1}{2(n + 1)}.$$

Thus

$$\lim_{n \to \infty} \frac{t_n - t_{n-1}}{\phi_n} = 1.$$

Comparing this with eq. (8), we see that, when n is large, the parallel world lines in de Sitter space behave like those of the original Michelson-Morley experiment in Minkowskian space.

## REFERENCES

1. Coxeter, H. S. M.: A Geometrical Background for de Sitter's World, Amer. Math. Monthly, **50**: 217-228 (1943).

2. Coxeter, H. S. M., The Real Projective Plane. Cambridge, England: Cambridge Univ. Press, 1961.

3. Coxeter, H. S. M., Non-Euclidean Geometry. Toronto: Univ. of Toronto Press, 1961.

4. Coxeter, H. S. M.: On Wigner's Problem of Reflected Light Signals in de Sitter Space, Proc. Roy. Soc., **A 261**: 435-442 (1961).

5. Coxeter, H. S. M., Projective Geometry. New York: Blaisdell Co., 1964.

6. Coxeter, H. S. M., Geometry, in vol. 3 of Lectures on Modern Mathematics, New York: John Wiley and Sons, 1965.*

7. Du Val, Patrick: Geometrical Note on de Sitter's World, Phil. Mag., **47** (6): 930-938 (1924).

8. Eddington, A. S., The Mathematical Theory of Relativity. Cambridge, England: Cambridge Univ. Press, 1924.

9. Robb, A. A., Geometry of Time and Space. Cambridge, England: Cambridge Univ. Press, 1936.

10. Synge, J. L., Relativity: The Special Theory. Amsterdam: North-Holland, 1965.

11. Veblen, Oswald and Hoffmann, Banesh: Projective Relativity, Phys. Rev., **36** (3): 810-822 (1930).

12. Wigner, E. P.: Relativistic Invariance and Quantum Phenomena, Rev. Mod. Phys., **29**: 255-268 (1957).

13. Wigner, E. P.: Measurement of The Curvature in a Two-dimensional Universe, Phys. Rev., **120**: 643 (1960).

14. Wigner, E. P.: Geometry of Light Paths Between Two Material Bodies, J. Math. Phys., **2**: 207-211 (1961).

*Reference 6 is Chapter 12 of this book.

# 12

# GEOMETRY

## INTRODUCTION

The subject of geometry is so vast that no adequate account of its recent developments could be given in two lectures. Felix Klein, David Hilbert, or Wilhelm Blaschke might have been able to give such an account in a whole course of lectures; but that would be far beyond my powers, because there are so many branches of the subject in which I am almost as ignorant as the proverbial man in the street. I must ask you to forgive me if I concentrate on my own favorite branches, and I must take the risk of offending various geometers who will ask why I have not dealt with algebraic geometry, differential geometry, symplectic geometry, continuous geometry, metric spaces [1],*Banach spaces, linear programming, and so on.

## EUCLIDEAN GEOMETRY

For most of us, our first contact with geometry was in the Euclidean plane with little problems about triangles and circles. Many such problems seem prohibitively difficult to solve until we happen

---

*For references, see pages 263–266.

to think of the right approach. A good example is the Steiner-Lehmus theorem, which says that, if two internal angle-bisectors of a triangle are equal in length, the triangle must be isosceles. In 1840, this was proposed by C. L. Lehmus and proved by Jacob Steiner. Since then many different proofs have been published. A pretty one, due to Lehmus himself (1848), was rediscovered 115 years later [34]. It might be argued that this is an *absolute* theorem (not involving Euclid's Fifth Postulate) and should have an absolute proof. Such a proof has been given by L. M. Kelly [20, p. 16, Ex. 4].

Kakeya's problem [5, pp. 99–101; 6] asks for the plane set of least area having the "Kakeya property," which means that a unit segment can be turned continuously in the set through 180° so as to return to its original position (reversed). Besicovitch proved that there is no least area: such a set with sufficiently many holes (like a lace tablecloth) can have as small an area as we please! In view of this complete solution, it is natural to modify the problem by asking for the smallest *convex* set, or *simply connected* set, that has the Kakeya property. It was proved by Julius Pál that the smallest convex set is the equilateral triangle of unit altitude (area $1/\sqrt{3}$). It is tempting to imagine that the smallest simply connected set is bounded by the deltoid or three-cusped hypocycloid (area $\pi/8$), within which the unit segment can be moved so as to remain a tangent while both its ends run along the curve. However, a curve enclosing a considerably smaller area was discovered independently by Melvin Bloom and I. J. Schoenberg in 1963. Schoenberg picturesquely describes this star-shaped curve as resembling the locus of the end of a Foucault pendulum swinging ten thousand times. The area is less than $\pi/11$. Defining the *Kakeya constant* $K$ to be the greatest lower bound of areas of simply connected regions having the Kakeya property, he raises the problem of computing $K$. So far, we know only that

$$K \leqq \frac{5 - 2\sqrt{2}}{24}\pi.$$

Fejes Tóth regards $K$ as the case $n = 2$ of the greatest lower bound of areas of simply connected regions within which a regular $n$-gon can be continuously turned through $360°/n$. We know only that this lies strictly between the area of the $n$-gon and the area of its circumcircle.

All three versions of Kakeya's problem can be extended to three-dimensional space, where we seek a set of least volume in which the unit segment can be moved from one direction to any other. Now even the case of the *convex* set is unsolved. Analogy suggests the regular tetrahedron of unit altitude, but Eggleston remarks that a smaller convex region will suffice.

## ORDERED GEOMETRY

The name *ordered geometry* has been given by Artin [2, p. 73] to the very simple geometry whose only primitive concepts are the *point* and the relation of *intermediacy* or "betweenness." If $B$ lies between $A$ and $C$, we write $[ABC]$. The axioms [20, pp. 177–178] tell us that the relation $[ABC]$ implies $[CBA]$ but not $[BCA]$, and so on. The *segment* $AB$ is the set of points $X$ such that $[AXB]$, the *ray* $A/B$ is the set of points $Y$ such that $[BAY]$, and the *line* $AB$ is the union of two points, two rays, and a segment:

$$A/B + A + AB + B + B/A.$$

Conceived by Pasch in 1882, this geometry was developed by Peano (1889), Hilbert (1899) and Veblen [65, Chap. I]. A model for it is easily obtained by considering the interior of a convex region in real affine or projective space. Conversely, any ordered space can be extended to a real projective space (of the same number of dimensions) by the adjunction of *ideal elements* such as "points at infinity." The best treatment of this extension is by Whitehead [62]. He, following Russell [49, p. 382], called ordered geometry "descriptive geometry," thus inviting confusion with Cayley's famous dictum "Descriptive geometry is all geometry," which really meant "Projective geometry is all geometry."

Any geometry worthy of the name has to justify itself by exhibiting interesting theorems. The first such theorem in ordered geometry is the Sylvester-Gallai theorem:

*If n points are not all collinear, there is at least one line containing exactly two of them.*

This was conjectured by Sylvester in 1893, revived by Karamata and Erdös in 1933, and proved (in a complicated manner) by T.

Gallai. It is always rather exciting when a problem is solved after remaining a challenge for as long as forty years. My own small contribution was to adapt Gallai's Euclidean proof to ordered geometry [20, pp. 181–182]. L. M. Kelly [20, pp. 65–66] discovered an "absolute" proof, so simple that anyone who has seen it once can always remember it.

The Sylvester-Gallai theorem and its extensions play a significant role in the part of ordered geometry that deals with *convexity* [25].

Hadwiger [35a] once asked whether every convex polyhedron can be transformed, by a finite sequence of truncations (cutting off a corner by means of a plane), into a polyhedron such that each face has a number of vertices that is a multiple of 3. The answer is yes or no according as the Four-Color Conjecture (for maps on a sphere) is true or false! This conclusion follows easily from Heawood's congruences [38, pp. 277–278; 5, pp. 230–231].

In $n$ dimensions, a *convex polytope* can be defined *either* as the convex hull of a finite set of points *or* as a bounded set which is the intersection of finitely many half-spaces. Supporting hyperplanes of an $n$-dimensional convex polytope $\Pi_n$ contain $k$-dimensional *elements* $\Pi_k$ for $k = 0, 1, \ldots, n - 1$. These are called *vertices*, *edges*, . . . , and *cells* (or *facets*). It is convenient to include also a unique $n$-dimensional element, the $\Pi_n$ itself, and a unique $(-1)$-dimensional element $\Pi_{-1}$, the null set, which belongs to all the other elements. If an element $\Pi_h$ belongs to an element $\Pi_k$, we say that $\Pi_h$ and $\Pi_k$ are *incident*. In particular, every edge $\Pi_1$ is incident with two vertices $\Pi_0$, and every $\Pi_{n-2}$ is incident with two cells $\Pi_{n-1}$. More generally, whenever a $\Pi_{k-1}$ and a $\Pi_{k+1}$ are incident, there are just two $\Pi_k$'s incident with both (for $0 \leq k \leq n - 1$). In 1852 Schläfli proved that, for a polytope having $N_0$ vertices, $N_1$ edges, and so on,

$$N_0 - N_1 + N_2 - \cdots + (-1)^n N_n = 1$$

or, if you prefer,

$$-N_{-1} + N_0 - N_1 + \cdots + (-1)^n N_n = 0$$

[50, p. 190; 22, Chap. IX].

A two-dimensional polytope $\Pi_2$ is simply a polygon, which has equally many vertices and sides ($N_0 - N_1 = 0$). It follows that, in any $\Pi_n$, whenever a $\Pi_{k-2}$ and a $\Pi_{k+1}$ are incident, there are equally many $\Pi_{k-1}$'s and $\Pi_k$'s incident with both (for $1 \leq k \leq n - 1$).

If this number $q_k$ depends only on $k$ (that is, if it is the same for every occurrence of an incident $\Pi_{k-2}$ and $\Pi_{k+1}$), the polytope is said to be (combinatorially) *regular*, and is denoted by the "Schläfli symbol"

$$\{q_1, q_2, \ldots, q_{n-1}\}.$$

For instance, $\{q\}$ is a $q$-gon, and the ordinary cube has the Schläfli symbol $\{4, 3\}$ because each face has four vertices (and four sides) and each vertex belongs to three edges (and three faces). Schläfli's formula (connecting the numbers $N_k$) is easily verified in the cases of the *simplex*

$$\{3, 3, \ldots, 3, 3\}, \quad \text{for which} \quad N_k = \binom{n+1}{k+1}$$

and the *cross polytope* (or "$n$-dimensional octahedron")

$$\{3, 3, \ldots, 3, 4\}, \quad \text{for which}$$

$$N_k = 2^{k+1} \binom{n}{k+1} \quad (k < n).$$

This discussion is the counterpart, in ordered geometry, of the metrical definition of a regular polytope [22, p. 288; 30, p. 134] as having a center from which all the $\Pi_k$'s are equidistant, for each $k < n$.

An $n$-dimensional convex polytope is said to be $m$-neighborly if every $m$ vertices lie in a supporting hyperplane. Thus every convex polytope is 1-neighborly, and a simplex is $m$-neighborly for every $m \leq n$. The possibility of other neighborly polytopes (in four or more dimensions) was recognized long ago by Brückner [8a] and Carathéodory [10a]. A 2-neighborly polytope may be described simply as having no diagonals; every two vertices are joined by an edge. Carathéodory considered a *cyclic* polytope: the convex hull of $N_0$ points (in affine $2m$-space) on the rational normal curve which is the locus of the point with barycentric coordinates

$$(1, t, t^2, \ldots, t^{2m}).$$

David Gale [32] has given a charmingly simple proof that such a polytope is $m$-neighborly and has

$$N_{2m-1} = \frac{N_0}{N_0 - m} \binom{N_0 - m}{m}$$

cells (simplexes). There is an unpublished proof by Martin

Fieldhouse that every $m$-neighborly polytope (with $N_0$ vertices in $2m$ dimensions) has this number of cells. In 1957 T. S. Motzkin conjectured that every $m$-neighborly polytope is combinatorially equivalent to a cyclic polytope. Gale established this conjecture for $N_0 \leq 2m + 3$. Grünbaum has pointed out that this inequality is "best possible," as one of Brückner's polytopes is the dual of a noncyclic 2-neighborly polytope having eight vertices (and twenty tetrahedral cells).

When $N_{n-1}$ is very large, the boundary of an $n$-dimensional polytope resembles an $(n-1)$-dimensional tessellation or *honeycomb*: an infinite collection of cells fitting together so as to fill and cover the $(n-1)$-space, each $\Pi_{n-2}$ belonging to two $\Pi_{n-1}$'s. If the $\Pi_{n-1}$'s are regular and alike, the honeycomb is said to be *regular;* it evidently has a Schläfli symbol. For instance, $\{6, 3\}$ is the familiar pattern of regular hexagons filling the Euclidean plane.


## SPHERE PACKING

It seems desirable to say something about the problem of packing equal spheres in Euclidean $n$-space, not only because of its intrinsic interest and its connection with the geometry of numbers, but also because it has been found to have a practical application in the theory of communication.

Let $M_n$ denote *the maximum number of spheres* (balls) *that can touch another of the same size without overlapping, in Euclidean $n$-space.* Everyone knows that $M_2 = 6$: seven silver dollars can be placed on a table so that one is surrounded by a ring of six. Moreover, this pattern can be extended to a packing of infinitely many circles (disks), each touching six others, and the centers of all the circles are the points of a lattice. One row of the lattice yields the analogous arrangement with $n = 1$. The one-dimensional "ball" is a line segment, and in a row of line segments each touches two others: $M_1 = 2$.

When $n = 3$, we are dealing with solid spheres, ordinary balls. In 1694 Sir Isaac Newton told David Gregory that $M_3 = 12$, but Gregory asserted that $M_3 = 13$. R. Hoppe, 180 years later, proved that Newton's value is correct. (For a fuller account of this story, see [21].) Although there is only one way of surrounding a circle with six equal circles, there are many ways of surrounding a sphere with twelve equal spheres. (This may even be done in such a way that no two of the twelve touch each other.) Two of the possible arrangements can be continued systematically so that *each* sphere touches twelve others [41; 13].   In one of these (described by Kepler in 1611), the centers form a lattice, namely the face-centered cubic lattice [20, pp. 225, 333, 407].   We naturally call this arrangement a *lattice packing* [48a].

Let $L_n$ denote *the number of spheres that touch each one in the densest n-dimensional lattice packing.*   We know that $L_1 = M_1 = 2$, $L_2 = M_2 = 6$, $L_3 = M_3 = 12$, and $L_n \leq M_n$ always.   Following Gauss [33, Vol. 1, p. 307; Vol. 2, p. 192] we can associate with each lattice a class of equivalent positive definite quadratic forms.   In fact if the lattice is generated by $n$ vectors $\mathbf{e}_1, \ldots, \mathbf{e}_n$, its general point has the position vector

$$\Sigma x^j \mathbf{e}_j,$$

where $x^1, \ldots, x^n$ are integers.   The square of the length of this vector is

$$\Sigma x^j \mathbf{e}_j \cdot \Sigma x^k \mathbf{e}_k = \Sigma\Sigma a_{jk} x^j x^k,$$

where $a_{jk} = \mathbf{e}_j \cdot \mathbf{e}_k$.   We are thus led to a positive definite quadratic form; let us denote it by $\phi$.   The smallest positive value attained by $\phi$ for integers $x^j$ is clearly equal to the square of the diameter of the largest equal spheres that can be centered at the lattice points so as to form a packing (that is, to avoid overlapping). Let us assume, for simplicity, that our spheres are of diameter 1, so that 1 is the minimal value of $\phi$.   Since $\det(a_{jk})$ is equal to the square of the content of the elementary cell of the lattice [20, p. 331], the densest packing occurs when this determinant is as small as possible.   It follows that the search for densest packings is equivalent to the search for absolutely extreme forms [15, p. 394].

Absolutely extreme forms for $n \leq 8$ have been known since 1935. They were established for $n \leq 5$ by Minkowski [42, p. 247] and for $n = 6, 7, 8$ by Blichfeldt [7].   Forms equivalent to those of Minkowski and Blichfeldt may be represented very simply by the graphs

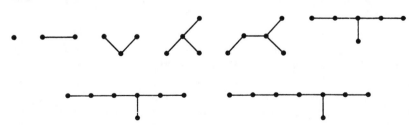

Each graph is a tree having $n$ nodes, one for each "square" term $(x^j)^2$, and $n - 1$ branches, one for each "product" term $-x^j x^k$; thus the first three trees represent the forms

$$(x^1)^2, \ (x^1)^2 - x^1 x^2 + (x^2)^2, \ (x^1)^2 - x^1 x^2 + (x^2)^2 - x^2 x^3 + (x^3)^2$$

[20, p. 337, Exercise, where $u^1 = x^1$, $u^2 = x^3$, $u^3 = -x^2$].

These trees will be recognized by anyone who followed the lectures of Kaplansky [40, pp. 125–126]; he used them as symbols for the Lie algebras $A_1$, $A_2$, $A_3$, $D_4$, $D_5$, $E_6$, $E_7$, $E_8$. This connection arises from the fact that every densest packing with $n \leq 8$ is symmetrical by reflection in the common tangent hyperplane at the point of contact of any two of the spheres that touch each other. Such reflections generate an infinite discrete group whose fundamental region is a simplex. When $n = 1$ the simplex is merely a line segment. The simplexes for $n = 2, \ldots, 8$ are conveniently represented by graphs

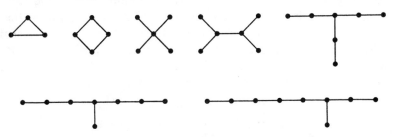

having $n + 1$ nodes, one for each bounding hyperplane (or mirror). Each branch joins two nodes that represent mirrors forming a dihedral angle of 60°. The remaining pairs of mirrors are orthogonal.

For instance, when $n = 2$ the packing consists of the incircles of the hexagons of the regular tessellation $\{6, 3\}$ (Figure 1), and the group generated by reflections in all the sides of these hexagons is adequately generated by reflections in the three sides of the shaded equilateral triangle. Two of these three sides are orthogonal to

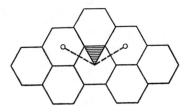

*Figure 1*

generating vectors of the lattice of centers. (These two vectors appear in the figure as broken lines.) Each graph in our second list can be derived from the corresponding tree in the first list by adding one extra node [15, pp. 405–412].

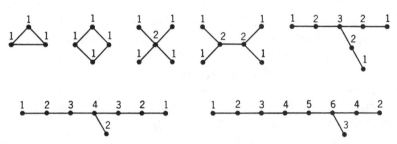

*Figure 2*

There is an amusingly simple way to derive the values of $L_n$ from these augmented graphs. We place the number 1 over the extra node; then we number the remaining nodes as in Figure 2, so that the number on each is half the sum of the numbers on the adjacent nodes. (In particular, the numbers along each "leg" are in arithmetic progression.) Steinberg [54; see also 22, p. 212] has proved that $L_n$ is equal to $n$ times the sum of these numbers:

| $n$ | 1 | 2 | 3 | 4 | 5 | 6 | 7 | 8 |
|---|---|---|---|---|---|---|---|---|
| $L_n$ | 2 | 6 | 12 | 24 | 40 | 72 | 126 | 240 |

C. Muses (in a letter of December 1963) epitomized these values in the simple formula

$$L_n = n\left(\left[\frac{2^{n-2}}{3}\right] + n + 1\right) \qquad (1 \le n \le 8).$$

Although $L_n$ is unknown for $n > 8$, we do know that the value given by this simple formula (for example, 468 for $L_9$) is too large, and that the densest lattice packing is no longer symmetrical by reflection in all the common tangent hyperplanes. It has been conjectured that $L_9 = 272$. In 1946 T. W. Chaundy believed he had a proof, but he withdrew his paper when he realized he had made the unjustifiable assumption that the densest $(n-1)$-dimensional packing must occur as a section of the densest $n$-dimensional packing.

Our knowledge of the numbers $M_n$ is even less satisfactory.

We know that $M_n = L_n$ for $1 \leq n \leq 3$, and we believe that $M_4 = 24$; but it is conceivable that $M_4$ is 25 or 26. Obviously $L_n$ provides a lower bound for $M_n$. An upper bound can be derived from Schläfli's work on the content of a simplex in spherical space (the analogue of the area of a spherical triangle) [50, p. 234; 21]. The conclusion is that

$$M_n \leq \left[ \frac{2f_{n-1}(n)}{f_n(n)} \right]$$

where $f_n(x)$ is defined recursively by $f_0(x) = f_1(x) = 1$,

$$f_n(x) = \frac{1}{\pi} \int_{n-1}^{x} \frac{f_{n-2}(x-2)}{x\sqrt{x^2-1}}\, dx.$$

With the aid of an electronic computer, John Leech obtained the following values:

| $n$ | 4 | 5 | 6 | 7 | 8 | 9 | 10 | 11 | 12 |
|---|---|---|---|---|---|---|---|---|---|
| Bound for $M_n$ | 26 | 48 | 85 | 146 | 244 | 401 | 648 | 1035 | 1637 |

The most remarkable of these is the case when $n = 8$, the range of possibilities being narrowed to

$$240 \leq M_8 \leq 244.$$

The next case explains why we can be sure that $L_9 \neq 468$. For large $n$, the upper bound is asymptotically

$$\frac{\sqrt{\pi}}{e}\, n^{3/2} 2^{(n-1)/2}.$$

## INTEGRAL QUATERNIONS AND INTEGRAL OCTAVES

The regularity of the arrangement when $n = 1$ or 2, and the near-regularity when $n = 4$ or 8, are connected with the rings of rational integers, Eisenstein integers ($a + b\rho$ where $\rho = e^{2\pi i/3}$), integral quaternions, and integral octaves. This application of geometry to the theory of numbers is, perhaps, worthy of a little more attention. Consider, for instance, the representation of the *quaternion*

$$x = x_0 + x_1 i + x_2 j + x_3 k$$

(where $i^2 = j^2 = k^2 = ijk = -1$) by the point $(x_0, x_1, x_2, x_3)$

in Euclidean four-space, or by the vector **x** that takes the origin to this point. In Hamilton's notation, the *conjugate* of $x$ is

$$\bar{x} = x_0 - x_1 i - x_2 j - x_3 k$$

and the *scalar part* of $x$ is $\mathrm{S}x = x_0 = \frac{1}{2}(x + \bar{x})$. The *norm*

$$\mathrm{N}x = x\bar{x} = \bar{x}x = x_0{}^2 + x_1{}^2 + x_2{}^2 + x_3{}^2$$

appears as the square of the length of the vector **x**. Multiplication is not always commutative. The conjugate of the product $xy$ is easily seen to be $\bar{y}\bar{x}$; therefore the conjugate of $x\bar{y}$ is $y\bar{x}$. The inner product of two vectors **x** and **y** is equal to the scalar part of the quaternion product $x\bar{y}$:

$$\mathbf{x} \cdot \mathbf{y} = x_0 y_0 + x_1 y_1 + x_2 y_2 + x_3 y_3 = \mathrm{S}(x\bar{y})$$

$$= \tfrac{1}{2}(x\bar{y} + y\bar{x}).$$

If **a** and **b** are two *unit* vectors, so that $\mathrm{N}a = \mathrm{N}b = 1$, we have

$$\mathrm{N}(a + b) = (a + b)(\bar{a} + \bar{b}) = a\bar{a} + b\bar{b} + a\bar{b} + b\bar{a}$$

$$= \mathrm{N}a + \mathrm{N}b + 2\mathbf{a} \cdot \mathbf{b} = 2 + 2 \cos (\mathbf{ab}),$$

where (**ab**) is the angle between the two vectors. $\mathrm{N}(a + b)$ takes the value 1 when (**ab**) = 120°, and 2 when (**ab**) = 90°.

Consider the reflection in the hyperplane through the origin orthogonal to the unit vector **a**. Since this reflection transforms any vector **x** into $\mathbf{x} - 2(\mathbf{x} \cdot \mathbf{a})\mathbf{a}$, it transforms the corresponding quaternion $x$ into

$$x - (x\bar{a} + a\bar{x})a = -a\bar{x}a$$

[64, p. 308].

Defining $g = \frac{1}{2}(1 + i + j + k)$, we can associate the quaternions 1, $ig$, $jg$, $kg$ with the nodes of the graph

in such a way that two of them, say $a$ and $b$, satisfy

$$N(a + b) = 1 \text{ or } 2$$

according as the nodes are adjacent or nonadjacent. These four quaternions generate, by addition and subtraction, the ring of integral quaternions $x$ whose constituents $x_0$, $x_1$, $x_2$, $x_3$ are either integers or (all of them) halves of odd integers. Also $ig$ and $jg$, satisfying

$$(ig)^3 = 1, \qquad (ig)(jg)(ig) = (jg)(ig)(jg),$$

generate by multiplication the 24 unities

$$\pm 1, \ \pm i, \ \pm j, \ \pm k, \qquad \tfrac{1}{2}(\pm 1 \pm i \pm j \pm k),$$

which constitute the binary tetrahedral group of order 24 [24 (6.69)]. Finally, the four reflections

$$x \to -a\bar{x}a,$$

where $a = 1$ or $ig$ or $jg$ or $kg$, generate the group $[3^{1,1,1}]$ of order 192 [22, p. 200], which shows that the points representing the 24 unities are the vertices of the regular 24-cell $0_{111}$ or $\{3, 4, 3\}$, and that the lattice of points representing all the integral quaternions are the vertices of the regular four-dimensional honeycomb $1_{111}$ or $\{3, 3, 4, 3\}$, whose cells are 16-cells (or cross polytopes) $1_{11}$ or $\{3, 3, 4\}$ [27a, p. 81].

This four-dimensional representation of integral quaternions has analogues both below (in two dimensions) and above (in eight dimensions). In the subgraph

$$\overset{\textstyle\bullet}{\underset{\textstyle 1}{\rule{0pt}{0pt}}}\!\!\!\!\rule[0.5ex]{5em}{0.4pt}\!\!\!\!\overset{\textstyle\bullet}{\underset{\textstyle ig}{\rule{0pt}{0pt}}}$$

we can replace the quaternion $ig$ (of period 3) by the complex number $\rho = e^{2\pi i/3}$. The two complex numbers 1 and $\rho$ are a basis for the ring of Eisenstein integers, which is represented in the obvious manner by the lattice of vertices of the regular tessellation $\{3, 6\}$ of equilateral triangles. Also $-\rho$ generates the cyclic group of six unities

$$\pm 1, \ \pm \rho, \ \pm \rho^2$$

represented by a regular hexagon $\{6\}$. The hexagon is symmetrical by the reflections

$$x \to -a\bar{x}a = -a^2\bar{x},$$

where $a = 1$ or $\rho$. These two reflections generate the dihedral group [3], of order 6.

*Octaves* (often called Cayley numbers, or Cayley-Graves numbers, or the Cayley-Dickson algebra) were discovered by Graves and

Cayley, and named by Hamilton. In Dickson's notation, the octave $q + Qe$ is derived from quaternions $q$ and $Q$ by adjoining a new unit $e$ which enters into (nonassociative) multiplication according to the rule

$$(q + Qe)(r + Re) = qr - \bar{R}Q + (Rq + Q\bar{r})e,$$

where the bars indicate quaternion conjugates.

The conjugate and norm of the octave

$$x = x_0 + x_1 i + x_2 j + x_3 k + x_4 e + x_5 ie + x_6 je + x_7 ke$$

are defined by

$$\bar{x} = x_0 - x_1 i - x_2 j - x_3 k - x_4 e - x_5 ie - x_6 je - x_7 ke$$

and

$$Nx = x\bar{x} = x_0{}^2 + x_1{}^2 + x_2{}^2 + x_3{}^2 + x_4{}^2 + x_5{}^2 + x_6{}^2 + x_7{}^2.$$

Since

$$\overline{q + Qe} = \bar{q} - Qe = \bar{q} - e\bar{Q} = \bar{q} + \bar{e}\bar{Q}$$

and $N(q + Qe) = (q + Qe)(\bar{q} - Qe) = q\bar{q} + \bar{Q}Q = Nq + NQ$, there is no conflict between our two uses of the bar and the N. Since $x_0 = \frac{1}{2}(x + \bar{x})$, the octave $x$ (like the quaternion $x$) satisfies the quadratic equation

$$x^2 - 2x_0 x + Nx = 0.$$

A ring of *integral* octaves can be selected in seven ways. One way uses the basis

$$1, j, e, ke, h, ih, jh, eh,$$

where $h = \frac{1}{2}(i + j + k + e)$ [12, p. 569]. Our rule about $N(a + b)$ applies again to the graph

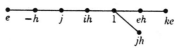

We represent the octave

$$x_0 + x_1 i + x_2 j + x_3 k + x_4 e + x_5 ie + x_6 je + x_7 ke$$

by the vector or point $(x_0, x_1, \ldots, x_7)$ in Euclidean eight-space. The reflection that reverses a unit vector **a** is still given by

$$x \rightarrow -a\bar{x}a.$$

(Although multiplication is not universally associative, we do have $ab \cdot a = a \cdot ba$, so that such an expression as $aba$ is unambiguous.)

The four octaves $i$, $j$, $e$, $h$ generate by multiplication the loop of 240 unities

$$\pm 1, \ \pm i, \ \pm j, \ \pm k, \ \pm e, \ \pm ie, \ \pm je, \ \pm ke,$$

$$\tfrac{1}{2}(\pm 1 \pm ie \pm je \pm ke), \qquad \tfrac{1}{2}(\pm i \pm j \pm k \pm e),$$

$$\tfrac{1}{2}(\pm 1 \pm j \pm k \pm ie), \qquad \tfrac{1}{2}(\pm i \pm e \pm je \pm ke), \text{ etc.}$$

(where "etc." means that we may cyclically permute $i$, $j$, $k$). Finally, the eight reflections

$$x \rightarrow -a\bar{x}a,$$

where $a$ takes the values $1$, $j$, $e$, $ke$, $h$, $ih$, $jh$, $eh$, generate the group $[3^{4,2,1}]$ of order $192 \cdot 10!$ [22, p. 204], which shows that the points representing the 240 unities are the vertices of the semiregular polytope $4_{21}$. (These numbers 4, 2, 1 indicate the lengths of the three legs of the graph.) Also the lattice of points representing all the integral octaves are the vertices of the eight-dimensional honeycomb $5_{21}$, whose cells are simplexes $5_{20}$ and cross polytopes $5_{11}$.

Since a cross polytope of edge 1 has circumradius $\sqrt{\tfrac{1}{2}}$ while a simplex has a smaller circumradius, every point in the space is within a distance $\sqrt{\tfrac{1}{2}}$ of some vertex of the honeycomb. It follows that, *for every octave $\xi$, there is an integral octave $x$ such that*

$$N(\xi - x) \leq \tfrac{1}{2}$$

[12, p. 577]. Similar considerations yield the same result for a quaternion $\xi$ and an integral quaternion $x$, and for a complex number $\xi$ and a Gaussian integer $x$. However, since the circumradius of an equilateral triangle is only $\sqrt{\tfrac{1}{3}}$, for every complex number $\xi$ there is an Eisenstein integer $x$ such that $N(\xi - x) \leq \tfrac{1}{3}$. (Here we are using the word "norm" in Hamilton's sense. Workers in Hilbert space use the same word for $|\xi - x|$, the square root of Hamilton's norm.) It is possible that this geometric approach to algebraic numbers could be further exploited.

## PROJECTIVE GEOMETRY

In projective geometry the primitive concepts are the *point*, the *line*, and the relation of *incidence*. The axioms [23, p. 15] tell

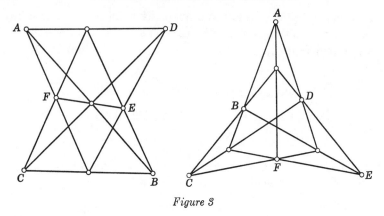

*Figure 3*

us that if $AB$ meets $CD$ (and the four points are all distinct) then $AC$ meets $BD$, and so on. If $C$ is not on $AB$, we can define the *plane ABC* as consisting of all the points on all the lines through $C$ that meet $AB$, and all the lines that join pairs of these points. If we include an axiom saying that, for any plane $ABC$, there is a point $D$ not in this plane (so that the number of dimensions is greater than 2), we can prove Desargues's theorem about perspective triangles and deduce the uniqueness of the fourth point of a harmonic set. If, on the other hand, we assume that all points lie in one plane (so that the number of dimensions is 2), we are free to assert or deny Desargues's theorem, that is, to work in a *Desarguesian* plane or a *non-Desarguesian* plane, respectively. A plane may be non-Desarguesian and still admit a unique fourth harmonic point for any three given collinear points; it is then called a Moufang plane [36, p. 370; 46, p. 137].

Another famous theorem that may be either taken as an axiom or denied is Pappus's theorem (Figure 3) which says, of the lines joining six coplanar points, that if $AB$, $CD$, $EF$ are concurrent and $DE$, $FA$, $BC$ are concurrent, then also $AD$, $BE$, $CF$ are concurrent [23, p. 90]. (G. Hessenberg's proof that Pappus implies Desargues has been published in many versions [e.g., 16, pp. 1, 46], but almost every author disregards at least one exceptional situation. Apparently the first complete proof is that of Pickert [47, pp. 144–148].) Since it is possible to deny Pappus without denying Desargues, geometries without Pappus can occur in any number of dimensions.

Pappus's theorem is equivalent to the Fundamental Theorem,

which says that a projectivity is determined by its effect on three collinear points. This plays an essential role in the classical treatment of polarities and conics [23, pp. 41–89]. In a projective plane satisfying Pappus's theorem, a polarity is an involutory correspondence between points and lines, such that each range corresponds to a projectively related pencil. A point $A$ (or $B$) and the corresponding line $a$ (or $b$) are called *pole* and *polar*. If $B$ lies on $a$, then $A$ lies on $b$, $A$ and $B$ are called *conjugate points*, $a$ and $b$ are called *conjugate lines*. If a self-conjugate point exists, it lies on a *conic* whose points and tangents are precisely all the self-conjugate points and self-conjugate lines. Other points are said to be *exterior* or *interior* according as they lie on two tangents or on no tangent. The polars of such points are *secants* or *nonsecants*, respectively. On a secant $AB$, any pair of conjugate points is a pair of harmonic conjugates with respect to $A$ and $B$. (These definitions and deductions, which are commonplace for the real projective plane, remain valid for the general Desarguesian plane, and can be adapted to non-Desarguesian planes [4].)

In the presence of a given polarity, a point $Q$ is said to be *accessible* from a point $P$ if $Q$ is the harmonic conjugate of $P$ with respect to some pair of distinct points which are conjugate for the polarity. The relation of accessibility is evidently reflexive and symmetric. By an application of Hesse's theorem, it is also transitive [23, pp. 101, 148], and the points of the plane can be distributed into one or more classes of mutually accessible points.

If the polarity admits self-conjugate points, their locus (the conic) constitutes a single class, and the exterior points constitute another single class. Thus the exterior points on a secant form a segment according to the ingenious definition of Pieri [16, p. 38]. Over some fields (such as real, rational, or finite) there are also interior points (lying on no tangent). We shall see later on that, in any real or finite plane, the interior points constitute a single class; but in the rational plane the interior points are distributed among infinitely many classes.

One of the most elegant developments in projective geometry has been the synthetic coordinatization of the general projective plane. This began in 1847 with von Staudt's calculus of "throws," and continued with Hilbert's remark that, in any Desarguesian plane, suitable definitions of addition and multiplication exhibit the points of a range as a division ring or "skew field" or *corpus*, which becomes a (commutative) *field* when Pappus's theorem holds. Further investigation by Oswald Veblen and J. H. M. Wedderburn inspired Marshall Hall [36, p. 123] to invent the *ternary ring* which enabled him to coordinatize a non-Desarguesian plane. This exciting story has been told many times, for example, by Pedoe [46, Chap. VI].

The most interesting case is that of the Moufang plane, in which the ternary ring reduces to an *alternative division ring:* addition is an Abelian group, both distributive laws hold, and each $a \neq 0$ has an inverse $a^{-1}$ satisfying

$$a^{-1}a = aa^{-1} = 1, \qquad a^{-1}(ab) = b = (ba)a^{-1},$$

whence $a(ab) = (aa)b$, $(ba)a = b(aa)$. It has been proved by Bruck and Kleinfeld [9] that every nonassociative ring of this kind is simply the ring of octaves over a field. (Earlier, we tacitly assumed this field to consist of the real numbers.) This aspect of the Moufang plane has been developed intensively by Freudenthal [31] and van der Blij [8].

In the algebraic approach to $n$-dimensional projective geometry over an arbitrary corpus [53, Part II], a point $(x)$ and a hyperplane $[X]$ are defined to be sets of "equivalent" symbols

$$(x_1, \ldots, x_{n+1}) \qquad \text{and} \qquad [X_1, \ldots, X_{n+1}],$$

where the $x$'s and $X$'s are not all zero, the rules for equivalence are that $(x_1, \ldots, x_{n+1})$ is equivalent to $(x_1\lambda, \ldots, x_{n+1}\lambda)$ $(\lambda \neq 0)$ and $[X_1, \ldots, X_{n+1}]$ to $[\mu X_1, \ldots, \mu X_{n+1}]$ $(\mu \neq 0)$, and the condition for incidence is $\{Xx\} = 0$, where

$$\{Xx\} = X_1x_1 + \cdots + X_{n+1}x_{n+1}.$$

It follows that the points collinear with $(x)$ and $(y)$ are $(x\lambda + y\mu)$, meaning $(x_1\lambda + y_1\mu, \ldots, x_{n+1}\lambda + y_{n+1}\mu)$. In particular, any three distinct collinear points may be expressed as

$$(x), (y), (x + y).$$

If a fourth point on the same line is $(x\lambda + y)$, the number $\lambda$ is called the *cross ratio* of the four collinear points [46, p. 170]. Dually, $\mu$ is the cross ratio of four "coaxial" hyperplanes

$$[X], [Y], [X + Y], [\mu X + Y].$$

If the four points are sections of the four hyperplanes, we have

$$\{Xx\} = 0, \qquad \{Yy\} = 0, \qquad \{Xy\} + \{Yx\} = 0$$

and

$$\mu\{Xy\} + \{Yx\}\lambda = 0, \qquad \text{whence} \qquad \mu\{Xy\} = \{Xy\}\lambda.$$

Thus we can conclude $\lambda = \mu$ if and only if the corpus is a field. In other words, the commutative law is logically equivalent to the invariance of cross ratio under projectivities, to the fundamental theorem of projective geometry, to Pappus's theorem [23, p. 38], to Möbius's theorem about mutually inscribed tetrahedra [20, p. 258] and to Gallucci's eight-line theorem [14, pp. 444–445]:

*If three skew lines all meet three other skew lines, any transversal to the first set of three meets any transversal of the second set.*

In the commutative case, let $\{AB, CD\}$ denote the cross ratio $\lambda$ of four collinear points $A$, $B$, $C$, $D$, so that, if $A$ is $(x)$ and $B$ is $(y)$ and $C$ is $(x + y)$, then $D$ is $(x\lambda + y)$, which is now the same as $(\lambda x + y)$. It follows easily [46, p. 171] that

$$\{AB, DC\} = \lambda^{-1} \qquad \text{and} \qquad \{AC, BD\} = 1 - \lambda.$$

Also, if $C$ and $D$ lie in hyperplanes $\alpha$ and $\beta$, respectively, we write $\{AB, CD\} = \{AB, \alpha\beta\}$, $\{AC, BD\} = \{A\alpha, B\beta\}$, and so on. If $\alpha$ is $[X]$ and $\beta$ is $[Y]$, so that $(x + y)$ lies on $[X]$, and $(x\lambda + y)$ on $[Y]$, we have $\{Xx\} + \{Xy\} = 0$ and $\{Yx\}\lambda + \{Yy\} = 0$, whence

$$\{AB, \alpha\beta\} = \frac{\{Xx\}\{Yy\}}{\{Yx\}\{Xy\}}$$

[39, pp. 120, 136].

According to a famous theorem of Wedderburn, every *finite* corpus is commutative (that is, every finite corpus is a Galois field). All known proofs, such as that of Ernst Witt [63], are deeply algebraic. Geometers are challenged by the thought that there may be a nonalgebraic proof of the equivalent statement that Pappus's theorem (or the fundamental theorem) holds in every finite Desarguesian plane. Such a thought was doubtless the motivation for the large section of Segre's book [53] that is devoted to line geometry over a general corpus.

If $a$, $b$, $c$ are three skew lines, the *regulus abc* has *directrices* which are their transversals, and *generators* which are transversals of triads of directrices. If the corpus is not a field, Gallucci's theorem fails, and so there are more generators than directrices. Such considerations provide a basis for the noncommutative theory of *conics*, which can be defined as sections of reguli.

One of the earliest instances of a non-Desarguesian plane was invented by Moulton [43]. However, that has an unpleasant air of artificiality. Far more satisfying is the example of Segre [51, p. 39], where the points and lines of a non-Desarguesian projective plane are represented by the points of the real projective plane and a certain two-parameter family of cubic curves, namely

$$(X_1{}^2 + X_2{}^2 + X_3{}^2)(X_1 x_1 + X_2 x_2 + X_3 x_3)(x_1{}^2 + x_2{}^2 + x_3{}^2)$$
$$+ \lambda X_3{}^3 x_3{}^3 = 0.$$

where the $X$'s are homogeneous parameters, varying from curve to curve, and $\lambda$ is a sufficiently small constant.

Before 1892, coordinates were always either real or complex. Then Fano [29] described an $n$-dimensional geometry in which the coordinates belong to the field of residue-classes modulo a prime $p$, so that the number of points on a line is $p + 1$. In 1906 Veblen and Bussey [61] gave this finite Projective Geometry the name $PG(n, p)$ and extended it to $PG(n, q)$ where the coordinates belong to the Galois field $GF(q)$, $q$ being any power of a prime. Many of the properties of this geometry depend only on the number $q + 1$ of points on a line. For instance [53, p. 27], only the most basic axioms of incidence are needed to prove that the number of $k$-dimensional subspaces $S_k$ in the $n$-space $S_n$ is

$$\frac{(q^{n+1} - 1)(q^n - 1) \ldots (q^{n+1-k} - 1)}{(q^{k+1} - 1)(q^k - 1) \ldots (q - 1)}$$
$$= \frac{(q^{n+1} - 1)(q^n - 1) \ldots (q^{k+2} - 1)}{(q^{n-k} - 1)(q^{n-k-1} - 1) \ldots (q - 1)}.$$

The former expression is appropriate when $k \leq \frac{1}{2}(n - 1)$, the latter when $k \geq \frac{1}{2}(n - 1)$. In particular, the finite plane contains $q^2 + q + 1$ points and (of course) the same number of lines.

In 1901 Dickson [26] represented many finite groups, especially simple groups, in finite projective spaces. About 1940 Brauer mentioned the desirability of finding a geometric representation for the Mathieu groups $M_{11}$, $M_{12}$, $M_{22}$, $M_{23}$, $M_{24}$. This remark inspired my work [18] which was continued by Todd [60], with the conclusion that the quintuply transitive groups $M_{12}$ and $M_{24}$ can be represented as collineation groups in $PG(5, 3)$ and $PG(11, 2)$, respectively.

## ELLIPTIC AND HYPERBOLIC POLARITIES

In the projective plane over a (commutative) field, a polarity is given by a bilinear form

$$(xy) = \Sigma \Sigma \, c_{jk} x_j y_k, \qquad c_{jk} = c_{kj}, \qquad \det(c_{jk}) \neq 0.$$

Two points $(x)$ and $(y)$, meaning $(x_1, x_2, x_3)$ and $(y_1, y_2, y_3)$, are conjugate if and only if $(xy) = 0$. Since the harmonic conjugates

$$(x \pm \mu y) = (x_1 \pm \mu y_1, \, x_2 \pm \mu y_2, \, x_3 \pm \mu y_3) \qquad (\mu \neq 0)$$

are conjugate (for the polarity) if

$$(xx) = \mu^2(yy),$$

this relation (for some nonzero square $\mu^2$) is the condition for $(x)$ to be accessible from $(y)$ [23, pp. 124, 153].

If the quadratic form $(xx)$ is indefinite, self-conjugate points exist, and their locus is the conic

$$(xx) = 0.$$

Multiplying the $c_{jk}$ (if necessary) by a suitable constant, we can ensure that, for some particular exterior point $(y)$, $(yy)$ is a square. Then, since the exterior points constitute a single class of mutually accessible points, $(xx)$ is a nonzero square for every exterior point $(x)$ and a nonsquare for every interior point $(x)$ [23, pp. 126, 155; 45, p. 201]. For instance, if

$$(xy) = x_1 y_1 + x_2 y_2 - x_3 y_3,$$

so that $(xx) = x_1{}^2 + x_2{}^2 - x_3{}^2$, the conic $(xx) = 0$ has tangents $x_2 \pm x_3 = 0$, and we can take $(y)$ to be their point of intersection $(1, 0, 0)$.

Over the real field, the statement "$(xx)$ is a nonzero square" becomes simply "$(xx) > 0$."

Over the complex field, every number is a square, so there are no interior points. Over the real field or any finite field, the product (or quotient) of two nonsquares is a square, and so *the interior points constitute a single class*. Over the rational field, we can find points $(x)$ for which $(xx)$ takes in turn various prime values; such points are interior but inaccessible from one another.

The same distinction between real and rational fields occurs when the polarity has no self-conjugate points, that is, when all the points of the plane are "interior." If such an "elliptic" polarity could exist over a finite field (of odd characteristic), that is, in a finite plane $PG(2, q)$ ($q$ odd), the number of points in each class of accessible points would be $\frac{1}{2}(q^2 + 1)$, which is not a divisor of $q^2 + q + 1$, the number of points in the whole plane [3, pp. 123–124; 23, pp. 101, 149]. Therefore *finite geometries* of two (or more) dimensions *admit no elliptic polarities* [2, p. 144; 52, p. 4].

This result is equivalent to the case $n = 3$ of the following theorem due to Chevalley [11]:

*Every Galois field is quasi-algebraically closed.*

This means that, over a finite field, if $F(x_1, \ldots, x_n)$ is any homogeneous polynomial of degree less than $n$, the equation $F = 0$ has a nontrivial solution.

CONICS IN THE REAL PLANE

The theory of accessibility shows that, in the *real* projective plane, two points on a conic decompose the secant joining them into two segments, consisting of exterior and interior points, respectively. The interior segment is naturally called a *chord*. Figure 4 shows two points $A$ and $B$ on a chord $MN$, so arranged that $A$ and $M$

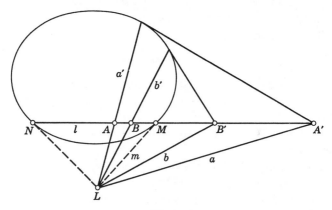

*Figure 4*

separate $B$ and $N$. The polars of $A$ and $B$, say $a$ and $b$, join $L$, the pole of the line $MN$, to points $A'$ and $B'$ on this line. These last points are the harmonic conjugates of $A$ and $B$ with respect to $M$ and $N$ [23, p. 73]. Hence, if $M, N, A, B$ are

$$(x), \quad (y), \quad (x + y), \quad (\lambda x + y),$$

respectively, then $A'$ is $(-x + y)$ and $B'$ is $(-\lambda x + y)$. Since the cross ratio of $(\kappa x + y)$, $(\lambda x + y)$, $(\mu x + y)$, $(\nu x + y)$ is

$$\frac{(\kappa - \mu)(\lambda - \nu)}{(\kappa - \nu)(\lambda - \mu)}$$

[23, pp. 119, 152], it follows that

$$\{AB, ba\} = \{AB, B'A'\} = \frac{(1 + \lambda)(\lambda + 1)}{(1 + 1)(\lambda + \lambda)} = \frac{(\lambda + 1)^2}{4\lambda},$$

where $\lambda = \{MN, AB\} = \{AB, MN\}$. This result extends easily to real projective $n$-space, with a nonruled quadric instead of a conic: if $\alpha$ and $\beta$ are the polar hyperplanes of two points $A$ and $B$ on a chord $MN$, then

$$\{AB, \beta\alpha\} = \frac{(\lambda + 1)^2}{4\lambda},$$

where $\lambda = \{AB, MN\}$.


CONICS AND $k$-ARCS IN A FINITE PLANE

In any *finite* projective plane, a set of $k$ points, no three collinear, is called a *k-arc*. A $k$-arc is said to be *complete* if it is not part of a $(k + 1)$-arc. Segre [53, pp. 270–294] has proved that, in a Desarguesian plane $PG(2, q)$ where $q$ is odd, every $(q + 1)$-arc is complete, the $(q + 1)$-arcs are just the conics, and there are no complete $q$-arcs. But complete $(q - 1)$-arcs occur when $q = 7$, 9, 11, 13; also complete 6-arcs and 7-arcs occur when $q = 9$.

In a plane $PG(2, q)$ where $q$ is even (namely a power of 2), the $q + 1$ tangents of a $(q + 1)$-arc concur in a point called the *nucleus* (or "center") of the $(q + 1)$-arc. The $(q + 1)$-arc is incomplete, but yields a complete $(q + 2)$-arc when its nucleus is added. Each point of the $(q + 2)$-arc is the nucleus of the $(q + 1)$-arc formed by the remaining points, but these $q + 1$ points usually *do not form a conic!* It has been proved with the aid of an electronic computer that, when $q = 16$, some $(q + 2)$-arcs contain no conic. The same situation arises when $q = 32$ or 128 or any higher power of 2, but the case when $q = 64$ remains undecided.

In virtue of Wedderburn's theorem, the only possible coordinates for a finite non-Desarguesian plane belong to a ternary ring which is not a corpus. Still denoting the number of points on a line by $q + 1$, we naturally ask what are the possible values for $q$. Although some progress has been made, this remains a challenging problem. The number $q$ is called the *order* of the plane.

The painstaking work of Tarry [59] on Euler's Officers' Problem [5, p. 190] proves the impossibility of $q = 6$. It is known [37] that the only planes of other orders less than 9 are the Desarguesian planes $PG(2, q)$; but non-Desarguesian planes are known for $q = 9$ and many greater values, including $p^r$ with $p$ odd and $r \geq 2$. In every known case, $q$ is a power of a prime [36, p. 394]. Bruck and Ryser [10] proved that there cannot be a plane with $q \equiv 1$ or 2 (mod 4) unless $q$ is expressible in the form $a^2 + b^2$. Since 10 is so expressible, but 6 not, this result includes Tarry's but leaves open the possibility of $q = 10$.

In $PG(2, q)$, where $q$ is a power of 2, every complete quadrangle has collinear diagonal points. Conversely, this property for every complete quadrangle in a finite plane is sufficient to make the plane Desarguesian [35]. In $PG(2, q)$, where $q$ is odd, every complete quadrangle has noncollinear diagonal points. Hanna Neumann [44] has conjectured that this property for every complete quadrangle suffices to make the finite plane Desarguesian.

## HYPERBOLIC GEOMETRY

When an ordered space is extended to a real projective space (that is, imbedded in a real projective space), each ordered line is extended to a projective line (on which the order of points is not serial but cyclic). On such a line there is either a single ideal point or an interval of ideal points. If every line has a single ideal point (its *point at infinity*) the ordered space is *affine* [20, Chap. 13]: through any point there is a unique line parallel to a given line, and we can compare distances along parallel lines. If every line has an interval of ideal points, the ordered space is *hyperbolic*, the ends of the ideal segments are called points at infinity (or simply *ends*), and their locus (in the projective space) is a nonruled quadric [17, p. 195] or, in two-dimensional case, a conic. This is Klein's approach to non-Euclidean geometry: the points and lines are the interior points and chords of a nonruled quadric (or conic).

When the points and lines of the hyperbolic plane are regarded as the interior points and chords of a conic in the real projective plane, the isometries of the former appear in the latter as collineations leaving the "absolute" conic invariant [17, pp. 201–203]. In the projective plane, a collineation of period 2 is a harmonic homology [23, p. 55], and this leaves a conic invariant if the center and axis are pole and polar. When applied to the absolute conic, this collineation appears as a reflection or a half-turn according to the nature of its axis. If the axis is a secant (so that its pole is exterior) the homology is the reflection in this line. If the axis is a nonsecant, so that its pole is interior, the homology is the half-turn about this point. It follows that *perpendicular* lines are conjugate with respect to the absolute conic. The various lines perpendicular to a given line are ultraparallel [55] to one another: their common point, being the pole of the given line, is exterior, and thus does not belong to the hyperbolic plane itself.

Any two *ends*, $M$ and $N$, determine a line $l = MN$ whose pole $L$ is the intersection of the tangents at $M$ and $N$ (Figure 5). For any point $P$, not on $l$, the segments $PM$ and $PN$ are the rays from $P$ *parallel* to $l$ (in the sense of Gauss, Bolyai, and Lobachevsky): they do not meet $l$ (in ordinary points), but every ray within the angle $NPM$ does meet $l$. This angle is bisected by the line $PL$, which, being perpendicular to $l$, acts as a mirror reflecting $PM$ into $PN$. This angle-bisecting technique, combined with continuity, soon yields a measurement of angles. In particular, $\angle LPM$ is the *angle of parallelism* for the point $P$ and line $l$.

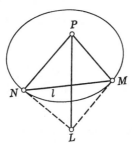

*Figure 5*

The measurement of *distance* can be developed from the observation that two directed segments, $AB$ and $A'B'$, on chords $MN$ and $M'N'$ of the absolute conic, are congruent if $ABMN \overline{\wedge} A'B'M'N'$, that is, if

$$\{AB, MN\} = \{A'B', M'N'\}.$$

The essential property of a directed distance is its additivity: if $[ABC]$, then

$$AB + BC = AC.$$

Taking into consideration the relation

$$\{AB, MN\}\{BC, MN\} = \{AC, MN\}$$

[17, p. 77], which holds for any five collinear points, we see that this requirement is neatly satisfied by defining

$$AB = \tfrac{1}{2}\log\{AB, MN\}.$$

(The $\tfrac{1}{2}$ is inserted for convenience [17, p. 201].)   Now let $a$ and $b$ be the absolute polars of $A$ and $B$, as in Figure 4.   Setting $\lambda = e^{2AB}$ in our formula

$$\{AB, ba\} = \frac{(\lambda + 1)^2}{4\lambda}, \quad \text{where} \quad \lambda = \{AB, MN\},$$

we deduce

$$\sqrt{\{AB, ba\}} = \frac{e^{2AB} + 1}{2e^{AB}} = \frac{e^{AB} + e^{-AB}}{2} = \cosh AB.$$

Thus the distance $AB$ is given by

$$\cosh AB = \sqrt{\{AB, ba\}}, \quad \operatorname{sech} AB = \sqrt{\{AB, ab\}},$$
$$\tanh AB = \sqrt{\{Aa, Bb\}}.$$

Similarly in $n$ dimensions, if $\alpha$ and $\beta$ are the absolute polar hyperplanes of $A$ and $B$,

$$\cosh AB = \sqrt{\{AB, \beta\alpha\}} \quad \text{and} \quad \tanh AB = \sqrt{\{A\alpha, B\beta\}}.$$

In terms of projective (homogeneous) coordinates, let the absolute quadric have the equation $(xx) = 0$, where

$$(xy) = \Sigma\Sigma c_{jk}x_j y_k, \quad c_{jk} = c_{kj}, \quad \det(c_{jk}) \neq 0.$$

Then the polar hyperplane of the point $(y)$ is $[Y]$, where

$$Y_j = \Sigma c_{jk} y_k.$$

Let $A$, $B$, $\alpha$, $\beta$ be $(x)$, $(y)$, $[X]$, $[Y]$, so that

$$\{Yx\} = \Sigma Y_j x_j = (xy).$$

Then

$$\cosh^2 AB = \{AB, \beta\alpha\} = \frac{\{Yx\}\{Xy\}}{\{Xx\}\{Yy\}} = \frac{(xy)^2}{(xx)(yy)}.$$

Somewhat similar considerations [17, p. 225] yield the formula

$$\cos^2 (\alpha\beta) = \frac{(xy)^2}{(xx)(yy)}$$

for the angle $(\alpha\beta)$ between two intersecting hyperplanes, $\alpha$ and $\beta$, whose poles are exterior points $(x)$ and $(y)$ on a nonsecant.

EXTERIOR-HYPERBOLIC GEOMETRY

As long ago as 1907, Study [55] described the geometry of the points *exterior* to a nonruled quadric $(xx) = 0$ in real projective $n$-space, that is, of the poles of the hyperplanes in hyperbolic $n$-space. He observed that the join of two such points may be either a secant or a tangent or a nonsecant. In the case of the secant, we can simply interchange the roles of the primed and unprimed letters in Figure 4, and conclude, as before, that

$$\cosh^2 AB = \frac{(xy)^2}{(xx)(yy)}.$$

In the case of the nonsecant, we borrow a trick from spherical trigonometry and measure the distance $AB$ by the angle $(\alpha\beta)$ between the polar hyperplanes, so that

$$\cos^2 AB = \frac{(xy)^2}{(xx)(yy)}.$$

In the intermediate case (when the line $AB$ is a tangent), we have

$$(xy)^2 = (xx)(yy)$$

[23, pp. 126, 154, Ex. 6]; so it is natural to define the length of $AB$ to be zero.

Dividing the coefficients $c_{jk}$ by the constant $(yy)$, where $(y)$ is a particular exterior point, we can contrive (as in our section on "Elliptic and Hyperbolic Polarities") to make $(xx)$ positive for *every* exterior point. Then, dividing the coordinates of each point $(x)$ by $\sqrt{(xx)}$, we can make $(xx) = 1$ for every exterior point [17, p. 281]. In other words, we replace our homogeneous coordinates by redundant nonhomogeneous coordinates satisfying $(xx) = 1$, with the pleasing result that

$$\cosh AB = |(xy)|, \qquad AB = 0, \qquad \text{or} \qquad \cos AB = |(xy)|$$

according as the line $AB$ is a secant, a tangent, or a nonsecant.

For instance, as we have already seen, $(1, 0, 0)$ is exterior to the conic

$$x_1{}^2 + x_2{}^2 - x_3{}^2 = 0,$$

so we can assume $x_1{}^2 + x_2{}^2 - x_3{}^2 = 1$ for every point $(x)$ in the exterior-hyperbolic plane determined by this conic. Study observed [55, p. 108] that in a triangle whose sides are lines of the first kind (secants), one side is *longer* than the sum of the other two!

## RELATIVITY

In its geometric aspect, the space-time of relativity theory is a diagram consisting of a four-dimensional space whose points represent events. As Synge remarks [57, pp. 5, 6]: "Anything that happens is an *event*, but we sharpen the concept to mean an occurrence that takes up no room and has no duration. . . . All possible events form a four-dimensional continuum." The locus of events in the "life" of any particular particle, being one-dimensional, is a curve in the four-space. Following Minkowski [42a, p. 55], we call this curve the *world line* of the particle. Synge's most charming and revealing description of world lines was given in a short paper in the *New Scientist* [56].

The passage of time appears in the diagram as a direction along the world line of each particle: thus the world line is a *directed* curve. Each event in the history of the particle appears as a point decomposing the curve into two parts: the past and the future. In particular, the world line of an unaccelerated particle appears as a straight line, on which the past and future are two opposite rays.

In the familiar spaces (Euclidean, affine, projective, non-Euclidean) all positions are alike and all directions are alike. In the four-dimensional space-time, all positions are alike (provided we ignore the "mountain ranges" that represent matter) but directions are of five essentially different kinds. For each event $O$, there are the directions of possible future events of a particle at $O$, the directions of possible past events of such a particle, and the directions of events $P$ which are neither before nor after $O$. These three sets of directions are separated by the two nappes of the *light cone* (or "null cone," or "isotropic cone") whose generators are the world lines of light signals (or photons) that could be emitted from $O$ or received at $O$ [57, p. 41]. This cone is, of course, a three-dimensional pseudo-manifold which joins its vertex $O$ by straight lines to all the points on a nonruled quadric surface like the familiar sphere.

The word *cone* is appropriate not merely because it gives a rough idea of the light rays associated with any event but because the most natural geometric spaces to use for our four-dimensional diagram are *real affine four-space* and *real projective four-space*, in both of which a quadric cone can be defined and a distinction can be made between three sets of lines through the vertex: generators that lie entirely on the cone, exterior lines on pairs of tangent hyperplanes, and interior lines that lie on no tangent hyperplanes. A generator is called a *null* line (or an *isotropic* line) because the measurement of any interval along it is zero. Other lines are said to be *timelike* or *spacelike* according as they are interior or exterior. The world line of a photon is a null line. The world lines of any other unaccelerated particle is timelike. The join $OP$ of two events that are essentially incapable of influencing each other is spacelike [57, p. 19; 58, p. 108].

Let $A$, $B$, $C$ be three unaccelerated particles, initially together and remaining collinear but parting with relative velocities $v_{BC}$, $v_{CA} = -v_{AC}$, and $v_{AB}$. In nonrelativistic kinematics we would have

$$v_{AC} = v_{AB} + v_{BC}.$$

The special theory of relativity changes this to

$$v_{AC} = (v_{AB} + v_{BC})/(1 + v_{AB}v_{BC}/c^2),$$

where $c$ is the velocity of light. The addition rule

$$\tanh (x + y) = (\tanh x + \tanh y)/(1 + \tanh x \tanh y)$$

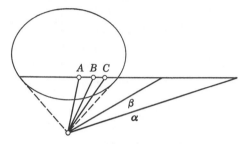

*Figure 6*

suggested to Robb the idea of using, instead of the velocity $v$, the *rapidity* $r$, defined by $v = c \tanh r$, so that

$$r_{AB} + r_{BC} = r_{AC}$$

[28, p. 22; 57, p. 126].

Robb showed in his Appendix [48, p. 405] that the timelike directions at any event $O$ can be represented by the points of a hyperbolic three-space in such a way that rapidities appear as distances. Thus the directions of the world lines of our three particles are represented by three collinear points $A$, $B$, $C$ (Figure 6) and the equation $r_{AB} + r_{BC} = r_{AC}$ becomes

$$AB + BC = AC.$$

It follows that the null vectors through $O$ are represented by the "points at infinity" of the hyperbolic space, and the light cone at $O$ is represented by the locus of points at infinity, which is the absolute quadric: a nonruled quadric in real projective three-space. From this it is a natural step to complete the picture by using the points outside the quadric to represent the spacelike directions from $O$.

Using the projective formula for a hyperbolic distance, we find that, if $\alpha$ and $\beta$ are the polar planes of $A$ and $B$, as in Figure 6, the relative rapidity $r_{AB} = AB$ of these two particles is given by

$$\tanh AB = \sqrt{\{A\alpha, B\beta\}}.$$

Therefore their relative *velocity* is simply

$$v_{AB} = c \tanh AB = c\sqrt{\{A\alpha, B\beta\}}:$$

a pretty connection between kinematics and projective geometry!

Although Robb [48, p. 22] was aware of the fact that this way of representing the directions of world lines is valid in the general theory of relativity, he was chiefly concerned with the special theory, according to which the underlying four-space is affine and the three kinds of directions from any event are *parallel* to those from any other.    He gave a sufficient set of axioms to describe a Minkowskian space-time which, from our standpoint, is simply a real affine four-space with a pseudo-Euclidean metric.    In its projective aspect, this is a projective four-space in which one hyperplane has been specialized and called the three-space at infinity. Two lines are parallel if and only if their points at infinity coincide. The metric is imposed by specializing a nonruled quadric surface $\Omega$ in this projective three-space.    Two lines are orthogonal if and only if their points at infinity are conjugate with respect to $\Omega$.    At any event $O$, the light cone joins $O$ to $\Omega$.    The timelike and spacelike lines join $O$ to the interior and exterior points.    In fact, this quadric $\Omega$, which is a common section of all light cones, serves as the absolute for Robb's hyperbolic space whose ordinary points represent the timelike vectors.

Using homogeneous coordinates $x_1, \ldots, x_5$ in the projective four-space, we are free to choose the simplex of reference and "unit point" so that the special hyperplane is $x_5 = 0$ and the quadric surface $\Omega$ is

$$x_1{}^2 + x_2{}^2 + x_3{}^2 - x_4{}^2 = 0, \qquad x_5 = 0.$$

Since the $x$'s are homogeneous, any point of the affine space $x_5 \neq 0$ may be expressed in the form $(x_1, x_2, x_3, x_4, 1)$; then the first four $x$'s serve as nonhomogeneous affine coordinates and can be identified with the familiar $x$, $y$, $z$, $ct$ of the special theory of relativity.    The light cone at any event $(a)$ is

$$(x_1 - a_1)^2 + (x_2 - a_2)^2 + (x_3 - a_3)^2 - (x_4 - a_4)^2 = 0.$$

The line joining two events, $A = (a)$ and $B = (b)$, is timelike, null, or spacelike according as the number

$$(a_1 - b_1)^2 + (a_2 - b_2)^2 + (a_3 - b_3)^2 - (a_4 - b_4)^2$$

is negative, zero, or positive; and the interval or *separation* $AB$ between these two events is defined to be the square root of the absolute magnitude of this number.

For comparison with Synge [57, p. 40], note that instead of $(x_1, x_2, x_3, x_4)$ we should strictly write $(x^1, x^2, x^3, x^4)$.    In fact,

Synge uses $(x_1, x_2, x_3, x_4)$ for the so-called Minkowskian coordinates, which do not concern us at all [57, p. 56].

As a diagram for relativistic cosmology in the large, Minkowski's world has been almost entirely superseded by de Sitter's, which, as Synge says, "opens up new vistas, introducing us to the idea that space (a slice of space-time) may be finite, and this seems to satisfy some mental need in us, for infinity is one of those things which we find difficulty in comprehending" [58, p. 257]. Timelike lines remain infinite, but most of us are willing to accept eternity.

In the same year that Eddington published the second edition of *The Mathematical Theory of Relativity* [28], Du Val [27] made a discovery that has been almost completely ignored, even by Robb and Synge. This discovery is that de Sitter's world, in its *polar* or *elliptic* form [58, p. 260] is exterior-hyperbolic four-space: the part of real projective four-space that lies outside a nonruled quadric hypersurface.

At any event $A$, the light cone envelops the absolute quadric; in other words, the null lines are tangents, the timelike lines are secants, and the spacelike lines are nonsecants (Figure 7). Study's observation of the "nontriangle inequality" for a triangle of timelike lines is thus revealed as an early version of Einstein's *twin paradox:* An astronaut flies away, at a speed $c - \epsilon$, to some distant planet, and returns, still in the prime of life, to find that his twin brother, who stayed at home, is now an old man.

The polar hyperplane $\alpha$ of $A$ intersects the quadric hypersurface in a quadric surface, and allows us to represent the lines of all three

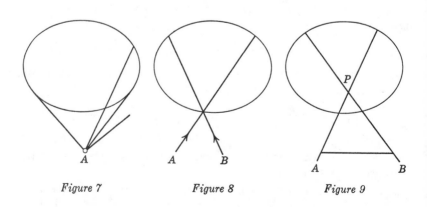

Figure 7          Figure 8          Figure 9

kinds through $A$ by the points that are their $\alpha$-sections, just like the hyperplane at infinity in Minkowski's world. It is only when we pass from $A$ to another "observer" $B$ that differences are seen: the null lines through $B$ are no longer "parallel" to those through $A$. But the adjective *parallel* can still be used for two timelike lines that have the same point at infinity, that is, two lines that meet on the quadric hypersurface (as in Figure 8), so that their extensions inside the quadric are parallel in the sense of classical hyperbolic geometry. If such lines are the world lines of two unaccelerated particles $A$ and $B$, the spacelike interval $AB$ does not remain constant but decreases asymptotically. On the other hand, two secants which have in common an interior point $P$ (as in Figure 9) are the world lines of unaccelerated particles which have a minimax interval on the polar hyperplane of $P$ and afterwards diverge.

This use of the word *minimax* has the following justification. For each position of $A$ on the first world line there is a corresponding position of $B$ on the second, namely the foot of the *perpendicular* from $A$ in the sense of the exterior-hyperbolic geometry. This is the point on the second world line that is *farthest* from $A$. (Two other positions of $B$, lying on tangents from $A$ to the quadric, are actually at *zero* distance from $A$.) This maximum distance $AB$ attains its minimum when $A$ is conjugate to $P$, so that $AB$ is perpendicular to both the world lines.

Similar remarks apply to *skew* world lines [19] which do not lie in a plane (but, of course, in a hyperplane). Thus two random particles (with an infinitesimal probability of having parallel world lines) will eventually diverge, and "a number of particles initially at rest will tend to scatter" [28, p. 161]. In fact, we can deduce the "expanding universe" from Du Val's projective diagram without invoking any calculation.

When we do wish to make calculations, we naturally take the absolute quadric to be

$$x_1{}^2 + x_2{}^2 + x_3{}^2 - x_4{}^2 + x_5{}^2 = 0.$$

Since the $x$'s are homogeneous, we can normalize the coordinates so that every event $(x)$ satisfies

$$x_1{}^2 + x_2{}^2 + x_3{}^2 - x_4{}^2 + x_5{}^2 = 1.$$

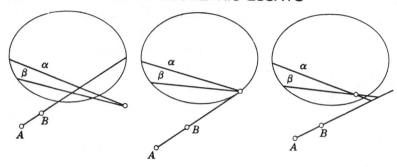

*Figure 10*

Then the light cone at any event $(a)$ is

$$x_1{}^2 + x_2{}^2 + x_3{}^2 - x_4{}^2 + x_5{}^2 - (a_1x_1 + a_2x_2 + a_3x_3 - a_4x_4 + a_5x_5)^2 = 0;$$

for example, the light cone at $(0, 0, 0, 0, 1)$ is

$$x_1{}^2 + x_2{}^2 + x_3{}^2 - x_4{}^2 = 0,$$

exactly as in the Minkowskian case. The line joining two events, $A = (a)$ and $B = (b)$, is timelike, null, or spacelike according as their polar hyperplanes $\alpha$ and $\beta$ (in the ordinary or "interior" hyperbolic space) are ultraparallel (having the line $AB$ as their common perpendicular), parallel, or intersecting (see Figure 10), that is [16, p. 193], according as the cross ratio

$$\{AB, \beta\alpha\} = (a_1b_1 + a_2b_2 + a_3b_3 - a_4b_4 + a_5b_5)^2 = (ab)^2$$

is greater than 1, equal to 1, or less than 1; and the separation $AB$ between these two events is given by

$$\cosh AB = |(ab)|, \qquad AB = 0, \qquad \text{or} \qquad \cos AB = |(ab)|,$$

respectively [17, p. 281]. It follows [17, p. 249] that the metric form $\Phi$ [58, p. 1] is

$$dx_1{}^2 + dx_2{}^2 + dx_3{}^2 - dx_4{}^2 + dx_5{}^2,$$

where $dx_4$ or $dx_5$ can be eliminated (if we so desire) by means of the relation

$$x_1 \, dx_1 + x_2 \, dx_2 + x_3 \, dx_3 - x_4 \, dx_4 + x_5 \, dx_5 = 0.$$

Synge [58, p. 262] described de Sitter's world as a space-time of constant positive curvature. We have simplified the discussion by choosing such units of measurement that this curvature is 1 and the velocity of light is 1. In other words, our unit of distance is "the radius of curvature of the universe" and our unit of time is the time light would take to travel this distance. It must be admitted that this projective approach yields only the nonorientable "polar" form of de Sitter's world, whereas Synge prefers its orientable covering manifold, the "antipodal" form. However, all the above formulae remain valid; the only difference is that two antipodal events, such as $(0, 0, 0, 0, \pm 1)$, are regarded as being distinct instead of coincident.

## CONCLUSION

We have considered projective geometries of two or more dimensions over various fields, and the possibility of replacing the field by a corpus or a ternary ring. We might just as well have considered affine, Euclidean, non-Euclidean, and pseudo-Euclidean geometries over the same fields. Thus there are many geometries, each describing another world: wonderlands and Utopias, refreshingly different from the world we live in.

I hope that these excursions, in a few of many possible directions, have helped to reveal the healthy state of development of this fascinating subject, including its interactions with other branches of pure and applied mathematics.

## REFERENCES

1. A. D. Alexandrow, *Die innere Geometrie der konvexen Flächen*, Berlin (1955) Akademie-Verlag.
2. Emil Artin, *Geometric algebra*, New York (1957) Interscience.
3. Friedrich Bachmann, *Aufbau der Geometrie aus dem Spiegelungsbegriff*, Berlin (1959) Springer.
4. Reinhold Baer, Polarities in finite projective planes, *Bull. Amer. Math. Soc.*, **52** (1946) 77–93.
5. W. W. Rouse Ball, *Mathematical recreations and essays*, London (1959) Macmillan.
6. A. S. Besicovitch, The Kakeya problem, *Amer. Math. Monthly*, **70** (1963) 697–706.
7. H. F. Blichfeldt, The minimum values of quadratic forms in six, seven and eight variables, *Math. Z.*, **39** (1935) 1–15.
8. F. van der Blij, History of the octaves, *Simon Stevin*, **34** (1961) 106–125.

8a. Max Brückner, Ueber die Ableitung der allgemeinen Polytope und die nach Isomorphismus verschiedenen Typen der allgemeinen Achtzelle (Oktatope), *Verk. Nederl. Akad. Wetensch.*, Sect. I, 10.1 (1909).

9. R. H. Bruck and E. Kleinfeld, The structure of alternative division rings, *Proc. Amer. Math. Soc.*, 2 (1951) 878–890.

10. R. H. Bruck and H. J. Ryser, The non-existence of certain finite projective planes, *Canad. J. Math.*, 1 (1949) 88–93.

10a. C. Carathéodory, Ueber den Variabilitätsbereich der Fourierschen Konstanten von positiven harmonischen Funktionen, *Rend. Circ. Mat. Palermo*, 32 (1911) 193–217.

11. Claude Chevalley, Démonstration d'une hypothèsis de M. Artin, *Abh. Math. Sem. Hamburg*, 11 (1936) 73–75.

12. H. S. M. Coxeter, Integral Cayley numbers, *Duke Math. J.*, 13 (1946) 561–578.*

13. H. S. M. Coxeter, Review of a paper by Melmore, *Math. Rev.*, 9 (1948) 53.

14. H. S. M. Coxeter, Self-dual configurations and regular graphs, *Bull. Amer. Math. Soc.*, 56 (1950) 413–455.*

15. H. S. M. Coxeter, Extreme forms, *Canad. J. Math.*, 3 (1951) 391–441.

16. H. S. M. Coxeter, *The real projective plane*, Cambridge, England (1955) Cambridge University Press.

17. H. S. M. Coxeter, *Non-Euclidean geometry*, Toronto (1961) University of Toronto Press.

18. H. S. M. Coxeter, Twelve points in PG(5, 3) with 95040 self-transformations, *Proc. Royal Soc.*, A 247 (1958) 279–293.*

19. H. S. M. Coxeter, Reflected light signals, in Hoffmann (Ed.), *Perspectives in Geometry and Relativity*, Bloomington (1966) Indiana University Press, 58–70.*

20. H. S. M. Coxeter, *Introduction to geometry*, New York (1961) Wiley.

21. H. S. M. Coxeter, An upper bound for the number of equal nonoverlapping spheres that can touch another of the same size, *Proc. Sympos. Pure Math.*, 7 (1963) 53–72 (Amer. Math. Soc., Providence, R.I.).*

22. H. S. M. Coxeter, *Regular polytopes*, New York (1963) Macmillan.

23. H. S. M. Coxeter, *Projective geometry*, New York (1964) Blaisdell.

24. H. S. M. Coxeter and W. O. J. Moser, *Generators and relations for discrete groups*, Berlin (1965) Springer's Ergeb. Math. N.F. 14.

25. Ludwig Danzer, Branko Grünbaum and Victor Klee, Helly's theorem and its relatives, *Proc. Sympos. Pure Math.*, 7 (1963) 101–180 (Amer. Math. Soc., Providence, R.I.).

26. L. E. Dickson, *Linear groups, with an exposition of the Galois field theory*, Leipzig (1901).

27. Patrick Du Val, Geometrical note on de Sitter's world, *Phil. Mag.* (6) 47 (1924) 930–938.

27a. Patrick Du Val, *Homographies, quaternions, and rotations*, Oxford, England (1964) Clarendon Press.

28. A. S. Eddington, *The mathematical theory of relativity*, Cambridge, England (1924) Cambridge University Press.

29. Gino Fano, Sui postulati fondamentali della geometria proiettiva in uno

*References 12, 14, 18, 19, 21 are Chapters 2, 6, 7, 11, 9 of this book.

spazio lineare a un numero qualunque di dimensioni, *Giorn. Mat.*, **30** (1892) 106–132.

30. László Fejes Tóth, *Regular figures*, Oxford, England (1964) Pergamon.

31. Hans Freudenthal, Zur ebenen Oktavengeometrie, *Proc. Kon. Akad. Wet. Amsterdam*, A **56** (1953) 195–200.

32. David Gale, Neighborly and cyclic polytopes, *Proc. Sympos. Pure Math.*, **7** (1963) 225–232 (Amer. Math. Soc., Providence, R.I.).

33. C. F. Gauss, *Werke*, Göttingen (1876).

34. A. Gilbert and D. MacDonnell, The Steiner-Lehmus theorem, *Amer. Math. Monthly*, **70** (1963) 79–80.

35. A. M. Gleason, Finite Fano planes, *Amer. J. Math.*, **78** (1956) 797–807.

35a. Hugo Hadwiger, Ungelöste Probleme, Nr. 17, *Elemente der Math.*, **12** (1957) 61–62.

36. Marshall Hall, Jr., *The theory of groups*, New York (1959) Macmillan.

37. Marshall Hall, Jr., J. D. Swift, and R. J. Walker, Uniqueness of the projective geometry of order eight, *Math. Tables Aids Comput.*, **10** (1956) 186–194.

38. P. J. Heawood, On the four-colour map theorem, *Quarterly J. Math.*, **29** (1897) 270–285.

39. L. Heffter and C. Koehler, *Lehrbuch der analytischen Geometrie* I, Karlsruhe (1927).

40. Irving Kaplansky, Lie algebras, in Saaty (Ed.), *Lectures on Modern Mathematics* I, New York (1963) Wiley, 115–132.

41. Sidney Melmore, Densest packing of equal spheres, *Nature*, **159** (1947) 817.

42. Hermann Minkowski, Diskontinuitätsbereiche für arithmetische Äquivalenz, *J. Reine Angew. Math.*, **129** (1905) 220–274.

42a. Hermann Minkowski, Raum und Zeit, in H. A. Lorentz, A. Einstein, and H. Minkowski, *Das Relativitätsprinzip*, Leipzig (1922) Teubner, pp. 54–71.

43. F. R. Moulton, A simple non-Desarguesian plane geometry, *Trans. Amer. Math. Soc.*, **3** (1902) 192–195.

44. Hanna Neumann, On some finite non-Desarguesian planes, *Archiv Math.*, **6** (1954) 36–40.

45. T. G. Ostrom, Separation, betweenness, and congruence in planes over non-ordered fields, *Archiv Math.*, **10** (1959) 200–205.

46. Daniel Pedoe, *An introduction to projective geometry*, New York (1963) Macmillan.

47. Günter Pickert, *Projektive Ebenen*, Berlin (1955) Springer.

48. A. A. Robb, *Geometry of time and space*, Cambridge, England (1936) Cambridge University Press.

48a. C. A. Rogers, *Packing and covering*, Cambridge, England (1964) Cambridge University Press.

49. Bertrand Russell, *The principles of mathematics*, London (1937).

50. Ludwig Schläfli, *Gesammelte mathematische Abhandlungen* I, Basel (1950) Birkhäuser.

51. Beniamino Segre, Plans graphiques algébriques réels non désarguesiens et correspondances crémoniennes topologiques, *Rev. Math. pures appl.*, **1.3** (1956) 35–50.

52. Beniamino Segre, Le geometrie di Galois, *Ann. di Mat.*, (4) **48** (1959) 1–96.

53. Beniamino Segre, *Lectures on modern geometry*, Rome (1961) Cremonese.
54. Robert Steinberg, Finite reflection groups, *Trans. Amer. Math. Soc.*, **91** (1959) 493–504.
55. Eduard Study, Beiträge zur nicht-euklidische Geometrie, *Amer. J. Math.*, **29** (1907) 101–167.
56. J. L. Synge, An introduction to space-time, *New Scientist*, May 1, 1958.
57. J. L. Synge, *Relativity: The special theory*, Amsterdam (1956) North-Holland.
58. J. L. Synge, *Relativity: The general theory*, Amsterdam (1960) North-Holland.
59. G. Tarry, Le problème des 36 officiers, *C.R. Assoc. Fr. Av. Sci.*, **2** (1901) 170–203.
60. J. A. Todd, On representations of the Mathieu groups as collineation groups, *J. London Math. Soc.*, **34** (1959) 406–416.
61. Oswald Veblen and W. H. Bussey, Finite projective geometries, *Trans. Amer. Math. Soc.*, **7** (1906) 241–259.
62. A. N. Whitehead, *The axioms of descriptive geometry*, Cambridge, England (1907) Cambridge University Press.
63. Ernst Witt, Über die Kommutativität endlicher Schiefkörper, *Abh. Math. Sem. Hamburg*, **8** (1931).
64. Ernst Witt, Spiegelungsgruppen und Aufzählung halbeinfacher Liescher Ringe, *Abh. Math. Sem. Hamburg*, **14** (1941) 289–322.
65. J. W. Young, *Monographs on topics of modern mathematics*, New York (1911) Longmans Green.

# INDEX

# INDEX

A CATALOG OF SELECTED
# DOVER BOOKS
IN SCIENCE AND MATHEMATICS

# A CATALOG OF SELECTED
# DOVER BOOKS
## IN SCIENCE AND MATHEMATICS

QUALITATIVE THEORY OF DIFFERENTIAL EQUATIONS, V.V. Nemytskii and V.V. Stepanov. Classic graduate-level text by two prominent Soviet mathematicians covers classical differential equations as well as topological dynamics and ergodic theory. Bibliographies. 523pp. 5⅜ x 8½. 65954-2 Pa. $14.95

MATRICES AND LINEAR ALGEBRA, Hans Schneider and George Phillip Barker. Basic textbook covers theory of matrices and its applications to systems of linear equations and related topics such as determinants, eigenvalues and differential equations. Numerous exercises. 432pp. 5⅜ x 8½. 66014-1 Pa. $12.95

QUANTUM THEORY, David Bohm. This advanced undergraduate-level text presents the quantum theory in terms of qualitative and imaginative concepts, followed by specific applications worked out in mathematical detail. Preface. Index. 655pp. 5⅜ x 8½. 65969-0 Pa. $15.95

ATOMIC PHYSICS (8th edition), Max Born. Nobel laureate's lucid treatment of kinetic theory of gases, elementary particles, nuclear atom, wave-corpuscles, atomic structure and spectral lines, much more. Over 40 appendices, bibliography. 495pp. 5⅜ x 8½. 65984-4 Pa. $13.95

ELECTRONIC STRUCTURE AND THE PROPERTIES OF SOLIDS: The Physics of the Chemical Bond, Walter A. Harrison. Innovative text offers basic understanding of the electronic structure of covalent and ionic solids, simple metals, transition metals and their compounds. Problems. 1980 edition. 582pp. 6⅛ x 9¼. 66021-4 Pa. $19.95

BOUNDARY VALUE PROBLEMS OF HEAT CONDUCTION, M. Necati Özisik. Systematic, comprehensive treatment of modern mathematical methods of solving problems in heat conduction and diffusion. Numerous examples and problems. Selected references. Appendices. 505pp. 5⅜ x 8½. 65990-9 Pa. $12.95

A SHORT HISTORY OF CHEMISTRY (3rd edition), J.R. Partington. Classic exposition explores origins of chemistry, alchemy, early medical chemistry, nature of atmosphere, theory of valency, laws and structure of atomic theory, much more. 428pp. 5⅜ x 8½. (Available in U.S. only) 65977-1 Pa. $12.95

A HISTORY OF ASTRONOMY, A. Pannekoek. Well-balanced, carefully reasoned study covers such topics as Ptolemaic theory, work of Copernicus, Kepler, Newton, Eddington's work on stars, much more. Illustrated. References. 521pp. 5⅜ x 8½. 65994-1 Pa. $12.95

PRINCIPLES OF METEOROLOGICAL ANALYSIS, Walter J. Saucier. Highly respected, abundantly illustrated classic reviews atmospheric variables, hydrostatics, static stability, various analyses (scalar, cross-section, isobaric, isentropic, more). For intermediate meteorology students. 454pp. 6⅛ x 9¼. 65979-8 Pa. $14.95

RELATIVITY, THERMODYNAMICS AND COSMOLOGY, Richard C. Tolman. Landmark study extends thermodynamics to special, general relativity; also applications of relativistic mechanics, thermodynamics to cosmological models. 501pp. 5⅜ x 8½. 65383-8 Pa. $15.95

APPLIED ANALYSIS, Cornelius Lanczos. Classic work on analysis and design of finite processes for approximating solution of analytical problems. Algebraic equations, matrices, harmonic analysis, quadrature methods, much more. 559pp. 5⅜ x 8½. 65656-X Pa. $16.95

INTRODUCTION TO ANALYSIS, Maxwell Rosenlicht. Unusually clear, accessible coverage of set theory, real number system, metric spaces, continuous functions, Riemann integration, multiple integrals, more. Wide range of problems. Undergraduate level. Bibliography. 254pp. 5⅜ x 8½. 65038-3 Pa. $9.95

INTRODUCTION TO QUANTUM MECHANICS With Applications to Chemistry, Linus Pauling & E. Bright Wilson, Jr. Classic undergraduate text by Nobel Prize winner applies quantum mechanics to chemical and physical problems. Numerous tables and figures enhance the text. Chapter bibliographies. Appendices. Index. 468pp. 5⅜ x 8½. 64871-0 Pa. $12.95

ASYMPTOTIC EXPANSIONS OF INTEGRALS, Norman Bleistein & Richard A. Handelsman. Best introduction to important field with applications in a variety of scientific disciplines. New preface. Problems. Diagrams. Tables. Bibliography. Index. 448pp. 5⅜ x 8½. 65082-0 Pa. $13.95

MATHEMATICS APPLIED TO CONTINUUM MECHANICS, Lee A. Segel. Analyzes models of fluid flow and solid deformation. For upper-level math, science and engineering students. 608pp. 5⅜ x 8½. 65369-2 Pa. $14.95

ELEMENTS OF REAL ANALYSIS, David A. Sprecher. Classic text covers fundamental concepts, real number system, point sets, functions of a real variable, Fourier series, much more. Over 500 exercises. 352pp. 5⅜ x 8½. 65385-4 Pa. $11.95

PHYSICAL PRINCIPLES OF THE QUANTUM THEORY, Werner Heisenberg. Nobel Laureate discusses quantum theory, uncertainty, wave mechanics, work of Dirac, Schroedinger, Compton, Wilson, Einstein, etc. 184pp. 5⅜ x 8½. 60113-7 Pa. $7.95

INTRODUCTORY REAL ANALYSIS, A.N. Kolmogorov, S.V. Fomin. Translated by Richard A. Silverman. Self-contained, evenly paced introduction to real and functional analysis. Some 350 problems. 403pp. 5⅜ x 8½. 61226-0 Pa. $11.95

PROBLEMS AND SOLUTIONS IN QUANTUM CHEMISTRY AND PHYSICS, Charles S. Johnson, Jr. and Lee G. Pedersen. Unusually varied problems, detailed solutions in coverage of quantum mechanics, wave mechanics, angular momentum, molecular spectroscopy, scattering theory, more. 280 problems plus 139 supplementary exercises. 430pp. 6½ x 9¼. 65236-X Pa. $14.95

ASYMPTOTIC METHODS IN ANALYSIS, N.G. de Bruijn. An inexpensive, comprehensive guide to asymptotic methods–the pioneering work that teaches by explaining worked examples in detail. Index. 224pp. 5⅜ x 8½. 64221-6 Pa. $7.95

OPTICAL RESONANCE AND TWO-LEVEL ATOMS, L. Allen and J. H. Eberly. Clear, comprehensive introduction to basic principles behind all quantum optical resonance phenomena. 53 illustrations. Preface. Index. 256pp. 5⅜ x 8½. 65533-4 Pa. $8.95

COMPLEX VARIABLES, Francis J. Flanigan. Unusual approach, delaying complex algebra till harmonic functions have been analyzed from real variable viewpoint. Includes problems with answers. 364pp. 5⅜ x 8½. 61388-7 Pa. $10.95

ATOMIC SPECTRA AND ATOMIC STRUCTURE, Gerhard Herzberg. One of best introductions; especially for specialist in other fields. Treatment is physical rather than mathematical. 80 illustrations. 257pp. 5⅜ x 8½. 60115-3 Pa. $7.95

APPLIED COMPLEX VARIABLES, John W. Dettman. Step-by-step coverage of fundamentals of analytic function theory–plus lucid exposition of five important applications: Potential Theory; Ordinary Differential Equations; Fourier Transforms; Laplace Transforms; Asymptotic Expansions. 66 figures. Exercises at chapter ends. 512pp. 5⅜ x 8½. 64670-X Pa. $14.95

ULTRASONIC ABSORPTION: An Introduction to the Theory of Sound Absorption and Dispersion in Gases, Liquids and Solids, A.B. Bhatia. Standard reference in the field provides a clear, systematically organized introductory review of fundamental concepts for advanced graduate students, research workers. Numerous diagrams. Bibliography. 440pp. 5⅜ x 8½. 64917-2 Pa. $11.95

UNBOUNDED LINEAR OPERATORS: Theory and Applications, Seymour Goldberg. Classic presents systematic treatment of the theory of unbounded linear operators in normed linear spaces with applications to differential equations. Bibliography. 199pp. 5⅜ x 8½. 64830-3 Pa. $7.95

LIGHT SCATTERING BY SMALL PARTICLES, H.C. van de Hulst. Comprehensive treatment including full range of useful approximation methods for researchers in chemistry, meteorology and astronomy. 44 illustrations. 470pp. 5⅜ x 8½. 64228-3 Pa. $12.95

CONFORMAL MAPPING ON RIEMANN SURFACES, Harvey Cohn. Lucid, insightful book presents ideal coverage of subject. 334 exercises make book perfect for self-study. 55 figures. 352pp. 5⅝ x 8¼. 64025-6 Pa. $11.95

OPTICKS, Sir Isaac Newton. Newton's own experiments with spectroscopy, colors, lenses, reflection, refraction, etc., in language the layman can follow. Foreword by Albert Einstein. 532pp. 5⅜ x 8½. 60205-2 Pa. $13.95

GENERALIZED INTEGRAL TRANSFORMATIONS, A.H. Zemanian. Graduate-level study of recent generalizations of the Laplace, Mellin, Hankel, K. Weierstrass, convolution and other simple transformations. Bibliography. 320pp. 5⅜ x 8½. 65375-7 Pa. $8.95

THE ELECTROMAGNETIC FIELD, Albert Shadowitz. Comprehensive undergraduate text covers basics of electric and magnetic fields, builds up to electromagnetic theory. Also related topics, including relativity. Over 900 problems. 768pp. 5⅜ x 8¼. 65660-8 Pa. $19.95

FOURIER SERIES, Georgi P. Tolstov. Translated by Richard A. Silverman. A valuable addition to the literature on the subject, moving clearly from subject to subject and theorem to theorem. 107 problems, answers. 336pp. 5⅜ x 8½. 63317-9 Pa. $10.95

THEORY OF ELECTROMAGNETIC WAVE PROPAGATION, Charles Herach Papas. Graduate-level study discusses the Maxwell field equations, radiation from wire antennas, the Doppler effect and more. xiii + 244pp. 5⅜ x 8½. 65678-0 Pa. $9.95

DISTRIBUTION THEORY AND TRANSFORM ANALYSIS: An Introduction to Generalized Functions, with Applications, A.H. Zemanian. Provides basics of distribution theory, describes generalized Fourier and Laplace transformations. Numerous problems. 384pp. 5⅜ x 8½. 65479-6 Pa. $13.95

THE PHYSICS OF WAVES, William C. Elmore and Mark A. Heald. Unique overview of classical wave theory. Acoustics, optics, electromagnetic radiation, more. Ideal as classroom text or for self-study. Problems. 477pp. 5⅜ x 8½.
64926-1 Pa. $14.95

CALCULUS OF VARIATIONS WITH APPLICATIONS, George M. Ewing. Applications-oriented introduction to variational theory develops insight and promotes understanding of specialized books, research papers. Suitable for advanced undergraduate/graduate students as primary, supplementary text. 352pp. 5⅜ x 8½.
64856-7 Pa. $9.95

A TREATISE ON ELECTRICITY AND MAGNETISM, James Clerk Maxwell. Important foundation work of modern physics. Brings to final form Maxwell's theory of electromagnetism and rigorously derives his general equations of field theory. 1,084pp. 5⅜ x 8½. 60636-8, 60637-6 Pa., Two-vol. set $27.90

AN INTRODUCTION TO THE CALCULUS OF VARIATIONS, Charles Fox. Graduate-level text covers variations of an integral, isoperimetrical problems, least action, special relativity, approximations, more. References. 279pp. 5⅜ x 8½.
65499-0 Pa. $8.95

HYDRODYNAMIC AND HYDROMAGNETIC STABILITY, S. Chandrasekhar. Lucid examination of the Rayleigh-Benard problem; clear coverage of the theory of instabilities causing convection. 704pp. 5⅜ x 8¼. 64071-X Pa. $17.95

CALCULUS OF VARIATIONS, Robert Weinstock. Basic introduction covering isoperimetric problems, theory of elasticity, quantum mechanics, electrostatics, etc. Exercises throughout. 326pp. 5⅜ x 8½. 63069-2 Pa. $9.95

DYNAMICS OF FLUIDS IN POROUS MEDIA, Jacob Bear. For advanced students of ground water hydrology, soil mechanics and physics, drainage and irrigation engineering and more. 335 illustrations. Exercises, with answers. 784pp. 6⅛ x 9¼.
65675-6 Pa. $19.95

NUMERICAL METHODS FOR SCIENTISTS AND ENGINEERS, Richard Hamming. Classic text stresses frequency approach in coverage of algorithms, polynomial approximation, Fourier approximation, exponential approximation, other topics. Revised and enlarged 2nd edition. 721pp. 5⅜ x 8½. 65241-6 Pa. $16.95

THEORETICAL SOLID STATE PHYSICS, Vol. 1: Perfect Lattices in Equilibrium; Vol. II: Non-Equilibrium and Disorder, William Jones and Norman H. March. Monumental reference work covers fundamental theory of equilibrium properties of perfect crystalline solids, non-equilibrium properties, defects and disordered systems. Appendices. Problems. Preface. Diagrams. Index. Bibliography. Total of 1,301pp. 5⅜ x 8½. Two volumes. Vol. I: 65015-4 Pa. $16.95
Vol. II: 65016-2 Pa. $16.95

OPTIMIZATION THEORY WITH APPLICATIONS, Donald A. Pierre. Broad spectrum approach to important topic. Classical theory of minima and maxima, calculus of variations, simplex technique and linear programming, more. Many problems, examples. 640pp. 5⅜ x 8½. 65205-X Pa. $17.95

THE CONTINUUM: A Critical Examination of the Foundation of Analysis, Hermann Weyl. Classic of 20th-century foundational research deals with the conceptual problem posed by the continuum. 156pp. 5⅜ x 8½. 67982-9 Pa. $6.95

ESSAYS ON THE THEORY OF NUMBERS, Richard Dedekind. Two classic essays by great German mathematician: on the theory of irrational numbers; and on transfinite numbers and properties of natural numbers. 115pp. 5⅜ x 8½.
21010-3 Pa. $6.95

THE FUNCTIONS OF MATHEMATICAL PHYSICS, Harry Hochstadt. Comprehensive treatment of orthogonal polynomials, hypergeometric functions, Hill's equation, much more. Bibliography. Index. 322pp. 5⅜ x 8½. 65214-9 Pa. $9.95

NUMBER THEORY AND ITS HISTORY, Oystein Ore. Unusually clear, accessible introduction covers counting, properties of numbers, prime numbers, much more. Bibliography. 380pp. 5⅜ x 8½. 65620-9 Pa. $10.95

THE VARIATIONAL PRINCIPLES OF MECHANICS, Cornelius Lanczos. Graduate level coverage of calculus of variations, equations of motion, relativistic mechanics, more. First inexpensive paperbound edition of classic treatise. Index. Bibliography. 418pp. 5⅜ x 8½. 65067-7 Pa. $14.95

COMBINATORIAL TOPOLOGY, P. S. Alexandrov. Clearly written, well-organized, three-part text begins by dealing with certain classic problems without using the formal techniques of homology theory and advances to the central concept, the Betti groups. Numerous detailed examples. 654pp. 5⅜ x 8½. 40179-0 Pa. $18.95

THEORETICAL PHYSICS, Georg Joos, with Ira M. Freeman. Classic overview covers essential math, mechanics, electromagnetic theory, thermodynamics, quantum mechanics, nuclear physics, other topics. First paperback edition. xxiii + 885pp. 5⅜ x 8½. 65227-0 Pa. $21.95

HANDBOOK OF MATHEMATICAL FUNCTIONS WITH FORMULAS, GRAPHS, AND MATHEMATICAL TABLES, edited by Milton Abramowitz and Irene A. Stegun. Vast compendium: 29 sets of tables, some to as high as 20 places. 1,046pp. 8 x 10½. 61272-4 Pa. $29.95

MATHEMATICAL METHODS IN PHYSICS AND ENGINEERING, John W. Dettman. Algebraically based approach to vectors, mapping, diffraction, other topics in applied math. Also generalized functions, analytic function theory, more. Exercises. 448pp. 5⅜ x 8¼. 65649-7 Pa. $12.95

A SURVEY OF NUMERICAL MATHEMATICS, David M. Young and Robert Todd Gregory. Broad self-contained coverage of computer-oriented numerical algorithms for solving various types of mathematical problems in linear algebra, ordinary and partial, differential equations, much more. Exercises. Total of 1,248pp. 5⅜ x 8½.
Two volumes. Vol. I: 65691-8 Pa. $16.95
Vol. II: 65692-6 Pa. $16.95

TENSOR ANALYSIS FOR PHYSICISTS, J.A. Schouten. Concise exposition of the mathematical basis of tensor analysis, integrated with well-chosen physical examples of the theory. Exercises. Index. Bibliography. 289pp. 5⅜ x 8½. 65582-2 Pa. $10.95

INTRODUCTION TO NUMERICAL ANALYSIS (2nd Edition), F.B. Hildebrand. Classic, fundamental treatment covers computation, approximation, interpolation, numerical differentiation and integration, other topics. 150 new problems. 669pp. 5⅜ x 8½. 65363-3 Pa. $16.95

INVESTIGATIONS ON THE THEORY OF THE BROWNIAN MOVEMENT, Albert Einstein. Five papers (1905–8) investigating dynamics of Brownian motion and evolving elementary theory. Notes by R. Fürth. 122pp. 5⅜ x 8½. 60304-0 Pa. $5.95

CATASTROPHE THEORY FOR SCIENTISTS AND ENGINEERS, Robert Gilmore. Advanced-level treatment describes mathematics of theory grounded in the work of Poincaré, R. Thom, other mathematicians. Also important applications to problems in mathematics, physics, chemistry and engineering. 1981 edition. References. 28 tables. 397 black-and-white illustrations. xvii + 666pp. 6⅛ x 9¼. 67539-4 Pa. $17.95

AN INTRODUCTION TO STATISTICAL THERMODYNAMICS, Terrell L. Hill. Excellent basic text offers wide-ranging coverage of quantum statistical mechanics, systems of interacting molecules, quantum statistics, more. 523pp. 5⅜ x 8½. 65242-4 Pa. $13.95

STATISTICAL PHYSICS, Gregory H. Wannier. Classic text combines thermodynamics, statistical mechanics and kinetic theory in one unified presentation of thermal physics. Problems with solutions. Bibliography. 532pp. 5⅜ x 8½. 65401-X Pa. $14.95

ORDINARY DIFFERENTIAL EQUATIONS, Morris Tenenbaum and Harry Pollard. Exhaustive survey of ordinary differential equations for undergraduates in mathematics, engineering, science. Thorough analysis of theorems. Diagrams. Bibliography. Index. 818pp. 5⅜ x 8½. 64940-7 Pa. $19.95

STATISTICAL MECHANICS: Principles and Applications, Terrell L. Hill. Standard text covers fundamentals of statistical mechanics, applications to fluctuation theory, imperfect gases, distribution functions, more. 448pp. 5⅜ x 8½.
65390-0 Pa. $14.95

ORDINARY DIFFERENTIAL EQUATIONS AND STABILITY THEORY: An Introduction, David A. Sánchez. Brief, modern treatment. Linear equation, stability theory for autonomous and nonautonomous systems, etc. 164pp. 5⅝ x 8¼.
63828-6 Pa. $6.95

THIRTY YEARS THAT SHOOK PHYSICS: The Story of Quantum Theory, George Gamow. Lucid, accessible introduction to influential theory of energy and matter. Careful explanations of Dirac's anti-particles, Bohr's model of the atom, much more. 12 plates. Numerous drawings. 240pp. 5⅜ x 8½. 24895-X Pa. $7.95

THEORY OF MATRICES, Sam Perlis. Outstanding text covering rank, nonsingularity and inverses in connection with the development of canonical matrices under the relation of equivalence, and without the intervention of determinants. Includes exercises. 237pp. 5⅜ x 8½. 66810-X Pa. $8.95

GREAT EXPERIMENTS IN PHYSICS: Firsthand Accounts from Galileo to Einstein, edited by Morris H. Shamos. 25 crucial discoveries: Newton's laws of motion, Chadwick's study of the neutron, Hertz on electromagnetic waves, more. Original accounts clearly annotated. 370pp. 5⅜ x 8½. 25346-5 Pa. $11.95

INTRODUCTION TO PARTIAL DIFFERENTIAL EQUATIONS WITH APPLICATIONS, E.C. Zachmanoglou and Dale W. Thoe. Essentials of partial differential equations applied to common problems in engineering and the physical sciences. Problems and answers. 416pp. 5⅜ x 8½. 65251-3 Pa. $11.95

BURNHAM'S CELESTIAL HANDBOOK, Robert Burnham, Jr. Thorough guide to the stars beyond our solar system. Exhaustive treatment. Alphabetical by constellation: Andromeda to Cetus in Vol. 1; Chamaeleon to Orion in Vol. 2; and Pavo to Vulpecula in Vol. 3. Hundreds of illustrations. Index in Vol. 3. 2,000pp. 6⅛ x 9¼.
23567-X, 23568-8, 23673-0 Pa., Three-vol. set $46.85

CHEMICAL MAGIC, Leonard A. Ford. Second Edition, Revised by E. Winston Grundmeier. Over 100 unusual stunts demonstrating cold fire, dust explosions, much more. Text explains scientific principles and stresses safety precautions. 128pp. 5⅜ x 8½. 67628-5 Pa. $5.95

AMATEUR ASTRONOMER'S HANDBOOK, J.B. Sidgwick. Timeless, comprehensive coverage of telescopes, mirrors, lenses, mountings, telescope drives, micrometers, spectroscopes, more. 189 illustrations. 576pp. 5⅝ x 8¼. (Available in U.S. only) 24034-7 Pa. $13.95

SPECIAL FUNCTIONS, N.N. Lebedev. Translated by Richard Silverman. Famous Russian work treating more important special functions, with applications to specific problems of physics and engineering. 38 figures. 308pp. 5⅜ x 8½. 60624-4 Pa. $9.95

THE EXTRATERRESTRIAL LIFE DEBATE, 1750–1900, Michael J. Crowe. First detailed, scholarly study in English of the many ideas that developed between 1750 and 1900 regarding the existence of intelligent extraterrestrial life. Examines ideas of Kant, Herschel, Voltaire, Percival Lowell, many other scientists and thinkers. 16 illustrations. 704pp. 5⅜ x 8½. 40675-X Pa. $19.95

INTEGRAL EQUATIONS, F.G. Tricomi. Authoritative, well-written treatment of extremely useful mathematical tool with wide applications. Volterra Equations, Fredholm Equations, much more. Advanced undergraduate to graduate level. Exercises. Bibliography. 238pp. 5⅜ x 8½. 64828-1 Pa. $8.95

POPULAR LECTURES ON MATHEMATICAL LOGIC, Hao Wang. Noted logician's lucid treatment of historical developments, set theory, model theory, recursion theory and constructivism, proof theory, more. 3 appendixes. Bibliography. 1981 edition. ix + 283pp. 5⅜ x 8½. 67632-3 Pa. $8.95

MODERN NONLINEAR EQUATIONS, Thomas L. Saaty. Emphasizes practical solution of problems; covers seven types of equations. ". . . a welcome contribution to the existing literature...."–*Math Reviews.* 490pp. 5⅜ x 8½. 64232-1 Pa. $13.95

FUNDAMENTALS OF ASTRODYNAMICS, Roger Bate et al. Modern approach developed by U.S. Air Force Academy. Designed as a first course. Problems, exercises. Numerous illustrations. 455pp. 5⅜ x 8½. 60061-0 Pa. $12.95

INTRODUCTION TO LINEAR ALGEBRA AND DIFFERENTIAL EQUATIONS, John W. Dettman. Excellent text covers complex numbers, determinants, orthonormal bases, Laplace transforms, much more. Exercises with solutions. Undergraduate level. 416pp. 5⅜ x 8½. 65191-6 Pa. $11.95

INCOMPRESSIBLE AERODYNAMICS, edited by Bryan Thwaites. Covers theoretical and experimental treatment of the uniform flow of air and viscous fluids past two-dimensional aerofoils and three-dimensional wings; many other topics. 654pp. 5⅜ x 8½. 65465-6 Pa. $16.95

INTRODUCTION TO DIFFERENCE EQUATIONS, Samuel Goldberg. Exceptionally clear exposition of important discipline with applications to sociology, psychology, economics. Many illustrative examples; over 250 problems. 260pp. 5⅜ x 8½. 65084-7 Pa. $8.95

THREE PEARLS OF NUMBER THEORY, A. Y. Khinchin. Three compelling puzzles require proof of a basic law governing the world of numbers. Challenges concern van der Waerden's theorem, the Landau-Schnirelmann hypothesis and Mann's theorem, and a solution to Waring's problem. Solutions included. 64pp. 5⅜ x 8½. 40026-3 Pa. $4.95

LECTURES ON CLASSICAL DIFFERENTIAL GEOMETRY, Second Edition, Dirk J. Struik. Excellent brief introduction covers curves, theory of surfaces, fundamental equations, geometry on a surface, conformal mapping, other topics. Problems. 240pp. 5⅜ x 8½. 65609-8 Pa. $9.95

ROTARY-WING AERODYNAMICS, W.Z. Stepniewski. Clear, concise text covers aerodynamic phenomena of the rotor and offers guidelines for helicopter performance evaluation. Originally prepared for NASA. 537 figures. 640pp. 6⅛ x 9¼.
64647-5 Pa. $16.95

DIFFERENTIAL GEOMETRY, Heinrich W. Guggenheimer. Local differential geometry as an application of advanced calculus and linear algebra. Curvature, transformation groups, surfaces, more. Exercises. 62 figures. 378pp. 5⅜ x 8½.
63433-7 Pa. $11.95

INTRODUCTION TO SPACE DYNAMICS, William Tyrrell Thomson. Comprehensive, classic introduction to space-flight engineering for advanced undergraduate and graduate students. Includes vector algebra, kinematics, transformation of coordinates. Bibliography. Index. 352pp. 5⅜ x 8½. 65113-4 Pa. $10.95

A SURVEY OF MINIMAL SURFACES, Robert Osserman. Up-to-date, in-depth discussion of the field for advanced students. Corrected and enlarged edition covers new developments. Includes numerous problems. 192pp. 5⅜ x 8½. 64998-9 Pa. $8.95

ANALYTICAL MECHANICS OF GEARS, Earle Buckingham. Indispensable reference for modern gear manufacture covers conjugate gear-tooth action, gear-tooth profiles of various gears, many other topics. 263 figures. 102 tables. 546pp. 5⅜ x 8½.
65712-4 Pa. $16.95

SET THEORY AND LOGIC, Robert R. Stoll. Lucid introduction to unified theory of mathematical concepts. Set theory and logic seen as tools for conceptual understanding of real number system. 496pp. 5⅜ x 8¼. 63829-4 Pa. $14.95

A HISTORY OF MECHANICS, René Dugas. Monumental study of mechanical principles from antiquity to quantum mechanics. Contributions of ancient Greeks, Galileo, Leonardo, Kepler, Lagrange, many others. 671pp. 5⅜ x 8½.
65632-2 Pa. $18.95

FAMOUS PROBLEMS OF GEOMETRY AND HOW TO SOLVE THEM, Benjamin Bold. Squaring the circle, trisecting the angle, duplicating the cube: learn their history, why they are impossible to solve, then solve them yourself. 128pp. 5⅜ x 8½.
24297-8 Pa. $5.95

MECHANICAL VIBRATIONS, J.P. Den Hartog. Classic textbook offers lucid explanations and illustrative models, applying theories of vibrations to a variety of practical industrial engineering problems. Numerous figures. 233 problems, solutions. Appendix. Index. Preface. 436pp. 5⅜ x 8½. 64785-4 Pa. $13.95

CURVATURE AND HOMOLOGY: Enlarged Edition, Samuel I. Goldberg. Revised edition examines topology of differentiable manifolds; curvature, homology of Riemannian manifolds; compact Lie groups; complex manifolds; curvature, homology of Kaehler manifolds. New Preface. Four new appendixes. 416pp. 5⅜ x 8½.
40207-X Pa. $14.95

HISTORY OF STRENGTH OF MATERIALS, Stephen P. Timoshenko. Excellent historical survey of the strength of materials with many references to the theories of elasticity and structure. 245 figures. 452pp. 5⅜ x 8½. 61187-6 Pa. $14.95

GEOMETRY OF COMPLEX NUMBERS, Hans Schwerdtfeger. Illuminating, widely praised book on analytic geometry of circles, the Moebius transformation, and two-dimensional non-Euclidean geometries. 200pp. 5⅜ x 8¼. 63830-8 Pa. $8.95

MECHANICS, J.P. Den Hartog. A classic introductory text or refresher. Hundreds of applications and design problems illuminate fundamentals of trusses, loaded beams and cables, etc. 334 answered problems. 462pp. 5⅜ x 8½. 60754-2 Pa. $12.95

TOPOLOGY, John G. Hocking and Gail S. Young. Superb one-year course in classical topology. Topological spaces and functions, point-set topology, much more. Examples and problems. Bibliography. Index. 384pp. 5⅜ x 8¼. 65676-4 Pa. $11.95

STRENGTH OF MATERIALS, J.P. Den Hartog. Full, clear treatment of basic material (tension, torsion, bending, etc.) plus advanced material on engineering methods, applications. 350 answered problems. 323pp. 5⅜ x 8½. 60755-0 Pa. $9.95

ELEMENTARY CONCEPTS OF TOPOLOGY, Paul Alexandroff. Elegant, intuitive approach to topology from set-theoretic topology to Betti groups; how concepts of topology are useful in math and physics. 25 figures. 57pp. 5⅜ x 8½.
60747-X Pa. $4.95

ADVANCED STRENGTH OF MATERIALS, J.P. Den Hartog. Superbly written advanced text covers torsion, rotating disks, membrane stresses in shells, much more. Many problems and answers. 388pp. 5⅜ x 8½. 65407-9 Pa. $11.95

COMPUTABILITY AND UNSOLVABILITY, Martin Davis. Classic graduate-level introduction to theory of computability, usually referred to as theory of recurrent functions. New preface and appendix. 288pp. 5⅜ x 8½. 61471-9 Pa. $8.95

GENERAL CHEMISTRY, Linus Pauling. Revised 3rd edition of classic first-year text by Nobel laureate. Atomic and molecular structure, quantum mechanics, statistical mechanics, thermodynamics correlated with descriptive chemistry. Problems. 992pp. 5⅜ x 8½. 65622-5 Pa. $19.95

AN INTRODUCTION TO MATRICES, SETS AND GROUPS FOR SCIENCE STUDENTS, G. Stephenson. Concise, readable text introduces sets, groups, and most importantly, matrices to undergraduate students of physics, chemistry, and engineering. Problems. 164pp. 5⅜ x 8½. 65077-4 Pa. $7.95

THE HISTORICAL BACKGROUND OF CHEMISTRY, Henry M. Leicester. Evolution of ideas, not individual biography. Concentrates on formulation of a coherent set of chemical laws. 260pp. 5⅜ x 8½. 61053-5 Pa. $8.95

THE PHILOSOPHY OF MATHEMATICS: An Introductory Essay, Stephan Körner. Surveys the views of Plato, Aristotle, Leibniz & Kant concerning propositions and theories of applied and pure mathematics. Introduction. Two appendices. Index. 198pp. 5⅜ x 8½. 25048-2 Pa. $8.95

THE DEVELOPMENT OF MODERN CHEMISTRY, Aaron J. Ihde. Authoritative history of chemistry from ancient Greek theory to 20th-century innovation. Covers major chemists and their discoveries. 209 illustrations. 14 tables. Bibliographies. Indices. Appendices. 851pp. 5⅜ x 8½. 64235-6 Pa. $18.95

DE RE METALLICA, Georgius Agricola. The famous Hoover translation of greatest treatise on technological chemistry, engineering, geology, mining of early modern times (1556). All 289 original woodcuts. 638pp. 6¾ x 11. 60006-8 Pa. $21.95

SOME THEORY OF SAMPLING, William Edwards Deming. Analysis of the problems, theory and design of sampling techniques for social scientists, industrial managers and others who find statistics increasingly important in their work. 61 tables. 90 figures. xvii + 602pp. 5⅜ x 8½. 64684-X Pa. $16.95

THE VARIOUS AND INGENIOUS MACHINES OF AGOSTINO RAMELLI: A Classic Sixteenth-Century Illustrated Treatise on Technology, Agostino Ramelli. One of the most widely known and copied works on machinery in the 16th century. 194 detailed plates of water pumps, grain mills, cranes, more. 608pp. 9 x 12. 28180-9 Pa. $24.95

LINEAR PROGRAMMING AND ECONOMIC ANALYSIS, Robert Dorfman, Paul A. Samuelson and Robert M. Solow. First comprehensive treatment of linear programming in standard economic analysis. Game theory, modern welfare economics, Leontief input-output, more. 525pp. 5⅜ x 8½. 65491-5 Pa. $17.95

ELEMENTARY DECISION THEORY, Herman Chernoff and Lincoln E. Moses. Clear introduction to statistics and statistical theory covers data processing, probability and random variables, testing hypotheses, much more. Exercises. 364pp. 5⅜ x 8½. 65218-1 Pa. $10.95

THE COMPLEAT STRATEGYST: Being a Primer on the Theory of Games of Strategy, J.D. Williams. Highly entertaining classic describes, with many illustrated examples, how to select best strategies in conflict situations. Prefaces. Appendices. 268pp. 5⅜ x 8½. 25101-2 Pa. $8.95

CONSTRUCTIONS AND COMBINATORIAL PROBLEMS IN DESIGN OF EXPERIMENTS, Damaraju Raghavarao. In-depth reference work examines orthogonal Latin squares, incomplete block designs, tactical configuration, partial geometry, much more. Abundant explanations, examples. 416pp. 5⅜ x 8¼. 65685-3 Pa. $10.95

THE ABSOLUTE DIFFERENTIAL CALCULUS (CALCULUS OF TENSORS), Tullio Levi-Civita. Great 20th-century mathematician's classic work on material necessary for mathematical grasp of theory of relativity. 452pp. 5⅜ x 8½. 63401-9 Pa. $11.95

VECTOR AND TENSOR ANALYSIS WITH APPLICATIONS, A.I. Borisenko and I.E. Tarapov. Concise introduction. Worked-out problems, solutions, exercises. 257pp. 5⅝ x 8¼. 63833-2 Pa. $9.95

THE FOUR-COLOR PROBLEM: Assaults and Conquest, Thomas L. Saaty and Paul G. Kainen. Engrossing, comprehensive account of the century-old combinatorial topological problem, its history and solution. Bibliographies. Index. 110 figures. 228pp. 5⅜ x 8½. 65092-8 Pa. $7.95

CATALYSIS IN CHEMISTRY AND ENZYMOLOGY, William P. Jencks. Exceptionally clear coverage of mechanisms for catalysis, forces in aqueous solution, carbonyl- and acyl-group reactions, practical kinetics, more. 864pp. 5⅜ x 8½.
65460-5 Pa. $19.95

PROBABILITY: An Introduction, Samuel Goldberg. Excellent basic text covers set theory, probability theory for finite sample spaces, binomial theorem, much more. 360 problems. Bibliographies. 322pp. 5⅜ x 8½. 65252-1 Pa. $10.95

LIGHTNING, Martin A. Uman. Revised, updated edition of classic work on the physics of lightning. Phenomena, terminology, measurement, photography, spectroscopy, thunder, more. Reviews recent research. Bibliography. Indices. 320pp. 5⅜ x 8¼. 64575-4 Pa. $8.95

PROBABILITY THEORY: A Concise Course, Y.A. Rozanov. Highly readable, self-contained introduction covers combination of events, dependent events, Bernoulli trials, etc. Translation by Richard Silverman. 148pp. 5⅜ x 8¼. 63544-9 Pa. $7.95

AN INTRODUCTION TO HAMILTONIAN OPTICS, H. A. Buchdahl. Detailed account of the Hamiltonian treatment of aberration theory in geometrical optics. Many classes of optical systems defined in terms of the symmetries they possess. Problems with detailed solutions. 1970 edition. xv + 360pp. 5⅜ x 8½.
67597-1 Pa. $10.95

STATISTICS MANUAL, Edwin L. Crow, et al. Comprehensive, practical collection of classical and modern methods prepared by U.S. Naval Ordnance Test Station. Stress on use. Basics of statistics assumed. 288pp. 5⅜ x 8½. 60599-X Pa. $8.95

DICTIONARY/OUTLINE OF BASIC STATISTICS, John E. Freund and Frank J. Williams. A clear concise dictionary of over 1,000 statistical terms and an outline of statistical formulas covering probability, nonparametric tests, much more. 208pp. 5⅜ x 8½. 66796-0 Pa. $7.95

STATISTICAL METHOD FROM THE VIEWPOINT OF QUALITY CONTROL, Walter A. Shewhart. Important text explains regulation of variables, uses of statistical control to achieve quality control in industry, agriculture, other areas. 192pp. 5⅜ x 8½. 65232-7 Pa. $8.95

METHODS OF THERMODYNAMICS, Howard Reiss. Outstanding text focuses on physical technique of thermodynamics, typical problem areas of understanding, and significance and use of thermodynamic potential. 1965 edition. 238pp. 5⅜ x 8½.
69445-3 Pa. $8.95

STATISTICAL ADJUSTMENT OF DATA, W. Edwards Deming. Introduction to basic concepts of statistics, curve fitting, least squares solution, conditions without parameter, conditions containing parameters. 26 exercises worked out. 271pp. 5⅜ x 8½.
64685-8 Pa. $9.95

TENSOR CALCULUS, J.L. Synge and A. Schild. Widely used introductory text covers spaces and tensors, basic operations in Riemannian space, non-Riemannian spaces, etc. 324pp. 5⅜ x 8¼. 63612-7 Pa. $11.95

A CONCISE HISTORY OF MATHEMATICS, Dirk J. Struik. The best brief history of mathematics. Stresses origins and covers every major figure from ancient Near East to 19th century. 41 illustrations. 195pp. 5⅜ x 8½. 60255-9 Pa. $8.95

A SHORT ACCOUNT OF THE HISTORY OF MATHEMATICS, W.W. Rouse Ball. One of clearest, most authoritative surveys from the Egyptians and Phoenicians through 19th-century figures such as Grassman, Galois, Riemann. Fourth edition. 522pp. 5⅜ x 8½. 20630-0 Pa. $13.95

HISTORY OF MATHEMATICS, David E. Smith. Nontechnical survey from ancient Greece and Orient to late 19th century; evolution of arithmetic, geometry, trigonometry, calculating devices, algebra, the calculus. 362 illustrations. 1,355pp. 5⅜ x 8½. 20429-4, 20430-8 Pa., Two-vol. set $27.90

THE GEOMETRY OF RENÉ DESCARTES, René Descartes. The great work founded analytical geometry. Original French text, Descartes' own diagrams, together with definitive Smith-Latham translation. 244pp. 5⅜ x 8½. 60068-8 Pa. $8.95

GAMES, GODS & GAMBLING: A History of Probability and Statistical Ideas, F. N. David. Episodes from the lives of Galileo, Fermat, Pascal, and others illustrate this fascinating account of the roots of mathematics. Features thought-provoking references to classics, archaeology, biography, poetry. 1962 edition. 304pp. 5⅜ x 8½. (USO) 40023-9 Pa. $9.95

THE HISTORY OF THE CALCULUS AND ITS CONCEPTUAL DEVELOPMENT, Carl B. Boyer. Origins in antiquity, medieval contributions, work of Newton, Leibniz, rigorous formulation. Treatment is verbal. 346pp. 5⅜ x 8½. 60509-4 Pa. $9.95

THE THIRTEEN BOOKS OF EUCLID'S ELEMENTS, translated with introduction and commentary by Sir Thomas L. Heath. Definitive edition. Textual and linguistic notes, mathematical analysis. 2,500 years of critical commentary. Not abridged. 1,414pp. 5⅜ x 8½. 60088-2, 60089-0, 60090-4 Pa., Three-vol. set $34.85

GAMES AND DECISIONS: Introduction and Critical Survey, R. Duncan Luce and Howard Raiffa. Superb nontechnical introduction to game theory, primarily applied to social sciences. Utility theory, zero-sum games, n-person games, decision-making, much more. Bibliography. 509pp. 5⅜ x 8½. 65943-7 Pa. $14.95

THE HISTORICAL ROOTS OF ELEMENTARY MATHEMATICS, Lucas N.H. Bunt, Phillip S. Jones, and Jack D. Bedient. Fundamental underpinnings of modern arithmetic, algebra, geometry and number systems derived from ancient civilizations. 320pp. 5⅜ x 8½. 25563-8 Pa. $8.95

CALCULUS REFRESHER FOR TECHNICAL PEOPLE, A. Albert Klaf. Covers important aspects of integral and differential calculus via 756 questions. 566 problems, most answered. 431pp. 5⅜ x 8½. 20370-0 Pa. $9.95

CHALLENGING MATHEMATICAL PROBLEMS WITH ELEMENTARY SOLUTIONS, A.M. Yaglom and I.M. Yaglom. Over 170 challenging problems on probability theory, combinatorial analysis, points and lines, topology, convex polygons, many other topics. Solutions. Total of 445pp. 5⅜ x 8½. Two-vol. set.

Vol. I: 65536-9 Pa. $8.95
Vol. II: 65537-7 Pa. $7.95

FIFTY CHALLENGING PROBLEMS IN PROBABILITY WITH SOLUTIONS, Frederick Mosteller. Remarkable puzzlers, graded in difficulty, illustrate elementary and advanced aspects of probability. Detailed solutions. 88pp. 5⅜ x 8½.

65355-2 Pa. $4.95

EXPERIMENTS IN TOPOLOGY, Stephen Barr. Classic, lively explanation of one of the byways of mathematics. Klein bottles, Moebius strips, projective planes, map coloring, problem of the Koenigsberg bridges, much more, described with clarity and wit. 43 figures. 210pp. 5⅜ x 8½.

25933-1 Pa. $6.95

RELATIVITY IN ILLUSTRATIONS, Jacob T. Schwartz. Clear nontechnical treatment makes relativity more accessible than ever before. Over 60 drawings illustrate concepts more clearly than text alone. Only high school geometry needed. Bibliography. 128pp. 6⅛ x 9¼.

25965-X Pa. $7.95

AN INTRODUCTION TO ORDINARY DIFFERENTIAL EQUATIONS, Earl A. Coddington. A thorough and systematic first course in elementary differential equations for undergraduates in mathematics and science, with many exercises and problems (with answers). Index. 304pp. 5⅜ x 8½.

65942-9 Pa. $9.95

FOURIER SERIES AND ORTHOGONAL FUNCTIONS, Harry F. Davis. An incisive text combining theory and practical example to introduce Fourier series, orthogonal functions and applications of the Fourier method to boundary-value problems. 570 exercises. Answers and notes. 416pp. 5⅜ x 8½.

65973-9 Pa. $13.95

AN INTRODUCTION TO ALGEBRAIC STRUCTURES, Joseph Landin. Superb self-contained text covers "abstract algebra": sets and numbers, theory of groups, theory of rings, much more. Numerous well-chosen examples, exercises. 247pp. 5⅜ x 8½.

65940-2 Pa. $8.95

STARS AND RELATIVITY, Ya. B. Zel'dovich and I. D. Novikov. Vol. 1 of *Relativistic Astrophysics* by famed Russian scientists. General relativity, properties of matter under astrophysical conditions, stars and stellar systems. Deep physical insights, clear presentation. 1971 edition. References. 544pp. 5⅜ x 8½.

69424-0 Pa. $14.95

---